To Doug,

with great admiration
and best wishes.

Bill

30 April 2002

What the Experts Say . . .

"This is a superb book on how to prevent and minimize technological disasters."

—P. Roy Vagelos, M.D.
Retired Chairman and CEO, Merck & Co., Inc.

"If you want to know how serious technological disasters can be, how poorly we tend to handle them, and what can be done to reduce or eliminate the dangers associated with them, *Minding the Machines* is the book for you."

—Russell L. Ackoff,
Professor Emeritus of Management Science at The Wharton School, University of Pennsylvania.

"A thorough compendium of technological disasters, complete with detailed descriptions, analyses of what happened, what went wrong, and why. This lucid book candidly addresses human and societal failings that need to be corrected if future disasters are to be prevented."

—Severo Ornstein,
Internet Pioneer and Founder of Computer Professionals for Social Responsibility.

"*Minding the Machines* provides us with insights that are greatly needed to cope with technological disasters that are endemic to our times."

—David A. Hounshell,
David M. Roderick Professor of Technology and Social Change, Carnegie Mellon University.

"An excellent, balanced, and highly readable book emphasizing human, social, and organizational elements universally present in technological disasters."

—Carver Mead,
Gordon and Betty Moore Professor Emeritus of Engineering and Applied Science at the California Institute of Technology, 1999 Lemelson—MIT Prize Winner.

"This book presents a systematic analysis of the root causes of technological disasters, accompanied by many riveting examples. More importantly, the authors provide an enlightening discussion on how we can prevent them."

—David J. Farber,
The Alfred Fitler Moore Professor of Telecommunication Systems in the School of Engineering and Applied Sciences and Professor of Business and Public Policy at The Wharton School, University of Pennsylvania.

"The authors of *Minding the Machines* teach us an invaluable lesson in social responsibility. Our continued reliance on complex technological systems will inevitably lead to disasters unless we take this lesson to heart."

—Brent H. Woodworth,
IBM Crisis Response Team Manager.

MINDING THE MACHINES
PREVENTING TECHNOLOGICAL DISASTERS

William M. Evan

*The Wharton School
and
School of Arts and Sciences
University of Pennsylvania*

Mark Manion

*Program in Philosophy
College of Arts and Sciences
and College of Engineering
Drexel University*

ISBN 0-13-065646-1

Prentice Hall PTR, Upper Saddle River, NJ 07458
www.phptr.com

Library of Congress Cataloging-in-Publication Data

Evan, William M
  Minding the machines : preventing technological disasters / William M.
Evan, Mark Manion.
    p. cm.
  Includes bioliographical references and index.
  ISBN 0-13-065646-1
  1. System failures (Engineering) 2. Disasters. I. Manion, Mark. II. Title.

TA169.5 .E93 2002
620.8—dc21   2001054867

Publisher: Bernard Goodwin
Editorial Assistant: Michelle Vincenti
Marketing Manager: Dan DePasquale
Manufacturing Manager: Alexis R. Heydt-Long
Cover Director: Jerry Votta
Cover Designer: Anthony Gemmellaro
Art Director: Gail Cocker-Bogusz
Production Coordinator: Anne R. Garcia
Composition/Production: Tiffany Kuehn, Carlisle Publishers Services

Cover photos courtesy of AP/World Wide Photos. Used by permission.

Prentice Hall books are widely used by corporations and government
agencies for training, marketing, and resale. For more information
regarding corporate and government bulk discounts please contact:

Corporate and Government Sales (800) 382-3419
or corpsales@pearsontechgroup.com

Printed in the United States of America

10 9 8 7 6 5 4 3 2 1

ISBN 0-13-065646-1

Pearson Education LTD.
Pearson Education Australia PTY, Limited
Pearson Education Singapore, Pte. Ltd.
Pearson Education North Asia Ltd.
Pearson Education Canada, Ltd.
Pearson Educación de Mexico, S.A. de C.V.
Pearson Education—Japan
Pearson Education Malaysia, Pte. Ltd.

For the next generation with our love:
Marc, Milena, Jason, Zachary, and Ali

Contents

LIST OF TABLES

LIST OF FIGURES

Preface

We live in an age of breathtaking technological innovation. Two developments of the 20th Century—the computer and the Internet—have revolutionized our everyday lives, transforming the way millions of people communicate, do their work, fall in love, and even buy birthday gifts. But while technological innovations have transformed and enhanced our lives in myriad ways, they have also created the potential for technological disasters of unimaginable consequences. We are vulnerable in ways we have never been vulnerable before. And yet, to turn the clock backward and eliminate machines from our lives is impossible. We are therefore faced with the challenge of minding the machines—of anticipating and preventing technological disasters. At the same time, we are faced with the challenge of seeing to it that technology's designers develop a stronger sense of social responsibility, a concern for human security and well-being. How do we evaluate our own risk assessment procedures? How do we ensure that policy makers, experts, and others involved in the risk assessment process act not only according to cost-benefit ratios but also with a commitment to social responsibility?

In the pages that follow, we present case studies of technological disasters that have occurred in all corners of the globe. Our purpose in examining these case studies is to develop an array of strategies—professional, organizational, legal, and political—that can help prevent technological disasters. What emerges from these case studies is both illuminating and deeply troubling. Hard as it is to believe, some of these case studies illustrate disasters that were *anticipated*. Potential problems were recognized long before lives were lost or property was damaged. For example, studies of the Challenger Shuttle tragedy identified memoranda written by engineers warning about the possible failure of the O-rings if the shuttle were launched in below-freezing temperature. In fact, the record shows that engineers working on the O-ring design were well aware of the problem a full year before the tragedy. During a teleconference the night before the scheduled launch, several engineers explicitly rec-ommended to management against launching the shuttle. What went wrong with the decision-making process of management that led to launching the Challenger shuttle? The answer has far less to do with technology and far more to do with the value judgments of the parties involved, the structure of organizations, and the inade-quacies of human communication. Why did the people making the decisions disregard the recommendations of their own engineers? What risk evaluation procedures were in place at the time? How can the lessons from the Challenger shuttle be applied to the future de-sign and assessment of technology?

Other case studies in our book focus upon technological disas-ters that were *unanticipated*. In these situations, there were inade-quate provisions for training workers to cope with crises, building-in fail-safe mechanisms to counter human errors, and preparing a well-thought-out emergency plan to meet hazardous and unex-pected developments. For example, the poison gas release at the Union Carbide plant in Bhopal, India, resulted in the death of thou-sands. A worker flushing some pipelines with water failed to insert a metal disk to seal valves in the pipeline leading to a storage tank. The tank contained a highly poisonous chemical, methyl isocyanate. Water from the pipeline leaking into the tank reacted violently with the methyl isocyanate, causing increased pressure and temperature in the tank. The ensuing chemical reaction caused the release of tons of toxic chemicals into the air surrounding the town. The result was catastrophic. In retrospect, certain questions haunt us: What design flaws and human errors made this catastrophe possible? What flaws in the risk assessment procedures made it possible for human beings *not* to consider that such problems might arise?

We distinguish between *anticipated* and *unanticipated* disas-ters. We study these tragic events in an effort to learn the lessons of history—and do everything we can to protect ourselves from similar events in the future.

As we move into the 21st Century, this is one of the greatest challenges that confront us. How do we design technology that will enhance human life? How do we live with the technology we have created? How do we mind the machines?

William M. Evan and Mark Manion
Philadelphia, Pennsylvania

Invitation to Our Readers

Greater public awareness of the risks of technological developments and increased public participation in technology policy decisions may indeed make a difference in creating safeguards to prevent technological disasters. Toward that end, we would welcome your comments and observations.

We hope to incorporate your responses in a future second edition of our book. Our e-mail addresses are:

William M. Evan: MindingTheMachines@wharton.upenn.edu

Mark Manion: MindingTheMachines@drexel.edu

Acknowledgments

A fter completing the manuscript for our book, we incurred many debts of gratitude when we turned to our colleagues, friends, and members of our families for their critical judgments. Their editorial and substantive comments greatly enhanced the quality of our book.

We wish to express our indebtedness to our colleagues at the University of Pennsylvania: Ivar Berg, Michael Cohen, Frits Dambrink, C. Nelson Dorny, Samuel Hughes, Samuel Klausner, Abba Krieger, Howard Kunreuther, Andrew Lamas, and Johannes Pennings. We are also deeply grateful to our colleagues at Drexel University: Lalit Aggarwal, Abby Goodrum, Christian Hunold, Moshe Kam, Andrew McCann, Eric Rau, and Amy Slayton.

We also wish to express our indebtedness to: Robert U. Evan (Kayne, Anderson Rudnick); Ezra G. Levin (Kramer Levin Naftalis and Frankel, LLP); Seymour Melman (Columbia University); Frances Portnoy (University of Massachusetts); Salvatore Tagliareni (Industry Consultant); and Gautham Venkat (Salomon Smith Barney).

For enlightening us about nuclear power reactors, we wish to thank Harold Feiveson (Princeton University), David Lochbaum (Union of Concerned Scientists), and Marvin Miller (M.I.T.).

We are especially grateful to Murray Eden of M.I.T. for reviewing the entire manuscript and giving us the benefit of his numerous creative editorial suggestions.

In addition, to each of the authors of the 12 case studies of technological disasters in Chapter 8, we wish to express our profound appreciation for granting us permission to reprint their exemplary essays.

At the eleventh hour of preparing the manuscript we were the recipients of sage advice from Raima Evan (Swarthmore College), for which we are immensely appreciative. We are also indebted to Craig Williamson (Swarthmore College) and to Sarah Evan, for their invaluable assistance.

Last but not least, we are deeply indebted to our publisher and editor, Bernard Goodwin, for originally recognizing the potential contribution of our book. Throughout the months of editing the manuscript, he exhibited a generosity of spirit, which greatly helped us surmount all of the expected and unexpected hurdles of transforming a manuscript into a book.

Needless to say, whatever shortcomings this book may have are the sole responsibility of the authors.

W. M. E. and M. M.

PART I

Introduction

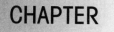

Technological Disasters:
An Overview

S pectacular technological successes worldwide have deflected attention from the high incidence of technological disasters. As technology becomes increasingly complex, technological disasters are likely to increase. Since the Industrial Revolution, we have created a cornucopia of technological systems. Sometimes these systems malfunction, and sometimes these malfunctions result in technological disasters. A *technological disaster* is one that brings on a major crisis, threatens the viability of a technological system, causes massive losses of life and property, and may endanger the social environment in which it occurs. By contrast, a *technological failure* is brought on by an unanticipated breakdown of one or more components of a technological system. It may impair the system's operation and may cause minor property damage, loss of life, or both. A technological failure, if not identified and corrected in time, can lead to a technological disaster.

In addition to the fact that technological disasters cause much devastation, they share another common feature: They are not "Acts of God." In other words, the causes of technological disasters do not stem from acts of uncontrollable nature; they stem from "Acts of Man." They are *human-made disasters*. Because human beings are responsible for such events, what often happens after a disaster is a public debate over how to assign blame for the failure. Such blame attribution is a natural response to the complex social disruptions and the impacts on individuals, communities, and the environment. Some technological disasters are global in nature, such as the Chernobyl nuclear power explosion. Some technological disasters, such as asbestos-related diseases, are not even bound by space and time.

DANGEROUS TECHNOLOGIES

Technological disasters are glaring symbols of our limited capacity to control the technology that human beings create. They confound our expectations about the safety of technology and our social institutions' abilities to manage technological risks. Computer networks are not supposed to crash; chemical production plants are not supposed to leak; airplanes are not supposed to crash, and nuclear power plants are not supposed to melt down. Hence, when they do, they confound their designers, operators, and owners who are usually unprepared to deal with such malfunctions. For many, the image of "runaway technology" is disconcerting, if not downright frightening, especially if one considers the catastrophic destruction that results from a technological disaster.

It is therefore high time to take a fresh look at the growing dependency of fallible human beings on dangerous technologies (Dumas, 1999: 12–13). One theme common to technological disasters is that, in each case, highly skilled professionals, ignoring conspicuous warnings of the dangers involved, pushed the technology they controlled beyond its limits. In almost all cases, previous successes with the technology had lulled designers, operators, managers, regulators, and society in general into overconfidence in the technology that, in turn, caused them to ignore the inherent limitations of the technology with which they were dealing. In many cases the results have been lethal. Technological disasters are not mere accidents—they are disasters directly linked to human action. They pose crises for the organizations and communities in which they occur. Henry Petroski, the renowned engineer, underscores, however, the critical role engineering failures can play in helping engineers recognize the limitations of their designs and helping them learn from their mistakes (Petroski, 1985: xii).

SELECTED EXAMPLES OF TECHNOLOGICAL DISASTERS

Asbestos-Related Illnesses

Although asbestos has not been used in products since the 1970s, class-action lawsuits continue to multiply. As of 1999, more than 275,000 asbestos lawsuits were either being settled or tried in the courts, up from 100,000 in 1993 (Kim, 2000: A4). Roughly 40,000 claims are filed each year, choking both state and federal courts. This prompted a U.S. Supreme Court ruling that called for a political solution to what many are calling a "litigation crisis." Justice David Souter put it aptly when he declared, "This case is a class action prompted by the elephantine mass of asbestos cases." This attitude was affirmed in a concurring opinion of Chief Justice William Rehnquist who charged that the asbestos case before the Supreme Court "cries out for a legislative solution" (Esteban Ortiz *et al., Petitioners v. Fiberboard Corporation,* 1999: sections I and VI).

An estimated 27 million workers have been exposed to asbestos, most of them more than 30 years ago. In a recent asbestos case, a Texas court awarded $115 million in damages, mostly punitive, to 21 plaintiffs who claimed to have succumbed to asbestosis (Robins, 2000: A21). In another such case, a New York court awarded $9.8 million to asbestosis victims (Troy, 2000: A 18). Crushing damage awards and litigation costs have driven at least 25 major asbestos manufacturing companies into bankruptcy (Kim, 2000: A4).

One reason for such high punitive damages is a court finding that large manufacturers of asbestos had known for decades about the negative effects of inhaling asbestos fibers. For example, the top management of Johns-Manville Corporation exhibited a long history of unethical behavior (Gini, 1996). This firm knew of the hazards of asbestos as early as the 1930s. In focusing on its bottom line in an effort to protect its shareholders, the corporation kept this information secret so that employees, customers, and government officials were unaware of the risks. Even when the evidence became irrefutable, Johns-Manville still tried to deny the dangers of asbestos (Broudeur, 1985). In 1973, a federal appellate court found that a number of large manufacturers of asbestos, including Johns-Manville, had known for decades about the negative effects of inhaling asbestos fibers (*Borel v. Fiberboard Paper Prods. Corp,* 1973). By deliberately suppressing information linking the inhalation of asbestos fibers to lung disease, corporations such as Johns-Manville endangered the health of many thousands of workers. Its behavior warranted not only civil liability actions but also criminal liability proceedings (Beck and O'Brien, 2000; Friedman, 2000; Koenig and Rustad, 1998). The end came when Johns-Manville, faced with more than 16,500 asbestos-related injury lawsuits, was forced into bankruptcy. The stockholders paid millions into two trusts set up to pay claims, eventually losing their company.

W.R. Grace, Inc., evidently followed Johns-Manville's deceptive practice by withholding information about the injurious effects of asbestos-laden fireproofing sprays (Moss and Appel, 2001: 1).

The Chernobyl Catastrophe

After the accident at Three Mile Island in 1979, some critics of nuclear power claimed that we had come within a hairsbreadth of radiation releases that could have left many thousands dead (Perrow, 1984). In spite of such fears, many advocates of nuclear power claimed that the worst conceivable accident had already occurred, and now everything was under control. Because the Three Mile Island accident was not regarded as a disaster, either by scientists or by the public, we thought we would never know which one of these statements was true. However, just seven years later, on April 26, 1986, the world got its answer when Unit 4 at Chernobyl, a nuclear power plant in the Ukraine, exploded, causing the reactor's 1,661 fuel rods to blast masses of radioactive material into the air, lighting up the night sky like destructive atomic roman candles. To date, the human toll of the disaster has been approximately 6,000 deaths and more than 30,000 injured. This disaster had the unintended effect of helping to strengthen the movement of *glasnost* (greater freedom of speech and information), as the Soviet Union was forced by the world community to release publications about Chernobyl (Johnson, 1986: 16). A final note to the Chernobyl disaster was the announcement by Russian officials that the entire Chernobyl nuclear power plant was officially shut down as of December 15, 2000. The total cost to shut down the plant was projected to be as high as $5 billion.

Bhopal Poison Gas Release

On December 3, 1984, a poisonous cloud of methyl isocyanate, a chemical compound used to make pesticides, escaped and passed over the town of Bhopal, India, eventually causing the deaths of 14,000 people. In addition, more than 30,000 permanent injuries (including blindness), 20,000 temporary injuries, and 150,000 minor injuries were reported. Lax governmental controls, incompetent operator training, and inadequate emergency preparedness and community education were causes of the disaster. In addition, Union Carbide Corporation's management's perception of the depreciated value of life in India resulted in negligent plant design, which did not include various fail-safe devices.

In October 1991, the Indian Supreme Court unanimously upheld a previous settlement requiring Union Carbide to pay $470 million to victims and their families for the death and destruction caused by tons of deadly gas that leaked from its plant in Bhopal. In addition, the Indian Supreme Court decision reinstated criminal charges against Union Carbide managers, including then-CEO Warren Anderson. In fact, it is reported that Anderson has been "on the run," supposedly hiding from attorneys who want to question him about the lawsuit charging him and Union Carbide with "culpable homicide" in the 1984 disaster (Elliott, 1987: 5).

The sentiments against Union Carbide and Warren Anderson remain intense in the community of Bhopal. One graffito on the walls of the old pesticide factory reads: "Take immediate steps to extradite Warren Anderson and the authorized representatives of Union Carbide from USA." Another reads simply: "Hang Anderson" (Pearl, 2001: A17).

The original judgment of 1989 found Union Carbide "absolutely liable" for damages from the accident, whether it was caused by negligence or sabotage, as the company has contended (Anonymous, 1988: 61). There is a chance that this decision will help globalize the application of absolute liability to an increasing range of hazardous business activities. Multinational corporations ought to take note. This ruling may well inspire multinational firms to be more circumspect in their global activities.

On the other hand, the Indian government, in effect, undermined the claims of the victims by making an out-of-court settlement with Union Carbide, which covered only a fraction of the actual damages. This ruling also limits the company's future liability. The $470 million settlement was paid to more than 550,000 claimants, including death and injury-related claims, about $855 per victim, hardly enough to cover medical and funeral expenses (Chakravarty, 1991: 10). Two prominent Indian Supreme Court judges condemned the settlement: "The multinational has won and the people have lost" (Cassels, 2001: 311).

The Bhopal disaster may not have happened if the refrigeration unit had not been disconnected; if all gauges had been properly working; if the proper safety steps had been taken upon the immediate detection of the deadly methyl isocyanate instead of waiting an hour or so to do anything about it; if the vent-scrubber had been in service; if the water sprays had been designed to shoot high enough to douse the emissions; if the flare tower had been of sufficient capacity. This is a long list of failures, but it is a common roster when chemical and other industrial mishaps occur. Such obvious violations of basic safety procedures were the unintended consequences of the decision to cut costs. Union Carbide had decided to drop the safety standards at the Bhopal plant well below those maintained at its nearly identical facility in Institute, West Virginia.

Nevertheless, the lessons of Bhopal seem to have fallen on deaf ears within Union Carbide. The safety record at Union Carbide, the 10th largest chemical corporation in the United States, has not really changed much since the Bhopal disaster. One need only look at the company's accident record since the 1984 explosion. Two accidents have already occurred at its West Virginia plant, and the company was fined $2.8 million after an explosion at its Seadrift, Texas, plant, which killed one and injured dozens. Occupational Safety and Health Administration (OSHA) fined the company after discovering that top management had ignored—and then attempted to cover up—a series of reports by its own safety engineers urging changes that might have prevented the accidents (Karliner, 1994: 726).

Nor did Union Carbide learn any lessons from a disaster, for which it was responsible, about 50 years prior to the Bhopal poison gas release. During 1930–1932 Union Carbide recruited several thousand predominantly poor, black workers to drill the Hawk's Nest Tunnel in West Virginia in order to divert water for a hydroelectric plant. Boring through the mountain, the workers inhaled silica dust, exposing them to what later became known as *silicosis,* a fatal disease that destroys the ability of the lungs to absorb oxygen.

. . . Union Carbide's management and engineers were mindful of the dangers associated with silica dust, and they wore facemasks or respirators for self-protection when they entered the tunnel for periodic inspections. The workers themselves, who spent eight to ten hours a day breathing the dust, were not told about the hazard, nor were they given face masks. . . . The company doctors were not allowed to tell the men what their trouble was, one of the doctors would testify later. If a worker complained of difficulty breathing, he would be told that his condition was pneumonia or "tunnellitis" (Rampton and Stauber, 2001: 76).

Constructing the Hawk's Nest Tunnel, America's worst industrial disaster, claimed the lives of approximately 764 workers who died of silicosis (Cherniack, 1986: 104).

DC-10 Crashes

On March 3, 1974, a Turkish Airlines DC-10 crashed outside of Paris, killing all 346 on board. The cause of the crash was traced to a defectively-designed rear cargo door. It blew open at an altitude of 12,000 feet, triggering rapid cargo cabin depressurization. Cabin depressurization, in turn, caused the floorboards, which separate the cargo cabin from the passenger cabin above, to tear apart. Unfortunately, the plane was designed to have all of the hydraulic and electrical control wires run along directly under the floorboard. Hence, when the floorboards ripped apart, so did the hydraulic and electrical systems that control the aircraft. With the crucial control systems destroyed, the pilots lost all control, and the plane crashed to earth with breakneck speed.

The Paris DC-10 case raises a host of organizational, professional, and ethical issues because top management at Convair, a subcontractor that designed the cargo door, had known about the faulty design from memoranda circulated by a senior vice president of engineering, warning of the likelihood of a crash. Likewise, management at McDonnell Douglas, the company that designed and manufactured the DC-10 jumbo jet, as well as managing directors at the Federal Aviation Administration, all knew of the potentially deadly problems associated with the rear cargo doors of the DC-10. They all became aware of the problem after an incident in 1972 over Windsor, Ontario, where a DC-10 had to make an emergency landing when its rear cargo door blew open (Fielder and Birsch, 1992: 3).

On May 25, 1979, an American Airlines McDonnell Douglas DC-10 jumbo jet crashed into a field shortly after taking off from Chicago O'Hare International Airport, killing 273 people on board. The cause of the crash was the result of a series of factors. During takeoff, the left engine broke loose, severing control and hydraulic cables housed in the body of the wing. The loss of those control cables made it impossible for the pilots to maneuver the wing slats, extensions of the wing that provide additional lift during takeoff and landing. Consequently, the left wing of the aircraft lost its ability to provide "lift" and it dipped low; at the same time, the right wing rose until the wings were perpendicular to the ground, causing the plane to crash. The National Transportation Safety Board (NTSB) reported that the engine broke loose of the large pylon holding it to the wing as a result of cracks in the pylon. According to the report, these cracks were overlooked because of an "improper maintenance procedure." The NTSB report chastised the Federal Aviation Administration's maintenance policies.

> Additional questions arose when it was learned that McDonnell Douglas knew about the improper maintenance techniques that led to the pylon cracks. Besides questions about the ethical adequacy of the FAA's maintenance policies, should McDonnell Douglas have informed the FAA that airlines were using a nonstandard method for removing the engine and pylon assembly? (Fielder and Birsch, 1992:8)

On July 19, 1989, a United Airlines DC-10 experienced a catastrophic loss of power in mid-flight due to a crack in the 370-pound fan disk assembly of the rear engine. The fan disk disintegrated from the pressure, causing flying debris to sever all of the airplane's hydraulic lines and causing the pilots to lose total control of the aircraft. The plane crashed while attempting an emergency landing at Sioux Gateway Airport, killing 111 people on board and injuring 174 others.

> The National Transportation Safety Board determines that the probable cause of this accident was the inadequate consideration given to human factors limitations in the inspection and quality control procedures used by United Airlines' engine overhaul facility which resulted in the failure to detect a fatigue crack originating from a previously undetected metallurgical defect located in a critical area of the stage 1 fan disk . . . (National Transportation Safety Board Report on the 1989 Sioux City, Iowa, crash, reprinted in Fielder and Birsch, 1992: 252)

Ford Pinto Rear-End Collisions

On September 13, 1978, the Ford Motor Company was indicted on three counts of reckless homicide and one count of criminal recklessness by an Elkhart County, Indiana, grand jury. The charge of reckless homicide had been brought under a 1977 revision of the Indiana Penal Code that allows a corporation to be treated as a person for the purposes of bringing criminal charges. Ford was charged with reckless homicide in the deaths of three teenage women whose Ford Pinto burst into flames after it was struck from behind by another vehicle on August 10, 1978. An eyewitness to the fiery crash testified during the trial that the car exploded "like a napalm bomb" when struck from behind by a van (Fritz, 1980: A9). The three women died of burns they suffered in the accident.

The charges held that Ford executives consciously allowed an "unsafe" car to be manufactured and sold to unsuspecting consumers. The controversy surrounded the design and placement of the gas tank. The Pinto gas tank was placed behind the rear axle of the car, where it was vulnerable to puncture if the car were struck from behind. In prototype testing, it was discovered that the Pinto had a propensity to burst into flames if struck in its rear end at speeds as low as 30–35 mph.

During the trial, attorneys for Ford Motor Company finally agreed to turn over internal Ford documents dealing with the design and production of the Ford Pinto to Indiana prosecutors, only after a Michigan judge denied attempts by the prosecutor in the case to subpoena Ford Chairman Henry Ford II and 29 other Ford executives (Kramer, 1979a: D8). Among other incriminating evidence, the documents showed that Ford executives knew in 1971 that an $11 part would significantly reduce fire risks in the Pinto, but postponed installing it until 1976 in order to save tens of millions of dollars. The prosecutor stated that the documents clearly show Ford management deciding to "sacrifice human life for private profit" (Kramer, 1979a: D8). The "smoking gun" in the Pinto case was a document that showed a cost-benefit analysis, concluding that it was more profitable for Ford management to settle burn victims' lawsuits than to fix the problem with the rear-end gas tank design.

According to Mark Dowie's Pulitzer Prize-winning exposé, published by *Mother Jones* in 1977, Ford management waited eight years to fix the defect "because its internal 'cost-benefit-analysis,' which places a dollar value on human life, said it wasn't profitable to make the changes sooner" (Dowie, 1977: 20). According to Dowie, between 500 and 900 victims died as a result of rear-end Pinto collisions. Dr. Leslie Ball, a retired NASA safety expert and founder of the International Society of Reliability Engineering studied the Pinto case and concluded: "The release into production of the Pinto was the most reprehensible decision in the history of American engineering" (cited in Dowie, 1977: 23). Byron Bloch, a Los Angeles auto safety expert had this to say: "It's a catastrophic blunder . . . Ford made an extremely irresponsible decision when they placed such a weak tank in such a ridiculous location in such a soft rear end. It's almost designed to blow up—premeditated" (cited in Dowie, 1977: 23).

As a result of the litigation against Ford, the Pinto was subject to one of the largest product recalls in history when the National Highway Traffic Safety Administration (NHTSA) determined that its gas tank had fuel leakage problems. Ford eventually decided to recall 1.5 million Pintos in an attempt to correct the problem (Kramer, 1979b: A1).

The importance of the 20-week trial in Elkhart, Indiana, charging the Ford Motor Company with criminal homicide goes beyond the deaths of the three victims. Although the jury found Ford not guilty of criminal homicide, the case has become a cautionary tale of the criminal liability of a corporation for the products it makes and sells. Here, too, as in the Bhopal catastrophe, corporations should take note of the often-costly liability damages that result from technological disasters.

Tenerife Runway Collision

On March 28, 1977, two jumbo jets—a Pan American 747 and a KLM Royal Dutch Airlines 747—collided on the runway as they were attempting to take off from Tenerife airport in the Canary Islands off the west coast of Africa. The Pan Am 747 had 378 passengers on board, and the KLM 747 was transporting 235 passengers. The death toll of 583 makes the Tenerife runway collision the worst accident in aviation history. The circumstances of the disaster suggest human failure, not technical failure, as the cause of the disaster. Both airliners were ready for takeoff at about the same time. The Pan Am jet was turning onto the main runway when the KLM jet, already beginning to lift off, slammed into it almost head-on.

Both airplanes were, in fact, diverted to the Tenerife airport because of a bomb threat at their intended destination of Las Palmas. The small airport at Tenerife was already crowded with other aircraft diverted from Las Palmas, and the arrival of the two 747s only complicated matters. Both aircraft were crowded with weary passengers and crews who wanted to proceed to their final destinations. The captain of the KLM jet was particularly concerned about time because he wished to complete his round-trip to Amsterdam before the number of hours he could legally fly between rest periods expired and he or his crew would be fined. According to the Netherlands Department of Civil Aviation, in an official report released by the *Subsecretaria de Aviacion Civil* in Spain, the probable cause of the disaster was as follows:

> The KLM aircraft had taken off without take-off clearance, in the absolute conviction that this clearance had been obtained, which was the result of a misunderstanding between the tower and the KLM aircraft. This misunderstanding had arisen from the mutual use of usual terminology, which, however, gave rise to misinterpretation. In combination with a number of other coinciding circumstances, the premature take-off of the KLM aircraft resulted in a collision with the Pan Am aircraft, because the latter was still on the runway since it had missed the correct intersection. (Spanish Report–Dutch Comments, 1977)

Hyatt Regency Walkway Collapse

On July 17, 1981, suspended walkways over the atrium lobby of the Kansas City Hyatt Regency hotel collapsed unexpectedly, killing 114 people and injuring at least 200 others (Roddis, 1993: 1546). The Hyatt-Regency walkway collapse is, so far, the worst structural disaster in U.S. engineering history, at least in terms of lives lost and persons injured. Three walkways were suspended from the ceiling by attaching long support rods, fastened to the ceiling, to each of the walkways. The three walkways transversed the atrium at the second, third, and fourth floors of the hotel. The third floor walkway was offset from the second floor and fourth floor walkways and hung from the ceiling by itself. The fourth floor walkway hung directly over the second floor walkway and was connected to it by steel rods.

Approximately 2,000 people had gathered that evening in the hotel atrium to watch and/or participate in a dance contest. As dancers and spectators populated the floor of the atrium, as well as the three state-of-the-art suspended walkways, or "skyways," the beams and bolts holding the fourth floor walkway broke, causing it to collapse onto the second floor walkway directly underneath and hurling both on top of the innocent spectators and dancers below.

The original design was to have a single steel rod running through both the second and fourth floor walkways. Each walkway would be fastened to the steel rod by a series of bolts. The single steel rod would, in turn, be fastened to the ceiling. However, prior to construction, the design was changed so that two steel rods were used. One rod would run from the ceiling to the top beam of the fourth floor walkway and be fastened by bolts. A second rod would be fastened to the bottom beam of the fourth floor walkway that, in turn, would be fastened to the second floor walkway beams. Unfortunately, this placed a greater stress on the single rod hanging from the bottom beam of the fourth walkway and connected to the top beam of the second floor walkway. The weight of the dancers and spectators eventually caused the beams to separate from the steel rods. This caused the disaster.

The design change was made by the steel company contracted to fabricate and erect the steel rods and beams needed for the walkways. Managers for the steel company sent drawings of the design changes to the engineers responsible for the original design. During the investigation, the design engineers denied receiving the shop drawings reflecting the design changes. For their part, the steel company presented the returned drawings they sent to the engineers, which included the engineering seals of two licensed engineers of the firm.

The Hyatt Regency walkway collapse highlighted the lack of established procedures for design changes as well as the lack of adequate professional communication between the various parties involved (Roddis, 1993). Responsibility for the cause of the disaster fell primarily on the engineers involved, although some researchers claim that the fabricators and installers of the steel used, as well as the owners of the hotel, should share in the blame. The legal process that ensued involved both civil and criminal charges.

After 20 months of investigation, the U.S. attorney and the Jackson County, MO, prosecutor found no evidence of criminality associated with the Hyatt failure. The attorney general of Missouri, on the other hand, charged the engineers with negligence. . . . However, a grand jury in Kansas City did not issue indictments for criminal negligence due to lack of evidence. . . . In 1984, the Missouri Board for Architects, Professional Engineers and Land Surveyors brought civil charges of gross negligence and misconduct against the structural engineering firm and the two engineers who were in charge of the structural design. . . . The decision found the firm and both engineers guilty of gross negligence, misconduct, and unprofessional conduct in the practice of engineering . . . (Roddis, 1993: 1550)

Eventually, the two engineers who affixed their seals to the drawings lost their license to practice engineering in the state of Missouri, and the company they worked for was forbidden to carry on as an engineering firm (Jenney, 1998).

Tylenol Poisoning

We readily associate acts of terrorism with acts of violence against governments. In recent years, however, corporations and individuals have also been the targets of terrorism. One dramatic act of terrorism is the Tylenol poisoning case. Johnson and Johnson's excellent reputation as a socially-responsible corporation was put to a test when six people died after ingesting Extra-Strength Tylenol capsules in three events, two in 1982 and one in 1986 (Shrivastava et. al., 1988: 295). Immediately after the 1982 incidents, James Burke, the CEO of Johnson and Johnson, issued a voluntary recall of more than 35 million bottles after it was discovered that tampered bottles of Tylenol were found with capsules laced with poisonous cyanide. The recall and subsequent production retooling, which included replacing all bottles with a new, tamper-proof lid, cost the company an estimated $150 million, but Johnson and Johnson deemed it well worth the expense to protect its reputation for integrity (Shrivastava, Mitroff, Miller, and Migliani, 1988: 295).

To enable the reader to fully appreciate the magnitude of the tragedies of the eight cases of technological disaster we have just outlined, Table 1–1 presents a summary of the property and damage costs, loss of life, and number of persons injured.

Table 1–1
Summary of effects of eight cases of technological disasters.

Selected Cases	Property and Damage Costs	Loss of Life	Injured
Asbestos-Related Illnesses	$10 billion+[1a]	259,000[1b]	27 million[1c]
Chernobyl Nuclear Catastrophe	$283–358 billion[2a]	6,000[2b]	30,000[2c]
Bhopal Poison Gas Release	$470 million[3a]	14,000+[3b]	200,000[3c]
DC-10 Crashes	$18.3 billion[4a]	730[4b]	174[4c]
Ford Pinto Rear-End Collisions	$137 million[5a]	500–900[5b]	?
Tenerife Runway Collision	$110 million[6a]	587[6b]	57[6c]
Hyatt Regency Walkway Collapse	$651 million[7a]	114[7b]	186[7c]
Tylenol Poisoning	$150 million[8a]	6[8b]	—

Sources:

[1a] Kim, Queena. (2000). "Asbestos claims continue to mount," *The Wall Street Journal*, December 7: A4.

[1b] Website of Roger Worthington, P.C. *www.mesothel.com/pages/deaths_pag.htm*. Accessed from the World Wide Web May, 10, 2001.

[1c] Twenty-seven million refers to the number of people exposed. Statistics compiled by Dr. Irving Selikoff, who, before his death, was one of the world's leading authorities on asbestos-related diseases. Selikoff, Irving (1982). *New York Times*, November 17 A13.

[2a] Hudson, Richard. (1990). "Cost of Chernobyl nuclear disaster soars in new study," *The Wall Street Journal*, March 29: A8.

[2b] Marples, David R. (1996). "Chernobyl's toll after ten years: 6,000 and counting," *The Bulletin of the Atomic Scientists*, 51 (3): 7–9. Estimates of Chernobyl casualties vary greatly depending upon the source of information. For example, the United Nations Scientific Committee on the Effects of Atomic Radiation (UNSCEAR) reported 31 deaths immediately after the explosion, whereas the Minister of Health for the Ukraine has estimated about 125,000 deaths by 1996. Understandably, pro-nuclear activists tend to underestimate casualties and anti-nuclear activists overestimate them.

[2c] Scherbak, Yuri M. (1996). "Ten years of the Chernobyl ear," *Scientific American*, 274 (4): 44–54.

[3a] Anonymous (a). (1993). "Retrying Bhopal claims ruled out," *New York Law Journal*, January 27: 1.

[3b] Anonymous (a). (1993). "Retrying Bhopal claims ruled out," *New York Law Journal*, January 27: 1.

[3c] Anonymous (b). (1986). "Union-Carbide-Bhopal," *Legal Times*, April 7: 12.

[4a] Anonymous (c). (1996). "Aerospace," *Los Angeles Times*, July 25: D2. According to this article, various lawsuits involving the 1989 Sioux City disaster were settled for $300 million in claims to the survivors. Given that there were 112 victims, the average compensation was $2.5 million. Using this average compensation per victim for the total of 730 victims of the three DC-10 crashes yields a total of $18.2 billion in damages. The value of the three DC-10s destroyed ($75 million) was obtained via an e-mail from the Aircraft Value Analysis Company in London.

[4b] Fielder, J., and Birsch, D. (1992). *The DC-10 Case: A study in applied ethics, technology, and society*. Albany, NY: State University of New York Press.

[4c] Fielder, J., and Birsch, D. (1992). *The DC-10 Case: A study in applied ethics, technology, and society*. Albany, NY: State University of New York Press.

(continued)

[5a] Dowie, Mark. (1977). "Pinto madness," *Mother Jones,* September/October. (Projected costs according to Ford's own cost-benefit analysis.)

[5b] Dowie, Mark. (1977). "Pinto madness," *Mother Jones,* September/October.

[6a] Anonymous (d). (1980). "Tenerife crash settlements," *The Washington Post,* March 25: B8; e-mail from Aircraft Value Analysis Company.

[6b] Anonymous (d). (1980). "Tenerife crash settlements," *The Washington Post,* March 25: B8.

[6c] Anonymous (d). (1980). "Tenerife crash settlements," *The Washington Post,* March 25: B8.

[7a] Since damage settlements are generally confidential, the only publicly available data on settlements are derived from the following four cases: *Kay Kenton v. Hyatt Hotels Corporation,* No. 66839, Supreme Court of Missouri, 693 S.W. 2d 83, 1985; *Sally Firestone v. Crown Center Redevelopment Corporation,* No. 66840, Supreme Court of Missouri, 693 S.W. 2d 99, 1985; *Occidental Fire and Casualty Co. of N.C. v. Hyatt Hotels Corporation,* WD No. 42,165, Court of Appeals of Missouri, Western District, 801 S.W. 2d 382, 1990; *Betty Wintz v. Hyatt Hotels Corporation,* No. WD35393, Court of Appeals of Missouri, Western District, 687 S.W. 2d 587, 1985. The total claims paid to the injured parties in these four cases were $14 million, or an average of $3.5 million per case. Using this average compensation per injured person for the total of 186 injured persons yields a total of $651 million in damages.

[7b] Daniel M. Duncan, Jack D. Gillum, and GCE International, Inc., *Appellants v. Missouri Board For Architects, Professional Engineers and Land Surveyors,* Respondant, No. 52655, Court of Appeals of Missouri, Eastern District, Division Three, 744 S.W.2d 524, 1988.

[7c] Daniel M. Duncan, Jack D. Gillum, and GCE International, Inc., *Appellants v. Missouri Board For Architects, Professional Engineers and Land Surveyors,* Respondant, No. 52655, Court of Appeals of Missouri, Eastern District, Division Three, 744 S.W.2d 524; 1988.

[8a] Shrivastava, Paul, Mitroff, Ian I., Miller, Danny, and Migliani, Anil. (1998). "Understanding industrial crises," *Journal of Management Studies,* July 25: 285–303.

[8b] Shrivastava, Paul, Mitroff, Ian I., Miller, Danny, and Migliani, Anil. (1998). "Understanding industrial crises," *Journal of Management Studies,* July 25: 285–303.

CAUSES OF TECHNOLOGICAL DISASTERS

When disaster strikes, there is an understandable urge to find a culprit or an explanation. Invariably, employees operating the failed technological system are assumed to be at fault. While there may be some truth to this ready assumption, oftentimes the causes are much more complex. If operator error is in fact the principal cause of disaster, then it points to the need for providing special training for personnel as part of the prevention program. Surely, the behavior of the pilot of the KLM 747 that caused the Tenerife runway collision is a case in point. In Figure 1–1 we characterize his behavior as an example of a human-factors type of problem. Likewise, the operators in Chernobyl exhibited gross negligence in handling that crisis.

Principal Causes of Disaster	Case	Explanation of Disaster
Human-Factors	Tenerife Runway Collision	Miscommunication among pilots and air flight controllers; inability of the pilots to control stress and anxiety
	Chernobyl Nuclear Catastrophe	Gross operator negligence in disabling key safety features, coupled with numerous faulty designs of the reactor as well as the absence of preparedness programs and evacuation procedures
Technical Design Factors	DC-10 Crashes	Defectively-designed rear cargo doors that were permitted to be installed on the DC-10; regulatory failures of the FAA to monitor and act on the knowledge of such design flaws; improper maintenance procedures for removing and replacing the engine and pylon assembly; human factors limitations in the inspection and quality control procedures that resulted in the failure to detect a fatigue crack originating from a previously undetected metallurgical defect
	Ford Pinto Rear-End Collisions	Placement of gas tank behind the rear axle exposes car to explosion upon rear-end collisions
	Hyatt Regency Walkway Collapse	Improper procedures for changing engineering designs
Organizational Systems Factors	Asbestos-Related Diseases	Refusal of management to acknowledge and report the medical evidence linking asbestos with lung disease
Socio-Cultural Factors	Bhopal Poison Gas Release	Multiple failures of design, coupled with gross managerial negligence in allowing key safety features to be compromised
Technological Terrorism	Tylenol Poisoning	Act of random terrorism, which prompted management to respond in a timely, ethical and courageous manner

Figure 1–1
Principal causes of eight cases of technological disaster.

One obvious factor that has received considerable attention is the technical design of a component of a technological system. Clearly, the three DC-10 crashes and the Ford Pinto rear-end collisions all involve technical design failures. In Figure 1–1 we classify such failures as problems in technical design.

Yet another potential cause of technological disasters is organizational in nature. The obstacles to effective communication between top management and rank-and-file employees—an all-too-frequent problem in large organizations—can deprive management of the information and knowledge necessary for rational and prudent decision making. In the case of asbestos-related diseases, it is likely that one or more industrial physicians employed at Johns-Manville and other asbestos manufacturing companies may have observed a relationship between exposure to asbestos and incipient problems of lung disease. It is also likely that this information was communicated to top management. Unfortunately, evidence points to top management's refusal to accept the dire implications of such information; covering up this information was deemed preferable. In Figure 1–1 we refer to this failure as involving an organizational systems problem.

The final root cause of technological disaster is what we call socio-cultural in nature. This refers to attitudes and values that are widely accepted by people in a society and that penetrate the attitudes and values of the corporate culture of various firms. In designing the Union Carbide plant in Bhopal, management could not help but know of the pervasive poverty in that city of approximately 2 million people. They also likely observed in the local government the absence of concern with, and resources for, protecting the health and safety of its citizens. Hence, when designing the plant, engineers and planners paid little heed to the importance of building in fail-safe devices to protect Bhopal residents against gas leaks. After all, life in Bhopal was deemed not quite as valuable as in Institute, West Virginia, where Union Carbide had already built a comparable plant, making sure to provide for an abundance of safety devices. Implicitly or explicitly, Union Carbide management accepted the Bhopal devaluation of human life and safety in arriving at its decisions on how to design the Bhopal plant (Shrivastava, 1995). In Figure 1–1, the Bhopal tragedy is characterized as a socio-cultural systems problem.

In short, to adequately explain any given technological disaster, one would need to consider all four factors we have just discussed: the role of the human operator, the role of technical design, the role of organizational systems, and the role of sociocultural factors.

Acts of technological terrorism are difficult to categorize into one of the four causes we have discussed. Such acts may cut across all four causes. Thus, in the case of the Tylenol poisoning, the terrorist was a malevolent person who exploited a weakness in the technical design of the Johnson and Johnson packaging system. He or she may conceivably be a member of a group or an organization determined to express resentment against a capitalist enterprise. In Figure 1–1 we classify such acts as technological terrorism.

STRATEGIES FOR PREVENTION

Can we draw some general lessons from the observed disasters in the eight cases, as summarized in Figure 1–1? Clearly, engineers, technologists, and applied scientists need to become more sensitive than they currently are to the problems of technological disasters. They also have to acknowledge the growing frequency of low-probability, high-consequence catastrophes. By carefully building-in a set of redundant fail-safe devices, engineers and corporate managers can reduce the incidence of such disasters. Implicit in such precautions is the imperative need to inculcate, by means of special training programs, a commitment to basic safety procedures in all levels of personnel in a firm. Such a commitment should be reflected in the provisions of a corporate code of conduct, with which all employees should be familiar.

Engineering schools are in a strategic position to transmit to the next generation of engineers a commitment to the value of safety in the design of technological systems. Such a commitment to responsible engineering design should be incorporated not only in engineering design courses and courses on engineering ethics, but should also be emphasized across the entire engineering curriculum. Engineering educators could thus convey the critical importance of designing technologies for the safety and well-being of potential users.

Some technological disasters have occurred in high-technology firms, mostly because of manifest or latent policies of corporate management. Government agencies, in regulating risky technologies of various industries, have also contributed to technological disasters through inadequate implementation procedures and lack of adequate resources to apply them. All organizations, in the public as well as in the private sector, have to develop effective mechanisms to ensure compliance with moral standards expressed in codes of conduct. They should also develop corporate social audits, which can identify negative social and environmental impacts of organizational actions. It is imperative that all employers implement standards for a safe work environment and that all employees have an understanding of the technologies employed in the firm to ensure safe operation.

In general, organizational crisis management provides a preventive strategy for the very survival of a firm. Almost all of the cases discussed in this chapter generated organizational crises. A main reason for such enormous loss of life and property associated with these events is the lack of coordination among the components of a firm and a lack of preparedness to cope with such crises. A crisis management and planning program must take into account the systemic nature of a crisis.

The legal system is often sluggish in responding to the risks and hazards of modern technological developments. Although extensive environmental legislation has been enacted in the past 30 years, legislators have been largely reactive in their response to the ills of technological developments. If enough community outrage over industrial pollution, chemical contamination, improper waste disposal, and technological failures is expressed, legislators are likely to respond. Legislators and government officials need to take a more proactive stance as leaders with respect to the risks and dangers of modern technology.

In addition to a heightened sense of corporate responsibility and government accountability for such disasters, as well as a call for more effective risk assessment from the technical and scientific communities, policymakers would do well to acknowledge that we are living in what social scientists call a "risk society" (Beck, Giddens, and Lasch, 1994). In producing products and services for the market economy we inadvertently generate technological disasters. According to such theorists as

Ulrich Beck, "the risk society designates a stage in industrial society in which the threats produced as a result of 'progress' begin to dominate" (Beck, 1994: 6). In other words, whereas the perceived benefits of technological developments used to be significant as compared with the associated costs, it is presently the case that the negative "side effects" or "externalities" of industrialization are beginning to take center stage.

WHO SHOULD BE CONCERNED?

Scientists and Engineers

Scientists and engineers working in private industry as well as those working for government agencies and in national laboratories (for example, Lawrence Livermore Laboratory, Los Alamos National Laboratory, Sandia, Oak Ridge, and Argonne) need to take far more heed of the potential for technological disasters than the cases suggest that they do.

In recent years, numerous engineers have faced ethical dilemmas in which engineering judgments ultimately led to technological disasters. What is more, these cases have received considerable attention and publicity. Examples abound: the Ford Pinto rear-end gas tank explosions, the various technical design flaws in the DC-10s, and technical miscalculations in the Hyatt Regency walkway collapse. All of these cases involved engineers and/or engineering decisions (Pletta, 1987). Accordingly, it is crucial that engineers and scientists become aware of the roles they play in determining the incidence of technological disasters and of the actions they can take to prevent and mitigate the harmful effects of such failures. Various organizational constraints operating on scientists and engineers have the effect of desensitizing them about problems that may result in technological disasters. Among such constraints are: rigid organizational hierarchies, compartmentalization of tasks, the prevalence of classified and proprietary information, and the implicit conflict between the need to enhance the firm's short-term profitability and the risk of overlooking a significant safety hazard.

Corporate Executives and Corporate R&D Managers

Apart from scientists and engineers working on the development of new technologies, corporate executives and technology management specialists need to be especially concerned about the possibility of hazards and disasters. Corporate executives whose firms may become vulnerable to liability lawsuits stemming from a technological disaster would do well to pay special attention to such cases as Bhopal, the DC-10 crashes, and the production and sale of asbestos and asbestos-related products. In addition, anyone responsible for introducing innovations in research and development needs to be aware of the pitfalls of extending known designs beyond their capacity. This is an issue that Petroski has addressed in his analysis of engineering failures, many of which resulted in technological disasters (Petroski, 1985; Petroski, 1994).

Government Agency Administrators

In addition to scientists and engineers working for such government agencies as the National Aeronautics and Space Administration (NASA), the Environmental Protection Agency (EPA), the Federal Aviation Administration (FAA), the Food and Drug Administration (FDA), and the Federal Trade Commission (FTC), top management of these agencies need to keep abreast of the rapidly growing number of technological failures, let alone disasters, that occur annually. Citizens turn to the EPA, FDA, FAA, and other government agencies for protection from the harmful consequences of technology. It behooves administrators of these agencies to be fully aware of the causes and effects of hazardous technologies.

Congress

The present Congress is the most scientifically sophisticated to date; 24 members of Congress have scientific or technology-based professional backgrounds. This, however, constitutes only 5 percent of federal lawmakers, who, more and more, are being asked to make decisions, craft legislation, and pass laws on technology-related issues ranging from nuclear power and biotechnology to the computerization of our critical infrastructure. These facts underscore the need for members of Congress and their staffs to be able to avail themselves of sound technical advice, something a previous Congress made more difficult for them by abolishing the Office of Technology Assessment (OTA) in 1995. Referred to as "Congress's own think tank," the OTA produced dozens of technology assessment reports to Congress. However, the OTA fell victim to budget cuts, and its demise left a gaping hole in the much-needed science and technology expertise available to Congress. Fortunately, some members of Congress are considering reviving the OTA in some form or another (Wakefield, 2001: 24).

The Academic Community

Several schools and departments of colleges and universities have, or should have, an abiding concern with problems of technological failure and disaster. Foremost among these are schools of engineering. In their efforts to educate the next generation of technologists with state-of-the-art knowledge and expertise, they expose their students to a range of problems and theories of engineering design. They also, in some measure, seek to sensitize their students to the dilemmas of engineering ethics. One such dilemma facing engineers is how to reconcile corporate demands for profit with requirements for safety. The latter falls squarely within the domain of the engineers' professional competencies.

Business schools, many of which offer courses in technology innovation and management, can address problems of technological disaster and inquire whether management's responses to such events promote or hamper organizational learning and adaptation.

Social science departments, such as history, economics, sociology, and political science, all focus on one or more facets of technology. Historians of science and technology consider the impact of culture on technological innovations; surely technological disasters become significant catalysts for innovation. Researchers who study science, technology, and society inquire into the positive and negative impacts technology has on society. Economists, concerned with problems of market failure and the role of government intervention to cope with them, would do well to calculate the annual costs of technological disasters that are borne by victims and consumers. Sociologists, who study the social impact of technological innovations, may wish to consider the impact of technological disasters on the family and the community. Finally, political scientists may wish to focus on the role of government regulatory agencies in preventing or, unwittingly, precipitating technological disasters.

The Citizenry

In the past 30 years, the citizenry has become painfully aware of the costs of technological disasters. Thousands have been killed and tens of thousands have been injured by human-made disasters. No longer can large corporations simply "write off" large-scale technological disasters as the "inevitable price to pay for progress." For many years, the citizenry has accepted this justification in the name of progress. Nowadays, people hold designers and decision makers accountable for technological harms and attempt to find out who was responsible for human-made disasters.

Technological disasters have forced large-scale high-tech industries to be held accountable to those who bear the greatest risks— employees and communities affected by such technologies—which may give rise to potential political mobilization at the grassroots level. Such protests emerged following the Bhopal tragedy. The antinuclear power plant movement that arose in the wake of Three Mile Island is also a case in point in which groups of citizens sponsored their own assessment of technologies that they feared may pose public harm, both directly or indirectly.

CONCLUSION

Some of the risks now confronting humanity are global in nature and cannot be mitigated by individual countries or even by regional communities. Unlike the risks of previous civilizations, technological disasters are "rooted in ecologically destructive industrialization and are global, pervasive, long-term, incalculable, and often unknown" (Shrivastava, 1995: 120). Radioactivity, chemical contamination, and the threat of biochemical and nuclear war are such risks. The risks stemming from the Chernobyl nuclear plant explosion, for example, spread across national boundaries, and the adverse consequences may affect future generations. Risks that are imperceptible but no less hazardous are global warming and ozone depletion. As the process of globalization continues—accompanied by the transfer of technologies from developed to developing economies—we may witness a rising incidence of technological disasters in developing countries.

As we have tried to show in this opening chapter, technological disasters arise as a result of human actions and judgments. Because of the human causes of technological disasters, it is essential to examine the larger political, economic, social, and historical contexts in which they occur. To further our understanding of technological failures and disasters, it is also crucial to compare and contrast them with the other major type of catastrophic event, natural disasters—to which we now turn.

References

Anonymous. (1988). "Absolutely liable: Bhopal disaster liability," *The Economist,* July 23; 308 (7560): 61.

Beck, M. and O' Brien, M. (2000). "Corporate criminal liability," *American Criminal Law Review* 37 (2): 361–389.

Beck, Ulrich. (1994). "The reinvention of politics: Towards a theory of reflexive modernization." In Ulrich, Beck, Anthony, Giddens, and Scott, Lash. (Eds.), *Reflexive modernization.* Stanford, CA: Stanford University Press: 1–55.

Beck, Ulrich, Giddens, Anthony, and Lash, Scott. (Eds.). (1994). *Reflexive modernization.* Stanford, CA: Stanford University Press.

Borel v. Fiberboard Paper Prods. Corp., 493 F.2d 1076, 5th Circuit, 1973.

Brodeur, Paul. (1985). *Outrageous misconduct: The asbestos industry on trial.* New York: Pantheon Books.

Cassels, Jamie. (2001). "Outlaws: Multinational corporations and catastrophic law," *Cumberland Law Review* 31: 311–335.

Chakravarty, Subrata. (1991). "Final act," *Forbes,* October 28; 148 (10): 10–12.

Cherniak, Martin. (1986). *The Hawk's Nest incident: America's worst industrial disaster.* New Haven, CT: Yale University Press.

Dowie, Mark. (1977). "Pinto madness," *Mother Jones,* September/October, 18–32.

Dumas, Lloyd. (1999). *Lethal arrogance: Human fallibility and dangerous technologies.* New York: St. Martin's Press.

Elliott, John. (1987). "Court likely to be long delayed," *Financial Times* (London), December 3: 5.

Esteban Ortiz et al., Petitioners v. Fiberboard Corporation. (1999). 527 U.S. 815; 119 S. Ct. 2295: sections I and VI.

Fielder, J., and Birsch, D. (Eds.). (1992). *The DC-10 case: A study in applied ethics, technology, and society.* Albany, NY: State University of New York Press.

Friedman, Lawrence. (2000). "In defense of corporate criminal liability," *Harvard Journal of Law and Public Policy* 23 (1): 833–858.

Fritz, Karen. (1980). "Pinto exploded like a big bomb, eyewitness says," *The Washington Post,* January 18: A9.

Gini, A. (1996). "Manville: The ethics of economic efficiency." In T. Donaldson and A. Gini (Eds.), *Case studies in business ethics* (4th ed.). Upper Saddle River, NJ: Prentice Hall: 58–66.

Jenny, Andrea. (1998). "The Hyatt Regency walkway collapse," available on the World Wide Web at: www.uoguelph.ca/~ajenny/webpage.htm

Johnson, Tom. (1986). "Lessons of Chernobyl: For pro-nukes, for anti-nukes," *The New Republic,* July 14; 195: 16–20.

Karliner, Joshua. (1994). "To Union Carbide, life is cheap: Bhopal—ten years later," *The Nation,* December 12; 259 (20): 726–730.

Kim, Queena. (2000). "Did broker of settlements unwittingly encourage more plaintiffs' suits?" *The Wall Street Journal,* December 12: A4, B6.

Koenig, Thomas, and Rustad, Michael. (1998). "Crimtorts as corporate just desserts," *University of Michigan Journal of Law and Reform* 31: 289–352.

Kramer, Larry. (1979a). "Ford to turn over Pinto papers," *The Washington Post,* December 26: D8.

Kramer, Larry. (1979b). "Recall asked of a million Ford autos," *The Washington Post,* April 28: A1.

Moss, Michael, and Appel, Adrianne. (2001). "Company's silence countered safety fears about asbestos," *The New York Times,* July 7: 1.

National Transportation Safety Board Report on the 1989 Sioux City Crash, reprinted in Fielder and Birsch (Eds.), 1992, *The DC-10 case.* Albany, NY: State University of New York Press: 247–265.

Pearl, Daniel. (2001). "An Indian city poisoned by Union Carbide forgets its past," *The New York Times,* February 12: A17–A18.

Perrow, C. (1984). *Normal accidents: Living with high-risk technologies.* New York: Basic Books.

Petroski, Henry. (1985) (1994). *To engineer is human: The role of failure in successful design.* New York: St. Martin's Press.

Petroski, Henry (1994). *Design paradigms: Case histories in error and judgment in engineering.* Cambridge, UK: Cambridge University Press.

Pletta, Dan. (1987). "'Uninvolved' professionals and technical disasters," *Journal of Professional Issues in Engineering* 113 (1): 23–31.

Rampton, S., and Stauber, J. (2001). *Trust us, we're experts!* New York: Penguin Putnam.

Robins, Mary. (2000). "Asbestos plaintiffs can sue again: Texas high court: Separate lawsuits allowed for injuries due to same exposure," *The National Law Journal*, December 25; 23 (18): A21.

Roddis, W. Kim. (1993). "Structural failures and engineering ethics," *Journal of Structural Engineering* (119) 5: 1539–1555.

Shrivastava, Paul. (1995). "Ecocentric management for a risk society," *Academy of Management Review* 20 (1): 118–138.

Shrivastava, Paul, Mitroff, Ian I., Miller, Danny, and Migliani, Anil. (1988). "Understanding industrial crises," *Journal of Management Studies*, July 25: 285–303.

Spanish Report–Dutch Comments, October 1978. Available on the World Wide Web at: http://aviation-safety.net/specials/tenerife/comments.htm

Troy, Tom. (2000). "A 'new wave' asbestos case gets a $9.8M verdict in N.Y.," *The National Law Journal*, July 10; 22 (46): A18.

Wakefield, Julie. (2001). "Flunking science," *The Washington Monthly*, January/February, 33(1/2): 23–27.

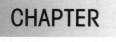

CHAPTER 2

Natural and Human-Made Disasters

"Acts of God or Acts of Man?"
—Wijkman & Timberlake

Natural disasters, or "Acts of God," according to conventional wisdom, are *unpredictable* and *unpreventable,* whereas human-made disasters, or "Acts of Man," are thought to be both *predictable* and *preventable.* Furthermore, "Acts of God" are perceived to be more catastrophic and destructive than "Acts of Man." These commonly accepted perceptions, along with our understanding of the relationship between "natural" and "human-made" disasters, deserve a critical reevaluation. Many disasters, although triggered by natural events such as floods and earthquakes, are increasingly human-made, and hence are both *predictable* and *preventable.* As for human-made disasters, they are, in principle, *predictable* and *preventable,* provided at least two conditions are fulfilled: first, the risks of new technologies are adequately assessed; second, individuals, organizations, and societies learn the proper lessons from past disasters.

According to statistics compiled by Swiss Reinsurance Company, a major firm that provides insurance against the losses of both types of disasters, the magnitude of natural and technological disasters has been increasing globally since 1970 (Swiss Reinsurance, 2001: 7). What preventive action can be taken to mitigate the losses of both types of disasters? We shall address these questions in this chapter by considering examples of natural and human-made disasters.

NATURAL DISASTERS

Natural, as opposed to human-made, disasters are triggered by climatic and geological variability. Such recurrent disasters exact an enormous human toll throughout the world. To achieve consistency in research on disasters, the Natural Hazard Research Group at the University of Colorado in 1969 formulated the following definition of *disaster:*

- more than $1 million in damage, or
- more than 100 people dead, or
- more than 100 people injured (Wijkman & Timberlake, 1984: 19)

One of the few longitudinal studies of major global natural disasters was initiated by Sheehan and Hewitt (1969) and updated by Thompson (1982). During the 35-year period from 1947 to 1981, the total loss of life recorded was 1,208,008, or an average of 34,574 deaths per year. Table 2–1 presents a total of 1,062 disasters ranked by type. The three most frequent types of disasters were floods (32 percent), hurricanes (20 percent), and earthquakes (15 percent). These three types accounted for two-thirds of all the disasters for the 35-year period. We will now consider each of these types of disasters in turn.

Floods

Of all natural disasters, floods are the most serious causes of both loss of life and damage to property. There are two types of floods: those caused by seawater inundating the land because of hurricanes, tsunamis, etc., and those caused by the overflow of inland rivers and lakes, sometimes called *riverine flooding* (Palm, 1990). The majority of floods occur unpredictably and abruptly after intense rain lasting anywhere from half an hour to several days.

During the 20th Century all but three floods claiming 10,000 or more lives occurred in Bangladesh. About 80 percent of this country is a broad, flat plain slightly higher than sea level, divided into thousands of islands at the banks of the Ganges and Brahma-

Table 2–1
Global disasters by type, 1947–81.

Agent	Number	Percentage
Floods	343	32
Hurricanes (typhoons, tropical cyclones)	211	20
Earthquakes	161	15
Tornadoes	127	12
Snowstorms	40	4
Thunderstorms	36	3
Landslides	29	3
Rainstorms	29	3
Heatwaves	22	2
Volcanoes	18	2
Coldwaves	17	2
Avalanches	12	1
Tsunamis	10	1
Fog	3	-
Frost	2	-
Sand and dust storms	2	-
	1,062	100

Source: Smith, Keith. (1992). *Environmental Hazards* © 1992. Reproduced by permission of Routledge, Inc., part of The Taylor & Francis Group.

putra rivers. In the last 35 years, Bangladesh has suffered at least seven major disasters (Zebrowski, 1997):

1963 22,000 deaths

1964 17,000 deaths

1965 30,000 deaths

1965 10,000 deaths

1970 300,000 deaths

1991 10,000 deaths

1991 131,000 deaths

However, as Wijkman and Timberlake (1984) report, disastrous inundation of waters on both the Ganges' and the Brahmaputras' floodplains is the direct result of massive deforestation—an "Act of Man."

Clearly, floods in Bangladesh are neither unanticipated nor unpreventable.[1] Unlike the Netherlands, this desperately poor, developing country does not have the resources to build dikes, sea gates, or dams. If such capital-intensive engineering projects were undertaken with the help of World Bank loans, flood-warning systems would save many lives, assuming Bangladesh had the capability not only to disseminate advance warning but also to evacuate potential victims.

Another example of the failure to cope with the ravages and devastation of floods is the present situation in China along the Yangtze River, where severe flooding occurs and causes devastating destruction (Eckholm, 1998: A1, A6).

The Chinese press has recognized the harmful effects of excessive deforestation and the human invasion of wetlands, as well as the negligent disregard for the upkeep of important dikes on the delicate ecosystem of the Yangtze floodplain. These failures of Chinese water-management technology result from a bias toward building dams and reservoirs to fight floods, despite their ineffectiveness. In recent years, economic incentives are guiding the choice of technology toward creating dams and reservoirs to generate more capital—in the form of electricity and municipal water supplies—rather than building an ecologically-safer system of dikes and locks.

The current controversial project to construct the Three Gorges Dam on the Yangtze River is a case in point (Ford, 2000: 7). The world's largest hydroelectric dam is being built in the scenic Three Gorges valley of the Yangtze River. By 2009, the reservoir—a serpentine 400-mile lake—will be filled. The dam will have 26 hydroelectric generators operating to distribute electricity as well as two massive locks and ship lifts. "The lake created by the dam . . . will submerge approximately 50,000 acres of land, 19 cities, 150 towns, and 4,500 villages. The relocation process for the 1.2 million people who will be forced out of their homes has already begun" (Kuo, 1998: 28).

One of the expected benefits of the project on the Yangtze River is the control of periodic flooding, which has killed tens of thousands of people in the 20th Century. On the other hand, one risk exists because the dam will be located over a seismically active fault line.

The weight of the water could trigger an earthquake, which would devastate the entire structure. Such a calamity would dwarf the 1975 collapse of 62 iron dams in the Henna province, which killed over 85,000 people. The floods also rendered millions homeless and triggered famine and disease. If the Three Gorges were to meet a similar fate, the result would be 40 times worse. (Kuo, 1998: 30)

Both the Yangtze and Ganges flooding problems indicate the linkage between human-made and natural disasters. If both the Ganges and the Yangtze continue to flood because of political and economic choices—that is, as the result of human choices that result in disaster—then the distinction between natural versus human-made disasters becomes, in at least some cases, fuzzy. It suggests to us that some disasters classified as natural can also be understood as human-made or technological disasters. This point has clearly been demonstrated by the Yangtze River flooding disasters. "... [B]ad policies and official neglect have worsened ... disastrous flooding in China, and the [Chinese] Government has made the unusual admission that its land use mistakes are partly to blame and announced sweeping policy changes" (Eckholm, 1998: A1).

Hurricanes

A hurricane is defined as a storm with wind speeds of 74 to 200 mph. The term *hurricane* comes from the Carib Indian *urican,* meaning "big wind," and is restricted to storms in the Atlantic Ocean. In the Pacific, the same phenomenon is known as a *typhoon* (from the Chinese *taifeng*). In the Indian Ocean and around Australia, such a storm is called a *tropical cyclone* (Robinson, 1993: 119–120). A tropical storm, on the other hand, is defined as a storm with wind speeds of more than 39 mph but less than 74 mph. In the United States, Tropical Storm Agnes resulted in $3.5 billion in property damages; because of an advanced warning system it was possible to evacuate hundreds of thousands of people, with no more than 118 lives lost. The average life of a hurricane is nine days, and the typical wind speed is 105 mph, although wind speeds of 150 to 175 mph were clocked in Hurricanes Camille in 1968 and Gilbert in 1988 (Palm, 1990: 5).

The most lethal hurricane in U.S. history hit Galveston, Texas, in 1900. It took 6,000 lives and destroyed one-half of the town. In 1992, when Hurricane Andrew hit Florida and Louisiana with winds of up to 160 mph, it caused up to $20 billion in property damage but only 20 fatalities. By this time, science and technology enabled the National Weather Service meteorologists, with the aid of a new Doppler radar system, to analyze distant wind patterns and "predict the storm's path . . . within thirty miles" (Tenner, 1996: 74). One of the worst hurricanes to date hit Central America during the fall of 1998. Hurricane Mitch caused the deaths of more than 19,000 people, many in Nicaragua. The cost of Hurricane Mitch has risen to more than $25 billion (Broder, 1999: A10).

After several decades of advances in weather satellites and computer forecasting models, forecasters have reduced errors in predicting the paths of hurricanes by only 14 percent (Robinson, 1993). However, coastal states exposed to hurricanes can take protective measures to limit property damages. Flood insurance, however costly, is also a rational protective measure (Kunreuther and Roth, 1998). In short, hurricanes are unanticipated events, but with the aid of meteorology and advances in communication they have become anticipatable. With evacuation planning and a rapid warning system many lives need not be lost.

Earthquakes

Earthquakes affect vast portions of the world, placing hundreds of millions of people at risk. An estimated 1.52 million deaths from earthquakes were officially reported from 1900 to the end of 1990. "Some of the largest recorded earthquakes have taken place in China, where an earthquake in 1556 in Shensi claimed 830,000 lives and another in 1976 claimed approximately 600,000 lives" (Palm, 1990: 6). Other countries that have suffered particularly destructive earthquakes in the past 50 years include Morocco, Chile, Peru, Nicaragua, Guatemala, Indonesia, The Philippines, Turkey, Iran, Algeria, Italy, and Japan. Earthquakes "take an average of ten thousand lives and cost $400 million a year worldwide" (Tenner, 1996: 77).

More than 95 percent of all deaths in earthquakes result from the collapse of buildings (Blaikie, Cannon, Davis, and Wisner, 1994: 169). The earthquake in Armenia in 1988, in which 25,000 people lost their lives, confirmed the long-standing seismological truth that "earthquakes don't kill people, buildings do" (Robinson, 1993: 60). The magnitude of the first shock was 6.9 on the Richter scale, but the earthquake itself and swarms of aftershocks leveled more than half the buildings in Leninakan, the result of both design deficiencies and faulty construction practices.

In 1999 there were major earthquakes in Taiwan, Turkey, and Greece, and in 2001 the death toll reached 20,000 in a major earthquake in India. A geologist seeking lessons from these disasters points to shoddy, illegal construction, buildings constructed on unstable ground or without adequate seismic resistance (Sieh, 1999: A29).

A Turkish journalist expressed outrage that his country was "unprepared for the expected," considering that Turkey lies on the great Anatolian fault line:

> Turkey's enforcement of building regulations is lax, and small-time contractors often economize by using less steel and iron and putting more sand in the concrete. The buildings that collapsed in a matter of seconds were the low-income housing projects and budget-priced summer vacation apartments. The ones that stand tall next to the wreckage are a testimony that tighter application of the building regulations could have saved lives. . . . For the killer was not just nature—it was contractors' greed and lawlessness and the government's ineptitude in preparing for, and responding to, a predictable natural disaster. (Aydintasbas, 1999: A10)

In many Indian cities there are no set standards for constructing a building higher than two stories. The brunt of destruction in Ahmedabad was borne by modern high-rise buildings. " 'Anybody can become an architect, anybody can become a building contractor—they don't need a license. And the contractors are not following the building codes,' says Vinod K. Sharma, academic head of the National Center for Disaster Management, a government agency. 'The contractors are just constructing graveyards for the people' " (Baldauf, 2001: 1).

In sharp contrast to the massive destruction and high death toll of earthquakes in Taiwan, Turkey, Greece, India, and Japan, the earthquake in Seattle, Washington, in March 2001, although 6.8 on the Richter scale, had relatively little impact. Only one person died and few buildings had structural damage. By contrast, the earthquake in Kobe, Japan, left more than 6,000 people dead and "turned the city inside out." Seattle's relative good fortune is due primarily to improved building codes and increased preparedness (Paton, 2001: 3).

An earthquake has been traditionally perceived as an "Act of God." In Europe, earthquakes were believed to be God's punishment for sin—at least according to the church. When a gigantic earthquake leveled Lisbon in November 1755, the Inquisition responded by roasting the survivors in the fires of the *auto-da-fé*. The pessimist Voltaire published first a poem and then, in 1759, his famous story *Candide*, attacking both the Pope, for attributing the earthquake to man's lack of faith in God, as well as the German philosopher and mathematician Leibnitz, who optimistically believed that God must have sent the earthquake as part of His plan for earth (Robinson, 1993: 49).

Although science and technology cannot currently accurately predict the occurrence of earthquakes, as in the case of floods, they *can* mitigate the loss of lives and property by means of protection and preparation. Seismology, soil mechanics, and engineering can provide various forms of protection by developing safer construction standards. Seismologists believe that warning systems, maps of fault lines, better building designs, careful zoning, stricter building codes, and, above all, enforcement of such codes can continue to reduce the risks posed by earthquakes (Kunreuther, 1998: 218–221). Kunreuther wisely advises us of the need for a program of insurance as a means for reducing disaster losses.

Seismologists also agree that there is a high probability that a major earthquake will strike northern and southern California by the year 2020 (Tenner, 1996: 77). Notwithstanding this gloomy assessment, the rate of migration to California has not noticeably diminished. Instead of being risk-averse, evidently many people are willing to live with risk while enjoying the benefits of living in California. In short, while earthquakes are still unanticipated events, they are nonetheless anticipatable.

The previous examples have shown that even though the triggering events of "natural" disasters such as floods, hurricanes, and earthquakes may be unanticipated "Acts of God," many of the disastrous results of these natural phenomena are anticipatable, and hence preventable, since "Acts of Man" either play a prominent role in their cause or have the potential to result in their successful mitigation. Global warming and the depletion of the ozone layer are telling examples of predictable and preventable natural disasters that are in the process of being created by human choices (Anonymous [a], 2000; Houlder, 2000; Kerkin, 2000).

For natural disasters to be anticipatable, many advances in science and technology may be required. For natural disasters to be preventable, a country has to allocate the necessary resources to use available funds of knowledge and to design appropriate public policies (Kunreuther and Roth, 1998).

HUMAN-MADE DISASTERS

Human-made disasters, or technological disasters, are legion: the collapse of a bridge or a dam, the escape of deadly radiation from a nuclear power plant or a toxin from a chemical plant, the crash of a train or an aircraft, oil and gas explosions. Such failures of technology are not as rare as we would like to think. Nor are technological disasters new phenomena. Technological innovation, a continuing feature of human history, has been accompanied by technological disasters. "Technological hazards are primarily seen as major 'man-made' accidents; that is, the initiating event in a disaster arises from a human, rather than a geophysical agency" (Smith, 1992: 271).

Smith has compiled a list of technological disasters that occurred before the end of World War I to underscore the point that technological disasters are not confined to the late 20th Century. Table 2–2 includes early examples of technological failures involving fires, the collapse of buildings and bridges, rail and sea accidents, and industrial explosions. Smith (1992: 26) notes that "the pace of change associated with the Industrial Revolution led to a marked increase in risk during the nineteenth century." Several industrial developments in the 20th Century, in turn, contributed to an increase in the frequency of technological disasters. For example, the rise of the nuclear industry and explorations for oil and gas deposits have been fraught with technological hazards and consequent disasters. Such accidents have been increasing in frequency in recent years. As Figure 2–1 illustrates, industrial accidents causing more than 50 deaths have been increasing in frequency throughout the 20th Century.

Table 2-2
Some early examples of technological accidents.

Structures (fire)

1666	Fire of London, England	13,200 houses burned
1772	Zaragosa theatre, Spain	27 dead
1863	Santiago church, Chile	2,000 dead
1871	Chicago fire, USA	250–300 dead, 18,000 houses burned
1881	Vienna theatre, Austria	850 dead

Structures (collapse)

Dam

1802	Puentes, Spain	608 dead
1864	Dale Dyke, England	250 dead
1889	South Fork, USA	2,000 dead

Building

1885	Palais de Justice, Theirs, France	30 dead

Bridge

1879	Tay Bridge, Scotland	75 dead

Public Transport

Air

1785	Hot air balloon, France	2 dead
1913	German airship LZ-18	28 dead

Sea

1912	<u>Titanic</u> liner, Atlantic Ocean	1,500 dead

Rail

1842	Versailles to Paris	60 dead
1903	Paris Metro, France	84 dead
1915	Quintinshill Junction, Scotland	227 dead

Industry

1769	San Nazzarro, Italy (gunpowder explosion)	3,000 dead
1858	London docks, England (boiler explosion)	2,000 dead
1906	Courrieres, France (coal mine explosion)	1,099 dead
1907	Pittsburgh steelworks, USA (explosion)	59 dead
1917	Halifax harbor, Canada (cargo explosion)	1,200 dead

Source: Smith, Keith. (1992). *Environmental Hazards* © 1992. Reproduced by permission of Routledge, Inc., part of The Taylor & Francis Group.

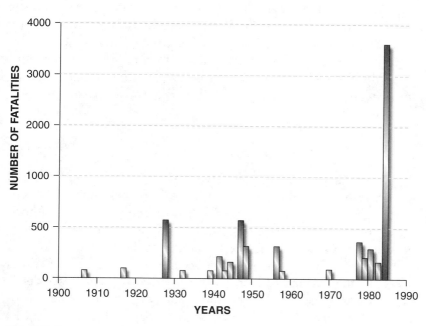

Figure 2–1

Annual numbers of deaths from industrial accidents causing more than 50 fatalities in the period 1900–1984.

Source: Smith, Keith. (1992). *Environmental Hazards* © 1992. Reproduced by permission of Routledge, Inc., part of The Taylor & Francis Group.

The year 1984 marked a watershed in the emergence of technological disasters. In that year three major failures caused approximately 15,000 deaths:

1. Cubatao, Brazil, 25 February: petroleum spillage and fire in a shanty town built illegally on the industrial company's land —500 deaths.

2. Mexico City, Mexico, 19 November: multiple explosions of liquefied petroleum gas in an industrial site in a heavily populated poor area—at least 452 deaths, 31,000 homeless, 300,000 evacuated.

3. Bhopal, India, 2–3 December: release of toxic gas from an urban factory—more than 14,000 deaths, 34,000 eye defects, 200,000 people voluntarily migrated (Smith, 1992: 278).

Two technological developments following the conclusion of World War II have left a continuing dangerous technological legacy: the nuclear arms race between the United States and the former Soviet Union resulted in the production of two gigantic nuclear arsenals, and the emergence of the nuclear power industry. As a result, for many years into the 21st Century, and possibly beyond, humanity will be confronted with two enormous technological hazards: nuclear reactor accidents and nuclear weapons accidents.

Nuclear Reactor Accidents

When nuclear power plants were being built in the 1950s and 1960s, technological optimism with regard to their safety and economic efficiency was rampant. The Rasmussen report, the authoritative nuclear reactor safety study commissioned by the U.S. government in 1975, concluded overall that the chances of anyone being killed as a result of a reactor accident are ". . . 100,000 times less than the chances of being killed in a motor vehicle accident" (Webb, 1976: 96). Nuclear power, argued the experts, would provide a clean, inexpensive source of electricity and free the United States from dependency on foreign oil. By 1979, there were approximately 80 commercial nuclear reactors in operation in the United States with another 60 either planned or under construction.

There was also the confident prediction that by the year 2000 the United States would have more than 500 nuclear plants in operation. This optimistic assessment encouraged 109 cities in the United States to become hosts to nuclear power plants. Approximately 40 countries, especially those lacking in oil resources such as Britain, Japan, Canada, and France, also decided to build nuclear power plants. As of December 1998, 437 nuclear power plants were operating worldwide (Dumas, 1999: 19–21).

This optimistic outlook was shattered on March 28, 1979, when the Three Mile Island (TMI) reactor accident occurred—the worst accident, as of that date, in the commercial nuclear power industry. Due primarily to human error, the Unit 2 reactor was allowed to lose cooling water, causing part of the core to melt and thus releasing millions of curies of radioactivity into the containment building. Some radioactive gases escaped from the containment building into the atmosphere; "the dose to the surrounding population was minimal" (Miller, 1994: 222).

Numerous evaluation studies have examined the carcinogenic effects of the TMI accident. One recent epidemiological study illustrates the long-lasting impact of the accident (Wing, Richardson, Armstrong, and Crawford-Brown, 1997: 52–57). A 10-mile area around TMI was divided into 69 study tracts, each assigned radiation dose estimates. Data on cancers from 1975 to 1985 was ascertained from hospital records and assigned to the study tracts. The findings support the hypothesis that radiation doses are related to increased cancer incidence, rates of leukemia, and lung cancer around TMI.

The TMI accident pales in significance when compared with the Chernobyl disaster. A violent explosion of a Russian graphite-moderated reactor released massive levels of fission products during the nine days the reactor burned. Radioactive fallout spread mainly north into Belarus, west into Ukraine, and into adjacent parts of Russia. In due time, the fallout spread into Eastern and Western Europe, eventually spreading over the entire Northern Hemisphere. The principal medical consequence of Chernobyl to date is a significant increase in childhood thyroid cancer (Bleuer, Averkin, and Abelin, 1997; Goldman, 1997; Lomat *et al.*, 1997).

> The annual incidence of thyroid cancer in children in Ukraine from 1981 to 1985 was 0.4–0.5 per million. It increased significantly to 2.2 cases per million in 1990, 1.8 in 1991, 3.9 in 1992, 3.5 in 1993, and at least 3.1 in 1994. This increase occurred mainly in northern Ukraine, where the highest radio iodine contaminations were found and more than 60% of childhood thyroid cancer cases were registered. (International Atomic Energy Agency, 1996: D46)

The Health Ministry of Ukraine acknowledged that children were the greatest sufferers of the Chernobyl disaster, with illnesses from radiation five times the level recorded before the explosion. "Chernobyl will be with us forever," said Health Minister Andriy Serdyuk. ". . . Our children will continue to be polluted by radiation" (Anonymous [b], 1998: A5).

Another medical consequence is an increase in cataracts in children from radiation exposure (Goldman, 1997: 1387). Yet another consequence of the accident is psychological stress due to "communication, miscommunication, and lack of communication" (Anonymous [b], 1998: A5). Exaggerated reports in the popular press precipitated fear and were accompanied by official pronouncements attempting to minimize the risk of radiation.

This contributed to a widespread phobia of radiation that led people to believe that many adverse health conditions stem from hidden radiation exposures (Goldman, 1997: 1387).

In addition to the two major nuclear reactor disasters, numerous minor accidents involving "radiological incidents" have occurred in various countries where more than 437 nuclear power plants are operating. These involve technological failures that were identified and corrected in a timely manner. The International Atomic Energy Agency, which devotes considerable attention to problems of nuclear reactor safety in its annual reports, does not compile statistics on nuclear reactor accidents throughout the world. As a consequence, we do not know the number of people in various countries exposed to the hazards of "radiological incidents."

In his survey of the nuclear power industry, Dresser (1993) enumerates approximately 200 instances in which nuclear power plants have experienced significant accidents, breakdowns, or technical failures. India's experience with nuclear power, to cite but one country's experience, is indeed startling. During 1992–93, there were 147 accidents: one every two and a half days. According to Praful Bidwai,

> India's 27-year long experience with atomic electricity generation is a story of accidents, flagrant violations of safety rules, avoidable exposure of workers and the public to radiation and toxic substances. . . . (Mian, 1995: 47)

In retrospect, the technological forecasts about nuclear power were erroneous in several respects. First, scientists and engineers underestimated the risks of radioactive emissions. Second, they overestimated how competitive nuclear energy is relative to electricity from other sources. Third, they virtually ignored the economic and technical problems concerning safe disposal of high-level radioactive waste produced by a nuclear reactor. Hudson, an official of an environmental education organization concerned with the problem of storing nuclear waste safely for hundreds of years in the future, asks:

> . . . how are we going to come up with a system of managing something for a minimum of 500 human generations? . . . The pyramids are only 150 human generations old, and we've got to make something to last at least 500. (Goldberg, 1998: A14)

Nuclear reactor accidents and problems associated with nuclear power were largely unanticipated though they were indeed anticipatable. Opponents of nuclear power would contend that they had indeed anticipated the safety problems associated with employing this source of energy, but that their arguments were dismissed (McLeod, 1995).

Nuclear Weapons Accidents

During the height of the Cold War, the United States and USSR each had nuclear arsenals consisting of 25,000 to 35,000 nuclear weapons, more than enough to obliterate the enemy as well as the entire human race. Military and political leaders on both sides subscribed to a deterrence theory, which held that since each superpower had more than enough nuclear weapons to survive a first strike by the enemy and mount a lethal retaliatory strike, this would prevent a deliberate nuclear war from breaking out. This strategy was known as "mutually assured destruction" (or MAD) and, circumstantially at least, it did prevent a nuclear confrontation between the superpowers during the 45 years of the Cold War. A perplexing problem confronted military and political leaders throughout the Cold War, namely, the potential failure of deterrence because of "broken arrows," false alarms or unauthorized use of nuclear weapons.

Broken Arrows

Technological failures involving nuclear weapons have occurred with surprising frequency since 1950. The Department of Defense classifies such accidents according to the degree of danger. The least dangerous is a "Dull Sword;" a significant incident is a "Bent Spear;" a substantial accident is a "Broken Arrow;" and the worst possible category is a "Nucflash," an accident that can be mistaken for a nuclear attack by the enemy (Cunningham and Fitzpatrick, 1983).

"Fail-safe" mechanisms are built into nuclear weapons to prevent accidental detonations. Fortunately, to date, there has never been an accidental detonation—or at least, none have ever been publicly reported. The Department of Defense does not rule out the possibility, but it insists the probability is very low. When a "Broken Arrow" does occur, the damage involves radioactive contamination from plutonium leakage or from other fissile material.

A substantial number of "Broken Arrows" have been acknowledged by the Department of Defense, including the following:

- In 1956, an unarmed B-47 crashed into a nuclear weapons stockpile at the Royal Air Force Lakenheath Base in East Anglia, Britain.
- In 1957, a B-47 accidentally dropped a 1-megaton bomb near Mars Bluff, South Carolina; the conventional explosive detonated, obliterating a farmhouse and killing several people.
- In 1968, a B-52 crashed and burned while landing at Thule Air Force Base, Greenland. All four nuclear bombs were destroyed in the fire. (Following this accident, B-52s with nuclear bombs were no longer kept on "airborne alert"). (Barash, 1987: 172)

Although not acknowledged by the Soviet government, the following Soviet Broken Arrows have been reported:

- In 1970, one nuclear-armed submarine collided with an Italian cruise liner, and another exploded and sank off the British coast.
- In 1981, a Soviet submarine, apparently on a spying mission and known to be armed with nuclear-tipped torpedoes, was grounded near the top-secret Swedish naval facility of Karlskrona.
- In 1984, a Soviet nuclear-armed submarine collided with the U.S. aircraft carrier Kitty Hawk in the Sea of Japan.
- Also in 1984, an errant submarine-launched cruise missile flew over Norwegian airspace and later crashed in Finland. (Barash, 1987: 172)

All of these Broken Arrows were unanticipated, but all were anticipatable.

False Alarms

The computerized warning system of North American Air Defense Command (NORAD) has generated a surprisingly large number of false alarms. During an 12-month period from June 1979 to June 1980, 3,804 such alarms were generated (Dietrich,

1984). Most of the false alarms were readily identified as false warnings caused by missile test firings, forest fires, or even a flock of geese (Dietrich, 1984: 16–17).

An example of a false alarm that occurred long after the Cold War ended will make clear the potential threat of these types of accidents. On January 25, 1995, Russian military technicians noticed a troubling blip on their radar screens. Realizing that a single missile from a U.S. nuclear submarine off the coast of Norway could launch eight nuclear bombs on Moscow in 15 minutes, the radar operators immediately alerted their supervisors. The message was transmitted to then Russian President Boris Yeltsin, who, in possession of the "nuclear briefcase," could order a nuclear missile response. For several anxious minutes, the radar crews tracked the trajectory of the mysterious object. After about eight minutes—just a few minutes short of the deadline for responding to an impending nuclear attack—Russian military officers ascertained that the rocket was heading out to sea and posed no threat to Russia. The mysterious "rocket," it turned out, was a U.S. scientific astronomical probe. Norwegians were informed of the probe weeks in advance. They, in turn, informed Russian authorities but word of the experiment failed to reach the ears of the proper Russian authorities (Blair, Feiveson, and Von Hippel, 1997).

False alarms are especially dangerous given the current "launch on warning" strategy, namely, the readiness of U.S. and Russian forces to fire nuclear missiles in retaliation when a radar signal indicates that an enemy attack is underway. Reliance on "launch on warning" greatly increases the risk of accidental nuclear war. Blair, Feiverson, and Von Hippel (1997: 79–81) have presented a persuasive argument for "de-alerting" our nuclear missiles now that the Cold War is over. Instead of keeping our nuclear warheads on hair-trigger alert, these authors recommend separating the nuclear warheads from the missiles to provide additional time to verify radar warnings and thus decrease the chances of an accidental nuclear war.

False alarms are not unanticipated events; they are indeed anticipated and anticipatable.

Unauthorized Use

In the event of a nuclear confrontation between the United States and Russia, very few individuals would have the authority to order the use of nuclear weapons. The exact number in both countries is kept secret, although it is acknowledged that both the President of the United States and the President of Russia have such authority. Following an attack on the United States or Russia, authority would be delegated to many different levels in the military hierarchy. It is also assumed that in the event of political and military "decapitation," there are procedures in both countries for delegating launch authority.

Although few individuals have the authority to launch nuclear weapons, numerous people are involved in the operations of various nuclear weapons systems. To reduce the chances of unauthorized use of nuclear weapons, several technological innovations have been introduced. In the case of intercontinental ballistic missiles (ICBMs), permissive action links (PALs) have been installed to ensure that the weapons cannot be fired until the missile silo personnel receive electronically coded authorization from a high level in the command structure (Barash, 1987: 173).

Strategic bomber crews operate under "positive control," meaning that after reaching a designated flight radius they are required to return to base unless they receive a specific "go-code" authorizing them to continue toward the target. Nuclear submarines, the most invulnerable of nuclear weapons systems, are also extremely difficult to communicate with by high levels of authority. A system known as ELF—extremely low-frequency electromagnetic waves—can provide some capability for transmitting messages to submarines. Because of limited capability to communicate with the commanders of strategic missile submarines, there is a danger that collisions at sea—not as rare an event as one might suppose—may be falsely interpreted as preemptive attacks. To reduce the likelihood that a commander of a U.S. or Russian submarine will fire a nuclear weapon in response to a false alarm, further technological innovations are required (Barash, 1987: 173).

The dangers of unauthorized use of nuclear weapons have been anticipated and are anticipatable.

Given that the United States and Russia possess 95 percent of the world's nuclear weapons, it is obviously urgent to reduce the threat of nuclear war between these two countries to zero. "Enough nuclear fire-power exists in each arsenal to kill tens of millions in each others' societies. . . . Any nuclear event of this kind would likely result in the worst catastrophe in human history. . . . Will we continue to sleep-walk into a nuclear catastrophe? And

	Unpredictable and Unanticipated	Unpreventable and Unanticipated	Preventable and Anticipated	Prevented and Anticipated
Natural Disasters	X	X		
Human-Made Disasters			X	X

Figure 2–2
Assumed distinctions between natural and human-made disasters.

will only such a catastrophe serve to move us toward a nuclear-weapons-free world? (McNamara and Blight, 2001: 169, 215).

The threat of nuclear war is clearly a potential human-made disaster that is both predictable and preventable.

We began our discussion in this chapter with an implicit hypothesis that natural and human-made disasters are distinct, as indicated in Figure 2–2.

Upon further consideration—and in the absence of a systematic body of comparative data, this hypothesis may not be confirmable. Natural disasters, owing to advances in science and technology, are becoming increasingly anticipatable, even if we are a long way from being able to predict such events.

COMPARISON OF NATURAL AND HUMAN-MADE DISASTERS

What do natural and human-made disasters have in common? First, they involve low-probability, high-consequence events (Weinberg, 1986: 10–14). Second, both types of hazards are commonly regarded as potential sources of grave physical, social, and psychological harm, which people and communities attempt to deal with through collective efforts and public policies. Comparing natural and technological disasters may help us develop effective mechanisms for dealing with the risks and dangers of technological disasters.

Especially informative in this regard are the statistical compilations by the Swiss Reinsurance Company on natural and human-made disasters. Figure 2–3 presents two rising curves of the frequency of both types of disasters from 1970 to 2000. The finding that the curve representing human-made disasters has been

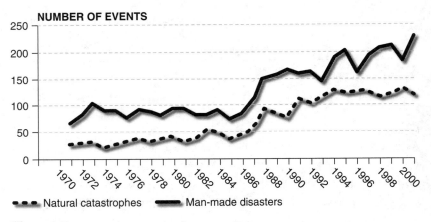

Figure 2–3

Number of natural catastrophes and man-made disasters 1970–2000.
Source: Swiss Reinsurance Company (2001). *Natural catastrophes and man-made disasters in 2000*, Sigma No. 2/2001: Zurich.

rising at a faster rate during the last decade of the 20th century than the curve for natural disasters is indeed noteworthy. The authors of a report by the Swiss Reinsurance Company present a clarifying observation about these two types of disasters:

> Since 1989 the number of disasters has averaged the high level of 120 natural catastrophes and 191 man-made losses per year. A large proportion of these are accounted for by the approximately 57 road-rail accidents and 39 shipping accidents which occur each year, followed by 35 major fires and 30 aviation and aerospace disasters. (Swiss Reinsurance Co., 2001: 8)

A related statistical comparison compiled by this reinsurance company pertains to the amount of insured losses in billions of dollars for 1970 to 2000. Table 2–3 summarizes the comparative losses for the three decades. What is striking about the finding in Table 2–3 is the marked trend for insurance losses to increase in both categories of disasters from 1970 to 2000, although the losses for natural disasters are clearly increasing at a faster rate, the staggering losses in 1990–2000 are due to devastating floods and earthquakes. Companies have traditionally insured against "Acts of God," whereas, they are evidently still reluctant to adequately insure against "Acts of Man." This may account for the relatively low rate of increase in insured losses during 1970–2000 for human-made disasters as compared to natural disasters.

Table 2–3
Insured losses (in billions of dollars) for natural and human-made disasters (adjusted for inflation).

	Natural Disasters	Human-Made Disasters
1970–1979	22.050	24.933
1980–1989	45.765	35.160
1990–2000	173.189	45.638

Source: Compilation of data from Swiss Reinsurance Company (2001). *Natural catastrophes and man-made disasters in 2000*, Figure 3, p. 9. Sigma No. 2/2001: Zurich.

The statistics in Table 2–3 are consistent with another report which states: "Disasters have cost the United States more than $500 billion in the past 20 years, and the annual toll is continuing to rise due to an increasingly complex society and more people moving to disaster-prone areas" (Anonymous [c], 1991: 303). This report also reiterates the point that human decisions and actions are often the cause of "natural disasters." "Human beings—not nature—are the cause of disaster losses, which stem from choices about where and how human development will proceed" (Anonymous [c], 1991: 303).

To gain a deeper understanding of how and why technology fails, it may be helpful to contrast technological disasters with natural disasters to identify characteristics unique to technological disasters. Studies comparing natural and technological disasters strongly suggest that technological disasters have longer-term effects, can affect people and communities far beyond the point of impact, and pose different and graver types of threat and harm than do natural disasters (Baum, Fleming, and Davidson, 1986; Freudenburg and Pastor, 1992; Kasperson and Pijawka, 1985). Two key distinctions are often made when comparing natural and technological disasters: one rests on the time dimension of the disaster; the other has to do with the perceived cause of the disaster (Freudenburg and Pastor: 1992: 391–393). Whereas natural disasters advance in reasonably clear stages from order, to chaos, to reconstitution of order, quickly ending in social system reequilibration, technological disasters are far less linear, and their boundaries and effects are much more difficult to measure (Couch and Kroll-Smith, 1991).

Selected Disaster Categories	Natural Disasters	Technological Disasters
Signification of the Disaster	Clear, unambiguous identification of the social crisis caused by the disaster	Disagreements over how to interpret the sudden crisis and its effects
Immediate Reaction to the Disaster	Uniform and concentrated effort to contain and control the external threat	Ambiguity of individual and organizational roles that hampers quick responses
Responsibility for the Disaster	Little community division over the question of blame	Much community division over assigning responsibility and apportioning justice for wrongdoing
Psychological Effects of the Disaster	Acute stress, anxiety, and trauma; relatively short-lived hysteria and confusion	Chronic anxiety, stress, and trauma; longer periods of hysteria and confusion
Overall Response to the Disaster	Little individual or public outcry	Often results in individual and public expressions of anger and outrage

Figure 2–4
Comparison of natural and technological disasters.
Sources: Data compiled from Couch, S. R. and Kroll-Smith, J. S. (1985). "The chronic technical disaster: Toward a social scientific perspective," *Social Science Quarterly* 66 (3) 564–575; Baum, A., Fleming, R., and Davidson, L. (1986). "Natural disaster and technological catastrophe," *Environment and Behavior* 15 (3):333–354.

Another key distinction is the perceived cause of the problem. Natural disasters are perceived to be "Acts of God" that are totally uncontrollable and unpreventable. Technological disasters, on the other hand, are "Acts of Man," or human-made disasters. The fact that humans are perceived as the cause of technological disasters creates a set of problems very different from those created by natural disasters. Researchers have identified some basic characteristics that make technological disasters different from natural disasters. These unique characteristics include how the disaster is individually and collectively interpreted and given social meaning; how the community responds to the threat; how the community seeks accountability and blame for the disaster; what psychological effects the disaster imposes; and, finally, how society responds to the disaster. These comparative characteristics are summarized in Figure 2–4.

It is safe to say that sudden, powerful, and devastating catastrophes such as the Chernobyl nuclear disaster and the Bhopal poison gas release shock and confound everyone affected by such events. Those affected include victims, local communities, corporate organizations in which the events transpire, and various government agencies. When such events happen, all those affected seek to make sense of the seemingly senseless carnage and destruction. How did it happen? How do we make sense of what happened? Answers to such questions are no more forthcoming from the corporations that manage the technology or from government agencies that regulate the technology than from the victims whose lives are shattered by such events.

The collective meaning and cultural interpretation of technological disasters are reflected in the ambiguity over what actions individuals and organizations should take in responding to the disaster. Because such events are totally unexpected, individuals, corporations, and government agencies are often caught totally unprepared. Who should respond, and why? Who should take responsibility for the technological disaster? In the case of the Ford Pinto, for example, management decided not to respond at all, after performing a cost-benefit analysis of the potential hazards of rear-end collisions.

Such uncertainty and confusion create a need to find out which individuals or organizations are responsible for the disaster. This may lead to community conflicts, especially when the disaster is perceived to emanate from human error or misjudgment. The more human ignorance, manipulation, or greed are thought to be the cause of the disaster, the more conflicts may arise over the question of assigning responsibility, blame, and accountability.

In the Bhopal poison gas release case, top management at Union Carbide initially attempted to avoid responsibility by blaming the disaster on a disgruntled employee (Lepkowski, 1988; Rich, 1986). Hundreds of victims filed lawsuits against Union Carbide in the Indian court system. Eventually, Union Carbide settled out of court—for a paltry $470 million—the lawsuits claiming billions of dollars of damages. One is inclined to compare such a figure with the $10 billion damages generated by asbestos-related lawsuits. As one reviewer put it, "by any measure, it's clear that for global corporations life is cheap in the developing world" (Karliner, 1994: 726). During the Chernobyl crisis, the Soviet government tried to hide major deficiencies in its ability to effectively manage and control its nuclear power plants by focusing exclusively on "operator error" as the only cause of the problem.

The disastrous effects of floods, hurricanes, and earthquakes often create tremendous stress and anxiety in the lives of affected individuals as they are forced to come to terms with the loss of loved ones and/or property. Studies of victims of massive floods and hurricanes have shown that they sometimes exhibit abnormal signs of stress and anxiety months after a catastrophe (Quarantelli, 1978; Quarantelli and Dynes, 1977). Research on the effects of natural disasters, as summarized in Figure 2–4, points to a tendency for community cohesion to develop to cope with the resulting stresses and dislocations, and efforts to avoid conflicts regarding blame. Common dangers and common fates tend to unite communities in the face of danger. The few studies that have been conducted comparing the overall effects of natural versus technological disasters on human communities (see Figure 2–4) suggest that the negative physical, psychological, and social impacts resulting from technological disasters are just as damaging, if not more so, than what people experience in dealing with natural disasters (Baum, Fleming, and Singer, 1983; Freudenburg, 1997; Gramling and Krogman, 1997; Kasperson and Pijawka, 1985). For one thing, technological disasters such as Chernobyl have caused widespread radiation panic lasting for years; and, in the case of Bhopal, collective hysteria seemed to last for many months after the event.

Immediately after the chemical contamination of their city, the residents of Bhopal marched in riotous demonstration on the Union Carbide factory, threatening to burn down the plant and calling for the immediate hanging of members of Union Carbide's top management, particularly the chairman of the corporation. More organized protest groups developed shortly thereafter and continued to stage protests on behalf of the victims, taking up such issues as social justice, public access to scientific information, medical care, and legal services (Sarangi, 1996: 102).

CONCLUSION

There is growing recognition that technological risks and disasters may begin to surpass natural disasters in terms of their threats, frequency, and damage to human life and property and environmental degradation. The extent of physical, psychological, social, and environmental damage of disasters such as Chernobyl, Bhopal, DC-10 crashes, and asbestos-related illnesses are greater than devastations caused by some floods, hurricanes, and earthquakes. And, as we have made clear, the continuing threat of nuclear war, not to mention the threat of biochemical war, are potential human-made disasters that will challenge the survival ingenuities of the human race.

As our ability to assess the impact of technology on society increases, there is reason to believe that human beings may yet develop the capacity to cope more adaptively with human-made disasters.

References

Anonymous (a). (2000). "Global warming," *The Economist,* November 18: 16–19.

Anonymous (b). (1998). "Ukraine tallies sharp rise in illnesses near Chernobyl," *The New York Times,* April 23: A5.

Anonymous (c). (1999). "Cost of natural disaster: $480 million a week," *The Christian Science Monitor,* May 20: 4.

Aydintasbas, Asla. (1999). "Unprepared for the expected," *The Wall Street Journal,* August 20: A10.

Baldauf, Scott. (2001). "Bad quake, worse buildings," *The Christian Science Monitor,* January 29: 1.

Barash, D.P. (1987). *The arms race and nuclear war.* Belmont, California: Wadsworth Publishing Company.

Baum, A., Fleming, R., and Davidson, L. (1986). "Natural disaster and technological catastrophe," *Environment and Behavior* 15 (3): 333–354.

Baum, A., Fleming, R., and Singer, J. (1983). "Coping with victimization by technological disaster," *Journal of Social Issues* 39 (2): 117–138.

Blaikie, P., Cannon, T., Davis, I., and Wisner, B. (1994). *At risk: Natural hazards, people's vulnerability and disasters.* New York: Routledge.

Blair, B.G., Feiveson, H.A., and von Hippel, F.N. (1997). "Taking nuclear weapons off hair-trigger alert," *Scientific American* 277: 75–81.

Bleuer, J.P., Averkin, Y.I., and Abelin, T. (1997). "Chernobyl-related thyroid cancer: What evidence for role of short-lived iodines?" *Environmental Health Perspectives* 105 (Suppl. 6): 1385–1391; 1483–1486.

Broder, J. (1999). "At Nicaragua mudslide site, Clinton's aid falls short of his goal," *The New York Times,* March 19: A10.

Couch, S.R., and Kroll-Smith, J.S. (1985). "The chronic technical disaster: Toward a social scientific perspective," *Social Science Quarterly* 66 (3): 564–575.

Couch, S.R., and Kroll-Smith, J.S. (Eds.). (1991). *Communities at risk: Collective response to technological hazards.* New York: Peter Lang Publishers.

Cunningham, A.M., and Fitzpatrick, M. (1983). *Future fire: Weapons for the apocalypse.* New York: Warner Books.

Dietrich, F. (1984). *Preventing war in the nuclear age.* Totowa, NJ: Rowman & Allanheld Publishers.

Dresser, P.D. (Ed.). (1993). *Nuclear power plants worldwide.* Detroit, MI: Gale Research Inc.

Dumas, Lloyd. (1999). *Lethal arrogance: Human fallibility and dangerous technologies.* New York: St. Martin's Press.

Eckholm, E. (1998). "China admits ecological sins played role in flood disaster," *The New York Times,* August 26: A1, A6.

Ford, Peter. (2000). "Dams generate a reservoir of controversy," *The Christian Science Monitor,* November 17: 7.

Freudenburg, William. (1997). "Contamination, corrosion and the social order: An overview," *Current Sociology* 45 (3): 19–39.

Freudenburg, William, and Pastor, Susan. (1992). "Public response to technological risks: Toward a sociological perspective," *The Sociological Quarterly* 33 (3): 389–412.

Goldberg, C. (1998). " In a post-nuclear town, some adjustments hurt," *The New York Times,* July 12: A14.

Goldman, M. (1997). "The Russian radiation legacy: Its integrated impact and lessons," *Environmental Health Perspectives* 105 (Suppl. 6): 1385–1391.

Gramling, R., and Krogman, N. (1997). "Communities, policy and chronic technological disasters," *Current Sociology* 45 (3): 41–57.

Houlder, V. (2000). "Hole in ozone layer could be closed within 50 years," *Financial Times,* December 4: 16.

International Atomic Energy Agency. (1996). *IAEA Yearbook.* Vienna, Austria: International Atomic Energy Agency.

Karliner, Joshua. (1994). "To Union Carbide, life is cheap: Bhopal— ten years later," *The Nation,* December 12; 259 (20): 726–730.

Kasperson, R.E., and Pijawka, K.D. (1985). "Societal response to hazards and major hazard events: Comparing natural and technological hazards," *Public Administration Review* 45: 7–18.

Kerkin, A. (2000). "Treaty talks fail to find consensus in global warming," *The New York Times,* November 26: 1.

Kunreuther, H. (1998). "A program for reducing disaster losses through insurance." In Howard Kunreuther and Richard Roth (Eds.), *Paying the price: The status and role of insurance against natural disasters in the United States.* Washington, DC: Joseph Henry Press.

Kunreuther, H., and Roth, R.J. (1998). *Paying the price: The status and role of insurance against natural disasters in the United States.* Washington, DC: Joseph Henry Press: 209–228.

Kuo, A. (1998). "Unbreaking the wall: China and the Three Gorges Dam," *Harvard International Review* (Summer): 28–31.

Lepkowski, W. (1988). "Union Carbide presses Bhopal sabotage theory," *Chemical Engineering News,* July 4; 66 (27): 8–12.

Lomat, L., Galburt, G., Quastel, M.R., Polyakov, S., Okeanov, A., and Rozin, S. (1997). "Incidence of childhood disease in Belarus associated with the Chernobyl accident," *Environmental Health Perspectives* 105 (Suppl. 6): 1529–1532.

McLeod, R. (1995). "Resistance to nuclear technology: Optimists, opportunists and opposition in Australia's nuclear history." In M. Bauer (Ed.), *Resistance to new technology.* New York: Cambridge University Press: 165–187.

McNamara, Robert S., and Blight, James G. (2001). *Wilson's ghost.* New York: Public Affairs.

Mian, Z. (1995). "The costs of nuclear security." In Z. Mian, (Ed.), *Pakistan's atomic bomb and the search for security.* Lahore, Pakistan: Gautham Publishers: 39–81.

Miller, K.L. (1994). "The nuclear reactor accident at Three-Mile Island," *Radio Graphics* 14 (January): 215–224.

Palm, R.I. (1990). *Natural hazards.* Baltimore, MD: Johns Hopkins University Press.

Paton, Dean. (2001). "Why Seattle's big quake had little impact," *The Christian Science Monitor,* March 2: 3.

Peterson, S. (2000). "Chernobyl closes, legacy endures," *The Christian Science Monitor,* December 8: 1.

Quarantelli, E.L. (Ed.). (1978). *Disasters: Theory and research.* Beverly Hills, CA: Sage Publications.

Quarantelli, E.L., and Dynes, R.R. (1977). "Response to social crisis and disaster," *Annual Review of Sociology* 3:23–49.

Rich, L. (1986). "Carbide realleges sabotage," *Chemical Week,* August 20, 139: 13–15.

Robinson, A. (1993). *Earth shock.* London: Thames and Hudson.

Sarangi, S. (1996). "The movement in Bhopal and its lessons," *Social Justice* (Winter); 23 (4): 100–109.

Sheehan, L., and Hewitt, K. (1969). *A pilot survey of global natural disasters of the past twenty years* (Working Paper/Tech. Rep. No. 11). Boulder, CO: University of Colorado, Boulder, Institute of Behavioral Science.

Sieh, Kerry. (1999). "Earthquake lessons," *The New York Times,* September 23: A29.

Smith, K. (1992). *Environmental hazards.* New York: Routledge.

Swiss Reinsurance Company. (2001). *Natural catastrophes and man-made disasters in 2000.* Sigma No. 2/2001. Zurich.

Tenner, E. (1996). *Why things bite back: Technology and the revenge of unintended consequences.* New York: Alfred A. Knopf.

Thompson, S.A. (1982). *Trends and developments in global natural disasters 1947–81.* (Working Paper/Tech. Rep. No. 45). Boulder, CO: University of Colorado, Boulder, Institute of Behavioral Science.

Webb, R.E. (1976). *The accident hazards of nuclear power plants.* Amherst, MA: The University of Massachusetts Press.

Weinberg, A.M. (1986). "Science and its limits: The regulator's dilemma." In National Academy of Engineering, *Hazards: Technology and fairness.* Washington, DC: National Academy Press: 4–23.

Wijkman, A., and Timberlake, L. (1984). *Natural disasters: Acts of God or acts of man?* Washington, DC: Earthscan Paperback.

Wing, S., Richardson, I., Armstrong, D., and Crawford-Brown, D. (1997). "A reevaluation of cancer incidence near the Three-Mile Island nuclear plant: The collision of evidence and assumptions," *Environmental Health Perspectives* 105 (1): 52–57.

Zebrowski, Jr., E. (1997). *Perils of a restless planet.* New York: Cambridge University Press.

Endnote

1. Our discussion in this chapter of anticipated and unanticipated consequences of disasters is informed by the seminal article by Robert K. Merton (1936), "The unanticipated consequences of purposive social action," *American Sociological Review* 1 (December): 894–964.

PART **II**

The Prevalence of Technological Disasters

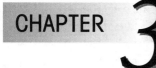
The Year 2000 (Y2K) Debacle: An Ironic Failure of Information Technology

"We leveraged the resources of the whole planet to squash an incredibly powerful problem."

—IBM Executive

*A*t the dawn of the computer age, when computing and information storage was time-consuming and expensive, computer programmers sought to economize on computer memory by abbreviating the number of digits pertaining to date representation. Instead of using an eight-digit date representation—month, day, and four digits for the year—programmers tended to represent dates with six digits: month, day, and two digits for the year. For example, April Fools' Day could be represented either as 04011999 or as 040199. By 1999, if the software and hardware running a computer were not fixed, April Fool's Day would be represented as 040199 instead of 04011999.

When the new millennium arrived, the last two digits of the six-digit representation would read only "00," and there was some uncertainty as to whether the computer would interpret the date as referring to the year 1900, the year 2000, or some other date. Because billions of lines of programming code—operating in millions of mainframe computers, personal computers, and embedded microprocessors—originally had date-fields that only represented six digits, when clocks rolled over from December 31, 1999, to January 1, 2000, experts were convinced that computers would not be able to recognize the correct date. Not being able to recognize the correct date or calculating the wrong date, computers could behave erratically, which would result in inaccuracies in date-related calculations and the generation of wrong data-processing results. In the worst-case scenario, entire computer systems and computer networks could fail completely. In short, computer programmers' decision to save the space taken up by the two digits eventually created the Y2K problem.

Among most information technology (IT) specialists, the threat of a serious crisis due to Y2K was real. A corporate preparedness survey published December 21, 1999, produced a number of "key findings" that illuminate the depth and breadth of concerns of corporations from September 1999 to December 1999. The key findings of this survey are summarized in Table 3–1.

All of the findings presented in Table 3–1 indicate that the problem as perceived by corporate officials was very real and hence account for the effort devoted to making all systems "Y2K compliant." Officials of state governments were likewise concerned about the Y2K problem. For example, New York State spent about $270 million on the problem. As one expert, Gary Davis, the New York State Year 2000 director, put it, "Had we not fixed our code and done the level of testing and gone out and got-

Table 3–1

Summary of year 2000 corporate preparedness survey.

	September 1999	December 1999
Incidence of Y2K-related failures	82%	88%
Overall causes of failures attributed to Y2K	41%	59%
Location of Y2K problems in premeditated systems	56%	44%
Supply-chain related failures	36%	24%
Percentage experiencing significant risk	12%	7%
Business-partner relations	48%	56%
Expectations of future failures	99%	84%

Source: Data compiled from Cap Gemini America Year 2000 Corporate Preparedness Survey, 1999, December 21. Accessed from the World Wide Web 12-28-2000 at: http://www.us.cgey.com/news/current_news.asp?ID=116

ten outside auditors, we would definitely have had failures" (Rata-jczak, 2000: H2). The fact that the U.S. government expected problems overseas was reflected in the decision to evacuate approximately 350 diplomatic personnel and family members from embassies in Russia and three other countries due to fears of Y2K-related problems (Urakami, 2000). The Department of Transportation closed U.S. commercial shipping at dozens of U.S. and foreign seaports during the New Year's weekend. The Coast Guard disclosed that about two dozen of the world's 16,000 cargo ships had been "red-flagged" and were barred from American ports during the New Year's weekend because their operators were unable to convince officials that they could operate safely. In addition, dozens of overseas ports, including some major oil terminals, were also closed during the January 1st weekend, from Angola to the United Arab Emirates. China also banned foreign shipping over New Year's weekend (Bridis, 1999: 3).

After the fateful date and only one week later, in a January 7 press release, John Koskinen, Y2K czar for the U.S. government's President's Council on Year 2000 Conversion Information Center, affirmed the reality of the threat and claimed that the overall relatively minor impact of the year 2000 turnover could be attributed to the conscientious and determined response from decision makers in government and industry. As Koskinen put it, "The fact that there continue to be date-change glitches remind us that the Y2K challenge was very real. The hard work of thousands of dedicated employees in the public and private sectors is the reason why what we have seen thus far are minor difficulties and not serious national problems." (www.y2k.gov)

Whereas Koskinen's remarks apply to work done in the United States, the dedicated efforts by countless Y2K workers had also produced similar success elsewhere in averting the potentially dangerous effects of the "millennium bug." In a press report on January 1, Bruce McConnell, Director of the International Y2K Cooperation Center (IY2KCC), a global Y2K monitoring agency established by the United Nations and funded by the World Bank, credited an "unprecedented level of international cooperation for keeping the world from experiencing serious Y2K problems. "The resources have been well spent," he said. The IY2KCC coordinated the efforts of more than 135 countries worldwide.

The fact that the world had no major difficulties with mission-critical infrastructures does not indicate the absence of the problem. It suggests, rather, that a coordinated effort by different organizations in the public and private sector managed to fix network systems and critical infrastructures dependent upon computerization.

THE OVERALL IMPACT OF Y2K

Although there were dozens of reported Y2K-related problems worldwide, some potentially serious and some merely quirky, thousands of IT professionals went to work to fix the problems before any serious damage could occur. The overall impact of Y2K reflects a position somewhere between the "I told you so" cynicism of the hoax mongers and the fears of Armageddon fostered by the millenarians and doomsayers. Given such wide discrepancies over the interpretation of the effects of Y2K, it is worth summarizing some of the major reported incidents in order to illustrate the potential seriousness of the problem. In addition, separating the reported glitches into those from developed and developing nations may throw light on how information technology is managed in different parts of the globe.

Incident reports from developed countries

- The Pentagon reported that a military reconnaissance system went down for several hours after it lost its satellite link. The failure involved a ground receiving system, not the satellite itself; backup systems were put in place to fill in the gap in reconnaissance. Deputy Defense Secretary John Hamre called it a "significant" Y2K failure.

- Nuclear plant incidents reported on January 1 occurred in Arizona, Arkansas, Connecticut, New York, Minnesota, and Massachusetts. Most involved communication links involving either weather instruments, which monitor and measure meteorological data, or control centers, which manage plant operations and building access.

- Eight U.S. utility companies reported Y2K-related problems with their geosynchronous timing devices, which failed on January 1, 2000. These synchronization devices are coordinated with the Global Positioning Satellite

(GPS) System. This caused the utilities to lose communications with satellites. All were quickly reset, officials said. An Arkansas power plant near Russellville experienced a Y2K failure with scanners for radiation monitors carried by plant workers.

- Airport systems were another sector reporting Y2K-related malfunctions. The Federal Aviation Association (FAA) said low-level wind-shear monitors at 10 airports suffered Y2K problems. All were brought back into operation within hours. One important failure reported in the aviation sector was the temporary breakdown of the communication system that distributes information to air personnel. As experts reminded us, efforts to fix the problem were rewarded because tests indicated that runway lights, security systems, and flight reservation computers would have failed at some airports without appropriate remedial work when the date became 2000.

- Amtrak reported a problem in its symbol tracking system that records the position of trains as they move up and down the network of tracks.

- Japan reported a total of 27 glitches, including weather monitoring and reporting systems at two nuclear power plants.

- In Sweden, 100,000 people were unable to access their bank accounts because the bank neglected to update its browser software.

- In Scotland, there were reports of problems associated with kidney dialysis machinery.

Sources: President's Council on Year 2000 Conversion Information Center, accessed from the World Wide Web on January 1–5, 2000, at: *www.y2k.gov*; Landers, 2000; Janofsky, 2000.

Incident reports from developing countries

- China witnessed numerous failures in small business financial systems and building control systems.

- Hong Kong reported that 18 "breathalyzer" tests used by police on suspected drunk drivers failed and remained inoperative for 10 days.

- Power outages that lasted six to eight hours were reported near the Pakistani city of Islamabad.
- In Malaysia, multiple failures, including medical systems and building automation systems, were reported. In the medical sector, failures occurred in everything from patient registration systems to ultrasound machines to blood-gas analyzers to bloodpressure monitoring devices.
- Medical operations systems failures were also reported in Sri Lanka, including certain E.C.G. monitoring devices, which were not compliant and were rendered inoperable. Nicaragua, Saudi Arabia, Taiwan, and Venezuela all reported similar failures in various sectors of the health care delivery industry.
- In Turkey, various glitches were reported in the health sector such as patient monitoring equipment, ultrasonic devices, tomography equipment, and kidney dialysis machines.
- In Africa, a major telecommunications glitch caused disruption of traffic and communication between Zambia and Malawi. Other major telecommunications disruptions occurred in Tanzania, Nigeria, and Namibia.

Source: The International Y2K Cooperation Center, accessed from the World Wide Web on Feb 15, 2000, at: *http://207.233.128.31/February2000Report.htm*

Clearly, there was no dearth of Y2K-related incidents, and these are not exhaustive lists. Nevertheless, none of the incidents reported caused serious infrastructural problems, and most system failures were fixed within days, many within hours. Moreover, there were no more important failures reported in the developing countries, which spent considerably less on the problem, than in the developed countries, which spent vast sums to make their systems compliant. Naturally, the question arises: how did countries that started so late—and appeared to do so little—manage to reach the year 2000 as smoothly as countries like the United States and the United Kingdom that spent so much money? It is a well-known fact that countries such as Italy, Venezuela, and Austria had been somewhat cavalier in addressing the problem. How did the countries that did so little get away so easily? Did countries like the United States and the United Kingdom overreact on the spending?

A World Bank survey published in January 1999 concluded that 54 of 139 developed countries had nationally coordinated Year 2000 programs, and only 21 were taking steps to prepare for the problem. Paraguay's Year 2000 coordinator was quoted as saying that his country would experience so many disruptions that their government would have to impose martial law (Sandberg, 2000).

In fact, it has even been suggested that one way to check if the problem was more a hoax than a real problem would be to compare the outcome in countries such as the United States and the United Kingdom, both of which spent vast sums of money, with what happened in countries such as Italy, Russia, and the developing countries that were more relaxed about the problem (Sandberg, 2000). This, however, would not be a meaningful comparison for many reasons. First, in the developing world, digital controls are not generally used to control infrastructural production and services. They do not control the production of power, and they do not control the production of telecommunications. In developing countries, "digital computers are primarily used, at least at present, simply to provide management information (*http://207.233.128.31/February2000Report.htm*)," said Bruce McConnell, director of International Y2K Cooperation Center.

In addition, experts also seemed to *overestimate* the world's dependence on computer technology. This is the second answer to the question of whether Y2K was a real problem or a hoax, given the major discrepancies between time and money devoted to the problem when comparing developed to developing countries. Many countries needed less Y2K preparation because they are far less computerized than the United States and European countries. In fact, compared to developed countries, computers in developing countries are much less likely to be tied together in complex networks and are often so old that they run on much simpler software, according to Louis Marcoccio, Year 2000 research director for the Gartner Group, a technology consulting firm (Feder, 2000: 1).

Once adjustments are made for a country's dependence on technology, the investment of the United States and others in Y2K preparations were more closely in line with those countries that started late. For example, Russia is estimated to have spent between $200 million and $1 billion to fix the problem (Feder, 2000). Even if Russia had spent only $400 million, it would have spent proportionately more than the United States, which spent $100 billion because the United States is estimated to be 300 times more reliant on computers, according to Louis Marcoccio (Feder, 2000: 2).

Furthermore, the U.S. government paid for many foreign nations to send representatives to the first United Nations meeting on Y2K in 1998 to jump-start lagging nations. The U.S. government also distributed hundreds of thousands of CD-ROMs in 10 languages, which provided background information as well as suggestions for how to organize Y2K compliance projects (Feder, 2000: 4). Of particular note is the effort devoted by the U.S. government to monitor Russian-designed nuclear systems, including regular on-site inspections and coordination of their efforts with Russian officials. In addition, the Defense Department sponsored programs to provide support to countries using approximately 65 Russian-designed nuclear power plants in order to increase the level of operational effectiveness of those plants (Feder, 2000: 4). Finally, the Defense Department set up an $8 million joint command post in Colorado so that the United States and Russia could be assured that miscommunication would not lead to missiles being launched erroneously (Janofsky, 2000).

The Leap Year Rollover Problem

In addition to the New Year rollover, computer experts were concerned with potential problems that might have arisen on February 29, when computers would have to deal with a leap year. As one expert explained:

> This so-called leap-year bug results from a peculiarity of the Gregorian calendar, which inserts an extra day at the end of February every four years, omits it on years that end a century, but *puts it back in when the year is divisible by 400.* Programmers who knew the second of these rules but not the third would have instructed their computers, erroneously, that 2000 is not a leap-year. This could result in all kinds of problems if the computer happens to be, say, keeping track of an oil tanker expected to arrive in port on the day after Feb. 28. 'We worked with 300 companies, and every single company had some leap-year problems,' says Nigel Martin-Jones, executive vice president of the high-tech consulting firm Data Dimensions. (Sandberg, 2000: 38–39)

Unlike the Y2K problem, where the use of a two-digit date field was a standard and accepted programming practice, the leap year problem resulted directly from erroneous programming. Experts projected that if left unfixed, erroneous leap year coding would affect primarily software applications, not hardware or operations systems that were the focus of so much attention during the millennium rollover period.

Although no major problems were reported, either in the developed countries or in the developing countries, numerous minor glitches did occur. In the United States, problems related to the leap year surfaced at the Department of Housing and Urban Development, the National Environmental Satellite Information Service, the Federal Communications Commission, and the Immigration and Naturalization Service. At the international level, the International Y2K Coordination Center monitored the leap year date change through six conference calls over two days with national and regional Y2K coordinators from every continent. Fewer than two dozen problems were reported, all of them minor. Problems included: incorrect date displays on cellular phones (Morocco); temporary interruptions in service at a few automatic teller machines (Japan); disruptions in the transmission of weather data to the media (Netherlands); inability to schedule medical appointments in some doctors' offices (United Kingdom); inability of a few merchants to verify credit card data (New Zealand); and the inability to enter correct expiration dates on new passports (Bulgaria). All problems were corrected within a few days.

John Koskinen, Director of the President's Council on Year 2000 Conversion Information Center, expressed confidence in handling the leapyear problem. As Koskinen put it, "As you look at the glitches reported internationally . . . and around the country in the U.S. . . . they basically reflect the kind of problems we expected, which is a system's inability to recognize the date—February 29th—causing it to reject data. So what you have is a transaction or information flow that stops. Again, none of those have turned out to interfere in any meaningful way with operations. . . ." Citing a comprehensive report of all problems related to the leap year in systems that were generally regarded as Y2K compliant, Koskinen was quick to report that "the bottom line is that there have been problems even when organizations fixed their systems." He noted that 98 percent of the organizations surveyed had taken Y2K precautions and still had minor glitches since January 1, 2000.

ANTICIPATION OF THE PROBLEM

Our dependence upon computer and communication systems is growing at a rapid rate, but as society becomes more dependent on computers, we also become more vulnerable to computer malfunctions. The Y2K problem, which posed a global threat, was in fact, *anticipated,* and hence was, in large measure, *preventable.* This particular example of technological disaster was not the result of so-called "unintended consequences" of technology (Tenner, 1996)—this problem was, in fact, *foreseen and fully anticipatable.*

As early as 1984, an article appeared in *Computerworld,* an IT trade journal, diagnosing the problem and calling for programmers and managers to take heed of the date conversion difficulties that would happen at the turn of the century (Gillin, 1984). Gillin, the editor of *Computerworld,* reported the findings of William Schoen, the programmer who identified the year 2000 problem. He discovered the problem in 1983 while working at one of the "Big Three" automakers. As Schoen attests, data processing professionals had known about the risk as early as the 1970s, but, as Schoen puts it, "It's just that no one thought their codes would last that long" (Gillin, 1984: 7).

Schoen even designed a programming solution to the predicament, calling it the "Charmar Correction," a cure for "the serious problem ignored by the entire data processing community." He then proceeded to create a consulting company, Charmar Enterprises, and went on a campaign to market his "correction" to the problem. However, he elicited only two sales for the Charmar Correction, and therefore dissolved Charmar Enterprises in 1984. The sale price for the Charmar Correction was a mere $995. This is indeed ironic, given the millions, even tens of millions, that many corporations and other organizations have since paid to "fix" the Y2K problem.

Schoen was not alone, however, in campaigning for attention to the date field encoding problem. As early as 1960, Greg Hallmuch of the U.S. Bureau of Standards was raising the issue (Munro, 1998). In 1967, Susan Jones, assistant director of the Department of Transportation, was urging Congress to address the date conversion situation (Munro, 1998).

In addition, the programming community claims that their pleas to top management of their corporations to address the potential problem were all ignored until the late 1990s. Why was management so shortsighted in the first place? Why, once management became aware of the problem, did major companies respond so lethargically—as if in denial?

THE CAUSES OF THE PROBLEM

The Y2K problem involved a failure on the part of corporate executives as well as a failure on the part of IT professionals.

The most common misconception about Y2K is that it was a single problem. Unfortunately, this perspective created a commonly held belief that the "problem" was trivial, although widespread, and that a single solution was possible. In reality, the causes and the solutions were multifarious and complex. As one computer guru put it:

> It [Y2K] . . . was *perpetrated* by people who decided that what we did yesterday was good enough for today and did not look out for tomorrow and evaluate the inevitable consequence of cutting corners. It was *exacerbated* by people who scoffed at warnings and were in denial and irresponsible. It was turned into a *crisis* by people who left it to the last minute. (DeJager, 1998)

Although the causes of the millennium bug were numerous, three factors were crucial: technical factors, programming factors, and managerial factors. The factors and associated causes are summarized in Figure 3–1.

Technical Factors

Lack of Internationally Accepted Date Standards

There exists no universal standard for date representation. Following the National Institute of Standards of Technology, the U.S. protocol is month-day-year, so January fourth, nineteen ninety-eight would be 01-04-98. Canadians and Britons reverse day and month so that the same day would be referenced as

Factors	Causes
Technical	Lack of internationally accepted date standards; high cost of computing
Programming	Unexpected tenacity of original programs; code reuse; systems compatibility
Managerial	Managerial accounting protocols; decisional inertia

Figure 3–1
Factors and causes of the Y2K problem.

04/01/98, which U.S. residents would identify as April Fool's Day. The Scandinavians use yet another system, putting the year first: 98.01.04. The International Organization for Standardization has as its standard: 1998:01:04. Notice that this standard includes the four-digit year. The lesson to learn is that standardization is crucial in the computer industry. The industry needs to universalize its standards of operation and standardize and keep extensive records on date-field labeling, programming documentation, and record keeping.

High Cost of Computing

Another technical factor relates to the primitive state of computing technology in the 1950s and 1960s. In the days of Hollerith cards, computer space was at a premium and computer memory was very expensive. Given that programmers wanted to conserve computer memory and storage space, which at the time was extremely expensive, they ended up encoding calendar dates in a six-digit format, mmddyy, rather than an eight-digit format. This entailed a 25 percent savings for relatively no loss of information. This may account for the original decision to use a six-digit date format.

Programming Factors

Unexpected Tenacity of Original Programs

Most programmers did not envision that the programs they wrote 30 years before would still be running at the end of the 20th Century. This permitted the development of psychological processes such as rationalization, dissociation, and other mechanisms of psychic numbing to avoid the cognitive dissonance between what they knew to be wrong and their insistence on continuing their defective practice.

Code Reuse

Virtually all new applications use algorithms from previous systems. The reuse of algorithms that have a hidden date-processing fault was one reason the Y2K problem was so extensive (Keough, 1997). Incidentally, this fact is what likens Y2K to a virus: faulty algorithms get used and reused, spreading their problematic payload to more and more systems. Since successive applications are often built on earlier programs or incorporate subroutines from other programs into their own structure, they can be inherently flawed.

Programs were still being written using computer code with six-digit date representations. Even the best and most modern code in the world could be hamstrung by faulty historical data. In fact, vendors and manufacturers continued to knowingly sell software "infected" with the millennium bug through the mid- and late-1990s (Miller, 1995; Condon, 1995).

Systems Compatibility

Systems operating software, as well as customized programs, have been designed to be compatible with older versions. Out-of-date programs have supported leading-edge replacement systems. Designing new systems to be compatible with older systems is generally a resourceful way to maximize efficiency. This feature may be consumer and producer friendly, but if this is done neglecting the values of quality, reliability, and science, the move to universal compatibility will inevitably lead to the design of faulty systems. Generally a good thing, systems compatibility actually enhanced the ability of the millennium bug to spread like a virus. Moreover, the date conversion problem was not addressed when designing systems for compatibility.

Managerial Factors

Managerial Accounting Protocols

One major cause of the Y2K problem stems from the fact that accounting procedures have treated software expenditure to be an expense in the period incurred. This means that spending money on maintaining software has been treated like a telephone bill. It gets paid regularly, but at the end of the day it is perceived as not increasing a corporation's net worth. This means that capital expenditure for fixing the problem is seen as coming off the bottom line. It was therefore difficult to convince a chief executive officer (CEO) or a chief financial officer (CFO) that a $5 million corporate expense to solve the Y2K problem in the early 1990s was a "good thing to do." This accounts for the management inertia on this issue (Garner, 1996). In other words, Y2K compliance had been a tough sell. How do you convince management to take on a multimillion-dollar project where the return on investment is zero? It reveals stubbornness, tenacity, and ignorance of other factors stemming from bureaucratic rationalism, efficiency, and capital accumulation (Meall, 1995). These kinds of factors are, in turn, the result of rigid organizational hierarchies.

Decisional Inertia

Another managerial cause of the problem was decisional inertia on the part of CEOs and CFOs. One cause of the indecisiveness is that many top managers had been deferring attention to the problem in the hope that a "silver bullet" would come along to solve the problem. Most experts, however, acknowledged that a silver bullet was unlikely to emerge (Newling, 1998). This is an example of the false optimism about technology—the belief that for every problem of technology, a technological "fix" can be found.

Yet another managerial cause of decisional inertia that led to failure to act in a timely manner is top management's general ignorance of information systems management. Corporate and governmental management certainly appreciated the benefits and results of computerization, but took little or no interest in understanding the complexity of information systems management. Hence one important lesson to learn from Y2K is that top management must come to a better understanding of the software and hardware on which their operations so crucially depend.

As recently as a few years ago, programs were being written that did not take into account date changes and date fields in data processing. In fact, various surveys report that only 20 percent of America's biggest companies had devised a full-fledged strategy to deal with the Y2K problem (Peters, 1997; Hicks, 1997; Feine,

1997). Timely planning depended on whether managers were alert to the issue. Most were not. Was the information technology industry in denial or simply negligent? Even when managers became aware of the problem, they also exhibited denial and neglect. A typical response from industry experts was as follows, "I won't be in this position or in this company in the year 2000. It's not *my* problem" (Furma, 1994: 70–74). For example, as recently as December 4, 1998, the *Wall Street Journal* quoted a corporate executive as stating, "This year 2000 stuff is *waayyyy* over done. It's complete, complete lunacy" (Binkley, 1998: 1).

Given the grave business, legal, and social risks and hazards caused by Y2K and given the large set of causes identified and discussed here, it is not difficult to conclude that accountability for safe, reliable, and beneficial information technologies had been greatly eroded in the Y2K case.

THE SCOPE OF Y2K

Three types of systems were affected by the date conversion problem: personal computers (PCs), mainframe computers plus the software running on them, and embedded microprocessors.

Hundreds of millions of PCs are owned and operated by individuals, mostly in industrialized and newly industrialized countries. If PCs had failed to correctly advance their internal clocks to the year 2000, innumerable problems would have resulted due to the many functions that depend on those clocks—e-mail, voice mail, banking transactions, data tracking, etc. Software companies that specialize in PC software, such as Microsoft, eventually began to respond to the problem by introducing upgrades or new software so that PCs would be Y2K compliant by the turn of the new millennium (Trott, 1998).

Payroll systems, sales-tracking systems, and mortgage and insurance data are typical examples of programs running on large mainframe computers. Tens and thousands of mainframe computers were in operation on December 31, 1999, and that did not include the AT&T, Unisys, Hewlett-Packard, Texas Instruments, Xerox, and other platforms that were also in use at the time (Eddy, 1998).

In terms of making mainframes Y2K compliant, the fact that both the mainframe platform and many mainframe applications are centrally located and controlled greatly eased the task of inventorying application components and simplified the challenge of information management carried out by most large organizations (Ulrich, 1997). The major problem, then, was the sheer number of lines of code that had to be checked, converted, and tested. Moreover, there was no "quick fix" answer to the problem, no black-box solution. As industry experts all agreed, the only sure way was to check every line of code in every program that was date dependent. For example, the IRS had 88,000 programs to debug on 80 mainframe computers. The Social Security Administration had more than 30 million lines of code to check. One expert claims that finding, fixing, and testing all Y2K-affected software required more than 700,000 person-years (Phillips, 1998).

The need for date information first appeared in computers used for commerce and finance many years ago, but its significance for embedded systems was not fully appreciated until the early 1990s. Estimates put the number of embedded chips in use at the turn of the 21st Century at more than 25 billion.

Embedded chips are found everywhere in:

1. manufacturing applications such as automated factories and bottling plants
2. process control systems such as those found in nuclear power plants, electric power grids, natural gas supply systems, water and sewage systems, and oil refineries
3. transportation industries such as airplanes and airlines, air traffic control systems, trains, buses, marine craft, automobiles, each of which may contain dozens of microprocessors, traffic control systems, and parking meters
4. telecommunications systems such as satellites, global positioning systems, telephone networks, and cable systems
5. finance applications such as automated teller systems, credit card systems, point of sale scanning, and cash systems

6. medical devices such as pacemakers, heart defibrillators, kidney dialysis machines, patient monitoring and information systems, medicine dispensing systems, and X-ray equipment

7. environmental, meteorological, and astronomical monitoring equipment

8. building maintenance systems such as electrical supply, fire control systems, backup generators, heating and ventilating systems, security systems, elevators and escalators, and door locks

9. office equipment such as telephones, faxes, and photo copiers

10. mission-critical weapons systems such as nuclear warheads, missile launch systems, air force bombers and fighter-jets, submarines, Sentinel Radars, the Air Battle Management Operations Center and the Forward Air Defense Command

There was much concern at the Department of Defense that the military's "launch-on-warning" system would fail as a result of the Y2K problem. As one reviewer put it, "There is a real possibility that command and control systems, including early warning systems, could fail. Such failures would leave the U.S. and Russia unable to track each other's moves, and therefore likely to misinterpret a computer failure as evidence of an attack, and thus a reason to launch a counterattack. The fact that each side has thousands of nuclear weapons targeting the other side only magnifies this risk" (Anonymous, 1999: 1).

It is safe to say, then, that the scope of the problem was immense, but the problems associated with embedded microchips were especially crucial, given their pervasive presence in most, if not all, of our sociotechnical critical infrastructure and safety-critical systems. Hence, the hazards associated with Y2K noncompliance were serious and far-reaching.

In at least four respects, the date conversion predicament in embedded chips differed from the problem as manifested in mainframe and personal computers. First, chips may be embedded deep within the system, i.e., in the platform on which the application software is based. Thus, even when the application software is date- and time-independent, the hardware/operating system platform on which it is based may not be. Second, the lifespan of embedded systems is often greater than that of commercial data processing systems, and hence they often remain in use longer, without the need to make alterations to their software. The fact that the software in embedded systems is generally older made them more vulnerable to date conversion problems. Third, embedded systems are often in continuous operation and so were liable to be in operation when the date changed from 123199 to 010100. Finally, time is often counted in intervals in systems using microchips rather than with specific dates. For example, the program may be set to react at 50-day intervals rather than on the fifth day of each month, or the fifth day of each week. Hence problems were expected to occur both before and for sometime after 01012000 and not at all on the Y2K date itself.

The two major aspects of the problem—that is, the mainframe problem and the embedded chip problem—demonstrated the complexity and seriousness of the problem. They demonstrated that Y2K was not just a financial management system problem for CFOs. It was a serious operational problem as well. If the dating systems were to fail, the real-time activity of the company would also fail. In response, everyone from governmental agencies such as the President's Council on Year 2000 Conversion to corporate boards knew since 1998 that they had to expand their compliance focus from financial systems to general operations areas such as power plants and telecommunications centers.

As one can see, the potential disruptive consequences of Y2K were enormous. The fact that corporations were reluctant to address the problem when it first surfaced eventually resulted in greatly ballooning the cost of fixing the problem.

THE COSTS OF Y2K

Estimates of overall costs to fix the problem in the United States alone vary from $100 billion to $200 billion. Moreover, worldwide costs reported were somewhere in the range of $400–600 billion, including the U.S. costs (Taylor, 2000). By any yardstick, this was a colossal global technological expense.

Many major governments, industries, and corporations started with an estimate of the projected costs of fixing the problem, but, by the time they were done, most estimates had at least doubled. For example, while the U.S. government estimate hovered around $2–3 billion, the final costs were closer to $8.5 billion. Nevertheless, critics suggested that 25 percent of the money was wastefully spent. In response, John Koskinen argued that, at most, 5–10 percent of the funds were unwisely spent. Even given the $8.5 billion price tag, Koskinen defended the costs. Without the expenditure, Koskinen said, $1.5 trillion in federal spending would have been disrupted, "which by itself could have caused a significant economic decline." The statistics in Table 3–2 illustrate the general costs to fix the problem for selected corporations.

Most experts agree that most of the money expended was well worth it. Much of the perceived overspending can be accounted for if one perceives it as money paid to insure against a potential problem. Consumers pay hundreds, even thousands of dollars a year to insurance companies to avert the costs of a potential undesirable event. If one is lucky enough to avoid serious illness in life, one might question the wisdom of spending all the insurance premiums for nothing, since it was never used for medical expenses. In reality, however, the money spent was, for the most part, worth it, given the astronomical costs of health care when one does fall seriously ill. Similar conditions apply to Y2K.

Table 3–2
General costs to fix the Y2K problem for selected corporations (as of December, 1997–1998).

Corporation	Y2K Budget	Lines of Code	People on Project
Atlantic Energy	$ 3.5 Million	25 Million	7
Canadian Imperial Bank of Commerce	$150 M	75–100 M	250–300
C. R. Bard	$ 11 M	8 M	10
Merrill-Lynch	$200 M	170 M	300
Nabisco	$ 22 M	17 M	50–60
Union Pacific	$ 44 M	72 M	104

Sources: Compiled from *Computerworld* 31 (51), December 22, 1997, pp. 2, 5, 6, 8 and *Computerworld* 32 (25), June 22, 1998, pp. 7, 8, 10.

Remember that serious infrastructure problems were at risk: nuclear power plants, electrical grids, telecommunications, transportation systems, and industrial operations. Given the potential gravity of the problem there was no choice but to err on the side of spending more rather than on spending less.

CONCLUSION

As society becomes more and more dependent on computing technology and information systems, it becomes increasingly vulnerable to harm when these systems fail. Nothing illustrates this better than the Y2K problem. In the long run, what ultimately counted was not whether all the minor Y2K bugs were fixed, but whether critical services were delivered. The world's strategy of devoting significant resources to tackling critical infrastructures first and saving less critical problems for later turned out to be successful.

The fact that the countries of the world have met the challenge does not diminish the magnitude of this information technology disaster and should not detract from the magnitude of the effort of all involved. Numerous government agencies, countless corporations, and thousands of computer professionals devoted enormous resources to solve the most significant technological challenge the world has faced in many decades. Perhaps the most reassuring aspect of the Y2K debacle is this: it demonstrated that human beings can, on occasion, organize themselves across political and cultural boundaries to cope with problems generated by the very technology they have created.

References

Anonymous. (1999). "Millennium may bug nuclear weapons," *PSR Reports* 20 (1): 1.

Binkley, Christina. (1998). "Millennium bugged: The big Y2K problem is the silly questions," *The Wall Street Journal,* December 4: 1.

Bridis, Ted. (1999). "U.S. fixes air traffic bug; ports to shut to fend off Y2K trouble," *The Philadelphia Inquirer,* December 31: A3.

Cap Gemini. (1999). "Cap Gemini America Year 2000 Corporate Preparedness Survey," December 21. Accessed from the World

Wide Web December 28, 1999 at:
http://www.us.cgey.com/news/current_news.asp?ID=116

Condon, Don. (1995). "Microsoft tries to rewrite programming," *Infoword* 24 (6): 7–12.

DeJager, Peter. (1998). "It's a people problem." Retrieved December 1999 from the World Wide Web:
www.year2000.com/archive/people.html

Eddy, David. (1998). "Is the mainframe a dead dinosaur?" Retrieved December 1999 from the World Wide Web:
www.comlinks.com/qow/30797.htm

Feder, B. (2000). "How computers eluded failure puzzles experts," *The New York Times*, January 9: 1–4.

Feine, Jacob. (1997). "Slow responses to year 2000 problem," *IEEE Software* 14 (3): 126–133.

Furma, Jeff, and Martola, Alberta. (1994). "Year 2000 denial," *Computerworld* 28 (43): 70–74.

Garner, M. (1996). "Why the year 2000 problem happened." Retrieved December 1999 from the World Wide Web:
www.is.ufl.edu/bawb080h.htm

Gillin, B. (1984). "The problem you may not know you have," *Computerworld*, February 13.

Hicks, John. (1997). "Many companies just starting to address the Y2K problem." *Byte* 21 (9): 24–28.

The International Y2K Cooperation Center. "Y2K: Starting the Century Right." Retrieved from the World Wide Web:
http://207.233.128.31/February2000Report.htm

Janofsky, M. (2000). "Monitors of missiles at year 2000 note relief," *The New York Times*, January 1: 10.

Keough, J. (1997). "Your safety net has big holes in it," Chapter 3 of *Solving the year 2000 problem*. Boston: AP Professional.

Landers, J. (2000). "Defense department experiences satellite glitch during Y2K switch," *The Dallas Morning News*, January 2: B1.

Meall, L. (1995). "The century's time bomb," *Accountancy* 116 (128): 52–57.

Miller, C.S. (1995). "Microsoft wakes up to the problem," *Infoworld* 20 (15): 1–3.

Munro, Neil. (1998). "The big glitch," *National Journal* 30 (25): 142–149.

Newling, R. (1998). "No magic bullet to save the laggards," *Financial Times*, December 2: 8

Peters, James. (1997). "If wishes were noises," *Computerworld*, December 22; 31 (51): 2.

Phillips, William. (1998). "Here comes the millennium bug," *Popular Science* 253 (4): 92.

President's Council on Year 2000 Conversion Information Center. Retrieved from the World Wide Web at: www.y2k.gov

Ratajczak, D. (2000). "Although overblown, Y2K preparation paid off," *The Atlanta Journal and Constitution*, January 23: H2.

Sandberg, Jared. (2000). "Why Y2K won't die," *Newsweek*, January 10: 38–39.

Taylor, Paul. (2000). "The surprise benefits of bug-free Y2K," *Financial Times* (London), January 5: B4.

Tenner, E. (1996). *Why things bite back: Technology and the revenge of unintended consequences*. New York: Vintage Books.

Trott, Bob. (1998). "Microsoft wakes up to the Y2K problem," *Infoworld*, 20 (15): 1.

Ulrich, William. (1997). *The Year 2000 software crisis: Challenge of the century*. Upper Saddle River, NJ: Yourdin Press.

Urakami, Leiji. (2000). "U.S. declares victory, defends huge repair costs," *Japan Economic Newswire*, January 3: A1.

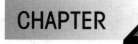
Theories of Technological Disasters

> *"Nothing is so practical as a sound theory"*
> —Kurt Lewin

*T*he pivotal concept of *technological disaster* seems to be self-evident. In reality, however, its definition is anything but self-evident. Four different theories—systems theory, normal accident theory, high reliability theory, and sociotechnical systems theory—are presented. Our task in this chapter is to clarify the meaning of technological disaster by analyzing these four theories.

A SYSTEMS APPROACH TO TECHNOLOGICAL DISASTERS

A *closed-systems analysis* treats any technological system as independent of external factors and tends to focus exclusively on internal structural properties and internal relations; analysis of parts of the system is ultimately subordinated to the analysis of the system as a whole. This approach, however, unnecessarily restricts the conception of a technological system because (1) it fails to consider the external environments impinging on the system, (2) it fails to consider the dynamic equilibrium that must be established between the internal and external environments of a system, and (3) it rarely considers changes of the system over time. Because of these defects, the closed-systems approach is not useful in explaining the conditions under which a technological system achieves a "steady state," namely, maintains a balance among opposing variables in a system. Nor does it explain the conditions under which the system errs or deviates in a destructive way from its intended purpose (Cherns, 1978). Because of the defects of a closed-systems approach, we will explore the implications of an *open-systems* approach for understanding technological disasters (von Bertalanffy, 1950; Miller, 1978).

Since all technological systems, from steamboats to petrochemical plants, are consciously designed to achieve a goal or set of goals, a systems definition of technological disaster is based on the observed discrepancies between the goals of the system and its overall performance. In this way, we can understand failures and disasters of technology as shortfalls between performance and standards, or gaps between promise and performance. An operational definition of this kind stresses goals and performance because failures and disasters in technological systems can only be measured against the intended purposes of a system (Turner, 1978).

In a complex technological system, failures and disasters are never due to a single cause. Therefore, in taking a systems approach to technological failure and disaster, it is necessary to take into account the functional context in which a system is operating (Bignell and Fortune, 1984). It is important to point out that a system's disaster is always a major failure of the goals and purposes of a system. Thus, the failure of a system cannot be separated from the intentions of the human beings who design, operate, or manage it. In other words, technological failures and disasters cannot be fully understood unless placed in the context of the social setting in which they occur. Moreover, these general observations apply to the whole continuum of technological systems. Any technological system, no matter how mundane—be it a thermostat, an automobile, a telecommunications satellite, or a large-scale technological system such as an electric utility or a nuclear reactor—exhibit various components functioning in a complex and integrated way as designed by human beings for a particular purpose.

A technological disaster can be understood as a radical departure from the pattern of expected inputs and outputs of a system. It almost always entails the failure of negative feedback to restore the system to a stable or steady state. Negative feedback involves the sensing of information regarding the current state of the system and using the information to achieve the desired balance between the input and output of the system. If there is a sufficient difference between the stated goal of a system and its actual performance, engineers or the mechanisms they have developed seek to reduce the difference. Negative feedback mechanisms, built into the system, are designed to detect and reduce deviations or errors in performance. By driving the controls in the direction opposite to the initial deviation or error in performance, negative feedback maintains the desired state of a system and reduces the discrepancy between the goals of a system and the performance of that system. When signals are fed back over the feedback channel in such a manner that they decrease the deviation of system output from a steady state, we observe negative feedback in operation (Rapoport, 1968).

The operation of a mundane thermostat illustrates the mechanism of negative feedback. The *desired* temperature setting is the goal, and the *recorded* temperature in the room is a measure of the performance of the heating system. If the difference between the desired temperature and the recorded temperature is less than zero, the furnace sends up heat. As soon as the difference in temperature readings reaches a value slightly greater than zero, the furnace is turned off.

The first task of a systems approach is to identify the relevant levels of analysis. Structures at the indicated level of analysis are called *systems*. Those at the level above a given system are called *suprasystems,* and those at the level below are *subsystems*. Complex systems may include several suprasuprasystems and subsubsystems (Miller, 1978). Applying a systems approach to our understanding of disaster involves at least two steps. First, we must identify all the components, subsubsystems, subsystems, systems, suprasystems, and suprasuprasystems thought to be relevant to the problem at hand. Second, we must identify breakdowns in the feedback mechanisms when the entire complex of systems, subsystems, and suprasystems have failed to function in an integrated manner.

These distinctions make it possible for us to make a preliminary distinction between a technological failure and a technological disaster. A failure involves a breakdown of one component and/or one subsystem of a technological system. By contrast, a disaster involves a breakdown of multiple components and multiple subsystems of a technological system, thereby threatening the viability of the entire system. Otherwise put, a technological failure is a partial breakdown of a technological system whereas a technological disaster is a complete breakdown.

FEEDBACK MECHANISMS AND THE DESIGN OF ENGINEERING SYSTEMS

The importance of feedback as a necessary mechanism in correcting errors is almost an axiom of engineering design. Negative feedback is crucial if components are to work as designed. If not, those components fail. Numerous components are kept near a target state, within a range of stability, by negative feedback controls. When these fail, the structures and processes of a system alter markedly, causing the system to fail, either partially or completely. A systems approach attempts to identify the different ways the components in a system affect one another and themselves. These interrelationships can be grouped into the categories of positive and negative feedback mechanisms.

Positive feedback in a system means that an increase in a given variable produces a further increase in that variable. Likewise, a decline in a given variable produces a further decline in that variable. Positive feedback tends to undermine the stability of a system, and a system in which many variables are governed by positive feedback can easily go haywire.

Negative feedback in a system means that an increase in one variable produces a decrease in another and vice versa. This kind of feedback tends to perpetuate the status quo. It maintains equilibrium in a system. As disturbances occur—and they are certain to occur—negative feedback works to return the system to equilibrium. Predator-prey relationships are a good example of negative feedback mechanisms. Such relationships work so that over time, the two sets of variables reach equilibrium near stable, average levels.

Negative feedback is frequently used in the design of technical systems to create stable situations, such as a constantly cool temperature in a refrigerator. A system incorporating many variables regulated by negative feedback is a well-buffered system. It can absorb a great many disturbances without becoming unstable.

Variable control in systems involves the recognition of at least four factors. First is the desired steady target value of an observed quantity. Second is the actual measured value of the system in operation. Third is the difference between the desired targeted value and the measured value. If the difference between these two values is unacceptable, this would constitute a systems error. Fourth is the action taken in response to the detected error rate between the target value and the real value of the system. If the action is directed so as to reduce error, then negative feedback is operating.

When designing a technological system, engineers build in redundant negative feedback mechanisms such as buffers and margins of safety that can reduce the deviation from the goal of system performance. Given the phenomena of technological failures and disasters, it may be that typical internal feedback mechanisms, such as "standards of safety" protocols used by engineers, may be inadequate, especially with highly complex, high-risk technologies. Also, due to their complexity and novelty, engineering designs may, paradoxically, not be grounded in sound scientific theory. For example, until recently computer programming has been more of an art than a science. Programming applications and, to a certain extent, programming languages are not always designed and tested according to scientific principles. Hence, some technological failures and disasters result from not grounding engineering designs in scientific theory, either due to the practical necessity of getting things done or because adequate theory is not yet available. The result is inadequate negative feedback mechanisms. "Limited knowledge implies that control . . . of systems will be limited" (Burns & Dietz, 1992: 211).

Another important factor in the failure of technological systems is the fact that management does not build–in organizational redundancies (Landau, 1969). An example of organizational redundancy can be found in matrix structures of organizations, which involve allocating employees to their functional specialization groups and to project groups. In such matrix structures, employees report to multiple supervisors of functional departments and of project teams. More generally, matrix structures of organizations are exemplified in multifunctional work teams, where members from different functional departments are brought together to work on complex projects and are managed and evaluated by more than one supervisor (Khandwalla, 1977).

All of these examples involve forms of organizational redundancy. Providing for organizational redundancies is one mechanism for countering destructive positive feedback, which should be distinguished from positive feedback effects of a beneficial and self-enhancing nature that amplify deviations in a positive direction (Rapoport, 1968; Evan, 1993). When a corporation that embarks on a new market strategy exceeds its projected growth rate by a significant percentage, it is enjoying positive feedback of a beneficial variety. It is important to point out, in addition, that destructive positive feedback can occur at *any* level of a system, subsystem, or suprasystem.

Systems failures clearly do not involve a concatenation of random events; they require time for destructive positive feedback to emerge. As Turner puts it:

> For each technological failure that emerges. . .there is an "incubation period" before the disaster that begins when the first of the ambiguous or unnoticed events, which will eventually accumulate to provoke the disaster, occurs. . .Large-scale disasters rarely develop instantaneously, and the incubation period provides time for resources of energy, materials, and manpower, which are to produce the disaster, to be covertly and inadvertently assembled. (Turner, 1978: 193)

An influential theory of failure and disaster of technological systems, that of Charles Perrow, takes an explicit systems approach.

PERROW'S THEORY OF "NORMAL ACCIDENTS" (NAT)

Perrow, an outstanding sociologist, analyzes the catastrophic potential of high-risk technologies through what he calls a "systems" analysis. Perrow begins by developing a hierarchy of failure. At the first level, an "incident" can occur, which usually pertains to failures of the smallest components of a system—the parts and units. Accidents tend to occur when an array of parts and units fail, causing multiple subsystems to fail, which, in turn, may cause the entire system to break down. This leads Perrow to define an accident as "a failure in a subsystem, or the system as a whole, that damages more than one unit and in doing so disrupts the ongoing or future output of the system" (Perrow, 1984: 67). Next, Perrow distinguishes between two types of accidents—"component failure accidents" and "systems accidents." These, in turn, are to be distinguished, says Perrow, "on the basis of whether any interactions of two or more failures is anticipated, expected, or comprehensible to the persons who designed the system and those who are adequately trained to operate it" (Perrow, 1984: 70–71). But, Perrow adds, "it is not the source of the accident that distinguishes the two types of accidents because all accidents start with a component failure—a valve or operator error. It is the presence or not of multiple failures that interact in unanticipated ways [. . .that are the distinguishing characteristics]" (p. 71). Perrow concludes that a systems accident is "an unanticipated interaction of multiple failures" (p. 70).

To account for an unanticipated interaction of multiple failures, Perrow distinguishes two critical dimensions along which systems can be classified. According to Perrow, systems can be either "linear" or "complex," or can be either "loosely coupled" or "tightly coupled."

The concept of coupling is well known in engineering and has been adopted as well in the social sciences. As Perrow defines it, *tight coupling* is "a mechanical term meaning there is no slack or buffer or give between two items [in a system]. What happens to one directly affects what happens in the other" (Perrow, 1984: 89–90). The opposite term, *loose coupling*, means the contrary—that there is a lot of slack between components, so when one fails, there is a buffer that allows the system to recover from a perturbation that has sent the system out of control. In a system exhibiting loose coupling, the relative insulation between subsystems slows the negative effects of a localized component mishap from spreading to other, larger units such as subsystems. Tightly coupled systems, on the other hand, need to have all the buffers and redundancies built in to the design and structure of the system itself.

The concept of *complexity* is a term of art used by Perrow to mean "baffling, hidden interactions" not anticipated in the original design that have the potential to "jump" from one subsystem to another (Perrow, 1984: 94). High-risk technologies are complex in that a single component often serves more than one function. To use an example from Perrow, a heat exchanger might be used to both absorb excess heat from a chemical reactor and to heat gas in a certain tank (Perrow, 1984: 79). Perrow suggests that when subsystems share pipes, valves, and feed-lines, and when feedback mechanisms automatically control key processes, accidents are to be expected, even inevitable—and hence "normal." Moreover, components in different subsystems are often in close operational proximity. If a component fails in one subsystem, the disruption might "jump over" into another subsystem, causing unplanned disruptive consequences. What Perrow is attempting to capture with his concept of complexity is what engineers call a *common-mode failure,* a phenomenon well studied in reliability engineering and engineering design science. Common-mode failures are relatively well known to design engineers and, more often than not, are included in safety and risk analysis techniques such as fault-tree and event-tree methods.

For Perrow, technical systems most prone to failure are complex, tightly coupled systems—that is, those technical systems that exhibit many potentials for common-mode failure. In his Normal Accident Theory (NAT), technological disasters are classified as "normal" due to the tight coupling and interactive complexity they exhibit, namely factors that make the chain of events leading to a disaster incomprehensible to the operators.

Perrow attempts to account for technological disasters without appealing to such factors as "operator error, faulty design or equipment, lack of attention to safety features, lack of operating experience, inadequately trained personnel, failure to use the most advanced technology, systems that are under financed, or poorly run" (Perrow, 1984: 63). As he states: "The analysis . . . focuses on the properties of systems themselves, rather than on the errors that owners, designers, and operators make in running them" (p. 63). Hence, human factors appear to play a relatively unimportant role in NAT. It is as if the technologies that concern Perrow have become autonomous—beyond the control of human beings.

It is a fact, however, that many of the accounts of technological disaster chronicled by Perrow in his book *Normal Accidents* concern negligent managers, incompetent operators, shortsighted owners, and disorganized social systems. As Hirschorn points out, Perrow seems to ignore his own evidence that technological disasters are caused by a complex interaction of technical systems, human factors systems, organizational systems, and sociocultural systems, and not as a result only of tightly coupled, complex technical and organizational systems (Hirschorn, 1985: 846). For example, many of the marine accidents discussed by Perrow were caused by the conscious refusal of captains to cooperate with one another when their ships were in danger of collision (Perrow, 1984: 174). Or, take, for example, what Perrow says in his narrative of the Flixborough chemical plant explosion, which killed 28 people, injured dozens of others, destroyed the plant, and damaged about 1,000 houses, shops, and factories in rural England. Perrow himself concludes that "fairly gross negligence and incompetence seem to account for this accident" (p. 111). To refer to the Flixborough chemical plant as an "accident" is clearly a misnomer. It was a technological disaster.

We can agree with Perrow that the common excuse of operator error is insufficient to account for technological disasters. Such excuses divert attention away from systems failures in Perrow's sense of the term. Such excuses also divert attention away from various human, organizational, and socio-cultural factors that are also at the root of technological failures and disasters. Interactions among operators, technology, and organizations are intrinsically complex; whatever failures occur cannot be blamed solely on operators. By declaring that operators must have failed, complex problems are covered up. As Perrow puts it:

> Finding the faulty designs responsible would entail enormous shutdown and retrofitting costs, finding that management was responsible would threaten those in charge, but finding the operators were responsible preserves the system, with some soporific injunctions about better training. (p. 133)

Perrow is unquestionably correct in this regard. One must see the operators as only one link in the chain. Systems failures should be seen as human-machine mismatches. The behavior of operators cannot alone account for failures because their behavior can only be understood in the context of the equipment they use, the design of the technology, and the administrative structure that controls operators. However, Perrow's marginalization of human, organizational, and socio-cultural factors surrounding the causes of technological disasters leads to an impoverished understanding of the complex dynamics of technological disasters. A comprehensive theory of technological disasters must focus on the complex effects of technical, human, organizational, and socio-cultural factors that cause technological disasters, in addition to such structural factors as interactive complexity and tight coupling.

It is noteworthy that, prior to the publication of his influential book *Normal Accidents*, Perrow himself outlined the rudiments of such a theory of technological disaster in an article

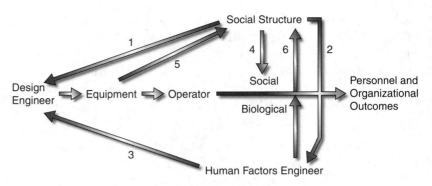

Figure 4–1
Perrow's "The Context of Design."
Source: Reprinted from "The organizational context of human factors engineering," by Charles Perrow published in *Administrative Science Quarterly* Volume 28, Number 4 (December 1983) by permission of Administrative Science Quarterly.

entitled "The Organizational Context of Human Factors Engineering" (Perrow, 1983). In this article Perrow accords a significant role to sets of interrelationships between operators, design engineers, and human factors engineers, as shown in his model, which is reproduced in Figure 4–1.

Perrow analyzes the various sets of interrelationships between the human factors engineer, the design engineer, the social structure, etc., as illustrated in lines 1 through 6.

In fact, Perrow identified four factors of organizational design that influence engineering design:

1. top management goals and perspectives
2. the reward structure of the organization
3. insulation of design engineers from the consequences of their design decisions
4. aspects of social structure

Perrow argues that goals such as speed, power, and maneuverability usually win out over considerations of easy maintenance, ease of operation, reliability, and sometimes safety when choosing among various designs (Perrow, 1983: 521). In addition, top management often fails to see the utility of a broader conception of technological design. The authority structure—

whether it is centralized or decentralized, or the span of control over operational procedures is tight or loose—also impinges on the design of hardware. As Perrow puts it:

> The design of systems, and the equipment that is used, is not entirely determined by technical or engineering criteria; designers have significant choices available to them that will foster some types of social structures and operator behaviors rather than others. Designers can choose, or they can implicitly accept, design rather than operating criteria, or the criteria implicitly preferred by top management. (1983: 534)

According to Perrow, design engineers have much to learn from human-factors engineers. Design engineers must pay attention to the way "things"—equipment, layout, ease of operation and maintenance—interact with human operators. As he puts it:

> This is especially necessary as the high-demand load of modern technologies threatens to exceed the physical, biological, and cognitive capacities of human operators. For example, passive-monitoring designs encourage de-skilling, tedium, and low system comprehension, which may lead to low morale, low output, and lack of skills to deal with operational problems. (1983: 522)

Hence, Perrow argues that the human-factors engineer "can bring to the design engineer knowledge about anthropometric limits, visual and motor sensitivity, response time, cognitive capacity and memory limits, and workload capacity for individual workers" (1983: 525), all of which may greatly reduce the friction between humans and machines operating within a system.

Perrow's general thesis regarding these relationships is that design engineers and top management fail to take into account relevant information that could be supplied by human-factors engineers regarding the physiological, affective, and cognitive properties of human operators, and their effects on the organization in which the technology is embedded.

According to Perrow, a major deficiency is the design engineer's inability to appreciate the vulnerabilities of human operators. For Perrow, design engineers have a tendency to see technical systems as "closed systems," that is, as systems composed of various hardware components, which are in turn designed as "expert" systems, with absolute rationality and error-free internal logic. In contrast, human beings exhibit "bounded rationality," they have substantial cognitive limits on rationality (Simon, 1957: 38–41). Yet the system design is not responsive to the "bounded" rationality of human operators. Psychological research (Tversky and Kahneman, 1974) and Perrow's own graphic account of accidents (Perrow, 1984) indicate that human operators exercise selective attention, engage in limited search behavior, and discount nonconfirming information when confronted with novel or emergency situations.

The operator is not simply a transfer device in the loop, but a creative interpreter of phenomena, a "bio-cognitive" creature with limited rationality. The design engineer (and the whole organization) should model creative operators in all their complexity more accurately, because the design employed will elicit some, rather than other, performance characteristics of the system. Perrow concludes that design engineers must pay attention to the way technological artifacts are shaped by the organizational structure and by top management interests, which, in turn, shape operator behavior.

Perrow's context of design model (Figure 4–1) is more complex than his later model, articulated in his book *Normal Accidents,* where human and organizational factors tend to be de-emphasized. However, the earlier model is still in need of refinement. For one thing, Perrow's model does not include socio-cultural factors, which consist of values and beliefs of the social institutions of the environing society. The engineers and managers who incorporate these values and beliefs exert an impact on the technological system and its management structure. An improved model would include the impact of human-factors engineering on operator behavior, as well as the impact of socio-cultural factors on managers and design engineers. In Figure 4–2, we present a modified version of Perrow's model.

Top managers, who have internalized socio-cultural factors, use the chain of command to direct the behavior of design engineers, human-factors engineers, and operators. Staff personnel, design engineers, and human-factors engineers do not have the authority to direct the behavior of operators; at best they instruct and advise the operators. If and when the suggestions of design and human-factors engineers are challenged by the operators, top management intervenes to resolve the conflict.

The hierarchical structure of a technological system provides "different contexts for understanding the system and different incentives to guide action" (Burns and Dietz, 1992: 212). The de-

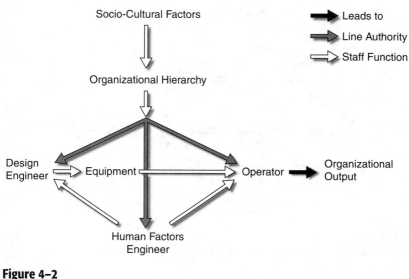

Figure 4–2
Evan and Manion's revised model of Perrow's "The Context of Design."

sign engineer is concerned principally with designing a functional, safe, and efficient system; his or her supervisor focuses on productivity and the cohesion of the work team, which may impede monitoring and auditing performance. In addition, top management is understandably oriented to the firm's profitability and shareholder returns on investment.

Notwithstanding our revised model of Perrow, as depicted in Figure 4–2, his overall theory of normal accidents tends to be pessimistic, maintaining that accidents in high technology industries are "normal," in the sense of "expectable," because complex technologies have "outrun our organizational abilities to manage and control them" (Bierly and Spender, 1995: 639). For Perrow, failures are inevitable because organizational powers are unable to handle the safe and reliable design, development, and management of high-risk technologies.

Perrow concludes that the bureaucracy model of organization is inadequate to handle the complexities of modern technology. But, rather than discard the bureaucracy model, which he argues is the best model for managing technology in complex systems, Perrow concludes that human beings cannot successfully manage certain high-risk technologies. But why? Perhaps we can develop

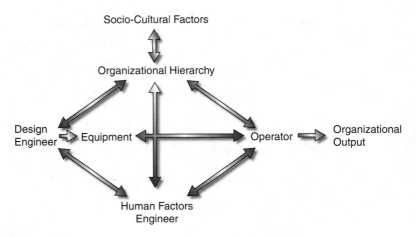

Figure 4–3
Evan and Manion's nonhierarchical consultative model of technological
systems.

a nonbureaucratic model of technology management. Figure 4–3
presents the outlines of such a model.

The two-directional arrows between staff and line underscore
the need for a *nonhierarchical and consultative relationship*, at
least in the planning stages and general operational processes.
Two-way flows of communication are especially essential in technological systems to maximize the sharing of information among
all personnel regardless of position in the organizational hierarchy.
Bureaucratic barriers to cooperation are particularly dysfunctional, given our limited understanding of technological systems
and our limited ability to control them. However, when a crisis
arises in the operations of a technological system, the command
model—namely, an hierarchical and single-directional mode of
communication—would supersede the nonhierarchical consultative model in an effort to contain the crisis and limit the damages.

Perrow's analysis of technological failures in *Normal Accidents* narrowly restricts analysis of the causes of technological
disaster to technical and organizational factors—for example, the
effects of complexity and coupling—and downplays human and
socio-cultural factors. We propose, instead, a *concentric* model of
technological systems that takes into account human and socio-
cultural factors as well as technical and organizational factors.
The difference between Perrow's model and ours is represented in
Figure 4–4.

Perrow's Model
of Technological Systems

Evan and Manion's Model
of Technological Systems

Figure 4–4
Perrow's model of technological systems versus Evan and Manion's model of
technological systems.

Organizational factors are often left implicit in NAT (Perrow, 1999: 368). For example, in analyzing the paradoxes of organizational design in high-risk systems, Perrow suggests that we have reached a cultural and organizational cul-de-sac (Perrow, 1984: 360). According to Perrow, we need centralized systems to ensure obedient responses in emergencies, but we also need decentralized systems so that workers can use their initiative to help solve unexpected problems. For Perrow, combining centralization and decentralization is impossible in tightly coupled, highly complex systems. This is because systems that exhibit tight coupling require increased centralization to facilitate rapid, authoritative, and correct decision making; whereas systems, which exhibit interactive complexity, require decentralization of decision-making authority. His underlying assumption is that our administrative capabilities are decidedly inadequate to manage high-risk technologies. However, models of organizational control can be both centralized and decentralized. In fact, as any proponent of the sociotechnical school of management will readily point out, many manufacturing organizations are designed according to "sociotechnical principles" in which close coordination is achieved between workers and work teams without resorting to hierarchy (Pasmore and Sherwood, 1978).

Perrow, however, is strongly committed to the superiority of the bureaucratic model, despite its limitations. In fact, he claims that:

> Bureaucracy is a form of organization superior to all others we know and can hope to afford in the near or middle future; the chances of doing away with it or changing it are probably

non-existent in the West in this century. Thus it is crucial to understand it and appreciate it. But it is also crucial to understand not only how it mobilizes social resources for desirable ends, but also how it inevitably concentrates those forces in the hands of a few who are prone to use them for ends we do not approve of, for ends we are not generally aware of, and more frightening still, for ends we are led to accept because we are not in a position to conceive alternative ones. (1979: 7)

Perrow finds himself on the horns of a dilemma primarily because he does not consider any other form of organizational control. The only model he considers is the standard bureaucratic structure. Since bureaucratic principles are ultimately unsuccessful in combining loose and tightly coupled structures, or centralized and decentralized authority, he concludes that high-risk technology cannot be safely and reliably developed and managed. Perrow does direct our attention to a significant problem that needs to be addressed, but other models of organizational design are available, such as those identified in the literature on high reliability organizations.

HIGH RELIABILITY THEORY (HRT)

In the closing chapter of his book *Normal Accidents,* Perrow classifies high-risk systems into three categories: (1) systems that should be abandoned because the risks outweigh any potential benefits; (2) systems that could be made less risky or those with potential benefits such that some risks should be tolerated; and (3) systems that are to some extent self-correcting and could be further improved. Technologies of the third category include mining, aircraft, dams, and chemical plants. Technologies in the second category include marine transport and genetic engineering. Nuclear weapons and nuclear power fall into his first category of technologies that cannot be tolerated because they are too prone to "systems" accidents; in other words, their failure is immanent and inevitable—"normal."

Perrow's prudent and circumspect conclusions regarding high-risk technological systems stimulated other researchers to study high-reliability organizations (HROs) such as nuclear submarines, nuclear-powered aircraft carriers, nuclear power plants, and air traffic control centers. These organizations are "characterized by both advanced technology . . . and a high degree of interdependence . . . " (Roberts, 1990 [a]: 161).

Roberts and her interdisciplinary team of researchers studied three HROs to ascertain "how they maintain safe and reliable operations under hazardous conditions" (Roberts, 1990 [b]: 102). The three organizations included the Pacific Gas and Electric Company (PG&E), the Federal Aviation Air Traffic Control system, and a U.S. Navy aircraft carrier. Each of these organizations involved technologies exhibiting Perrow's complex interaction and tight coupling characteristics. Nevertheless, their performance was highly reliable: PG&E, a utility serving 4 million customers in California, was 99.96 percent reliable in terms of outages; air traffic control centers had not had a single mid-air collision in the 10 years preceeding the study; the U.S. Navy aircraft carrier "crunch rate" (accidents of aircraft being moved on deck) in 1989 was one in 8,000 moves; and the fatality rate due to deck fire in 1989 on aircraft carriers was 2.97 for every 100,000 hours of flight time (Roberts, 1990 [b]: 102–103).

Another striking example of an HRO is the Diablo Canyon nuclear power plant operated by PG&E. This plant is capable of generating 2,190 megawatts (MW) of electricity. Plant personnel include operations, maintenance, engineering, administration, and/or technical, totaling 1,250 employees on site, plus more than 1,100 outside consultants.

In stark contrast, PG&E also operates a conventional fossil-fueled steam generating plant near Pittsburg, California. The plant produces roughly the same amount of electricity with one-fourth of the personnel of the Diablo Canyon plant. Its budget for operations and maintenance in 1990 was $30,318,000, whereas the budget of the Diablo Canyon facility was $187,000,000. The difference in operating costs of the two plants—one is six times more expensive than the other—is due to "the added complexity of nuclear technology, the current political and regulatory climate surrounding nuclear power generation, and specific organizational strategies adopted for the management of this technology at Diablo Canyon" (Schulman, 1993: 356).

The impressive performance record of HROs in this study was achieved by systematically applying strategies to counter the potential hazards associated with technological complexity and tight coupling. Among the organizational strategies and processes identified are continuous training of personnel, job redesign, and redundancy (Roberts, 1990 [a]). Another strategy is building-in a "culture of reliability," norms and values pertaining to safety, accountability, and responsibility. "Managers of HROs must confront the cost of potential catastrophe and build into their organizations strong norms for cultures of reliability" (Roberts, 1990 [b]: 112).

In their study of a nuclear submarine, Bierly and Spender (1995) underscore the contributions of Admiral Hyman Rickover, the creator of the U.S. nuclear navy, to the unique culture of this type of organization. Insisting on stringent recruitment standards, Rickover personally interviewed each potential officer. His principles of selection—apart from the highest level of professional competence—were commitment, trust, communication, and performance under pressure. Along with his unique approach to the selection and training of naval reactor officers, his "obsession with safety" affected the design, manufacture, and operation of the nuclear power plant. Rickover "forced contractors to approach their task with the same demanding zero-defect engineering standards that builders of supersonic fighters had used for years" (Bierly and Spender, 1995: 652).

As of 1979, when the Three Mile Island (TMI) accident occurred, Rickover had been managing 152 naval reactors, operating almost 30 years without a *single* accident resulting in radioactive emissions. In the course of the TMI inquiry, Rickover "insisted that this extraordinary reliability was not due to the military context and personnel, rather that it was due to careful selection of highly intelligent and motivated people who were thoroughly trained and then held personally accountable" (Bierly and Spender, 1995: 651). If all managers of nuclear power plants around the world had Rickover's dedication to reliability, safety, and professional discipline of their employees, the safety records of such plants would be greatly enhanced. Rickover's conception of a culture of reliability entailed the following components:

> . . .each individual's ownership of the task, responsibility, attention to detail, high professionalism, moral integrity, and mutual respect created the cultural context necessary

for high quality communications under high risk and high stress conditions. Communication and recommendations can flow upward from the crewmen to the officers as well as downward. Likewise communication about all kinds of mistakes, operational, technical or administrative, can flow rapidly through the system. Anyone making a mistake can feel free to report it immediately so that the watch officers can really understand what is happening to the system. Rickover believed that the real danger lay in concealing mistakes, for when this happens those in charge become disconnected and disoriented. This could be disastrous in the high-risk circumstances of a nuclear warship. (Bierly and Spender, 1995: 651)

Perrow's NAT has reified "technological systems" to the point that no human operators or agents can ever be held accountable or responsible for any system failure. This impersonal conception of systems fails to account for the fact that some high-risk systems are highly reliable, presumably because motivating human operators makes a difference in reducing the vulnerability of systems to failure. The research of Roberts and her colleagues in high-reliability organizations, as well as Rickover's experience with nuclear submarines, evidently do not fit Perrow's theory: *interactive complexity and tight coupling need not result in catastrophe.*

As Roberts and her colleagues have shown in their work on HRT, and as Rickover has dramatically established, the principal characteristics of high-reliability organizations are (1) top management's commitment to safety as an organizational goal, (2) the need for personnel redundancy as well as engineering redundancy, (3) the development of a culture of reliability, and (4) the valuation of organizational learning (Sagan, 1993). When these principles are implemented, they have the effect of countering the potentially catastrophic consequences of interactive complexity and tight coupling that Perrow's theory predicts.

Perrow's analysis reminds us that corporate elites have constructed our technologies and that we can abandon them if they become oppressive. They are indeed social constructions (Perrow, 1984: 285). Yet, he simultaneously evokes a picture of "transcendental technologies" that defy our capacities to organize and manage technology.

A SOCIOTECHNICAL SYSTEMS ANALYSIS OF TECHNOLOGICAL DISASTERS

As we have seen, Perrow's model suggests that the dynamics of technological failure and disaster are a result of interactive complexity and tightly coupled systems. However, the model we are advancing will demonstrate that, in addition to the structural features of interactive complexity and tight coupling, the dynamics of technological failure and disaster reflect fundamental problems of organization design, worker competence, management systems, and the socio-cultural context in which the technology develops.

On Perrow's account, design engineers are incapable of anticipating all of the possible effects of complex systems and, hence, failure is inevitable. But, technological systems are human constructions ultimately under the control of human beings. People *can* modify them to minimize and possibly eliminate errors if they are so motivated. As Perrow himself states, "ultimately, the issue is not risk, but power, the power to impose risks on the many for the benefit of the few" (Perrow, 1984, p. 306). Hence, the causes of the risks and harms posed by technology are more of an organizational structure issue than those associated with the abstract system itself.

In contrast to organizational designs that focus on either the social system or the technical system exclusively, the *sociotechnical systems approach* integrates the demands of both. Emery (1959) referred to this dual concern with the social and technical systems as *joint optimization*. A sociotechnical systems approach focuses attention on dynamic processes within organizations and between organizations and their environments.

For complex, hazardous technologies, one major cause of loss of control is the failure of both design engineers and organizational planners to incorporate adequate sociotechnical detection and monitoring systems that feed back observable deviations between the sociotechnical system and its external environment. According to E.L. Trist, sociotechnical systems breakdowns almost always "reflect the mutual permeation of a particular organization and its environment that is [often] the cause of such imbalance" (Trist, 1978: 45). Creating high-reliability, high-performance sociotechnical systems requires analyzing the *interdependencies* among the social, technical, and environmental subsystems that render each system unique. With this approach, the emphasis is more on interdependencies between the technical and social sub-systems rather than on isolated problems. The gap between the technical systems analysis and the social systems analysis leads to a frequent lack of coordination between the builders and design-ers of the systems (technical system) on the one hand and the op-erators and managers of the system (social system) on the other. This leads to "different people working in very different contexts and according to different rules with different constraints" (Burns and Dietz, 1992: 212).

Thomas Hughes, a distinguished historian of technology, ar-ticulates a systems approach in his writings. Hughes's analysis fo-cuses on what he calls "large-scale technological systems." This approach stresses the importance of paying attention to the dif-ferent but interlocking elements of physical artifacts, institutions, and their environments, thereby integrating the technical, eco-nomic, social, and political aspects of technological development. As Hughes puts it:

> Technological systems contain messy, complex, problem solving components. They are both socially constructed and socially shaping. Among the components in technological systems are physical artifacts, such as turbo generators, transformers, and transmission lines in electric light and power systems. Techno-logical systems also include organizations, such as manufactur-ing firms, utility companies, and investment banks, and they incorporate components usually labeled scientific, such as books, articles, and university teaching and research programs. Legisla-tive artifacts, such as regulatory laws, can also be part of tech-nological systems. Because they are socially constructed and adapted in order to function in systems, natural resources, such as coalmines, also qualify as system artifacts. (Hughes, 1994: 51)

The work of Hughes draws attention to the socially con-structed nature of technological systems. After all, sociotechnical systems *are* constructed by individuals and collectivities (Latour and Woolgar, 1986). Since different stakeholders or constituents have different, often conflicting interests and resources, they tend to differ in their views as to the proper structuring and design of technological systems. In addition, different stakeholders may have different and competing processes for determining what ought to be the proper operation of a technological system (Barnes, 1974). From this perspective, an explanation of technological failure and disaster depends on two factors. The first is the study of the socially constructed conditions of technological development. Because different stakeholders often have competing definitions as to the problems to be solved in determining the best design of a technological system, there needs to be much negotiation and interpretative flexibility in design decisions (Barnes, 1982).

Second, various stakeholders will decide differently not only about the definition of the "best" design, but also about the "steady state" of the system in operation. Accordingly, decision making concerning the steady state of a sociotechnical system must be explained by referring to the varied and often competing social and economic interests attributed to the various stakeholders concerned and their capacity to mobilize forces in the course of debate and controversy (Latour and Woolgar, 1986). Social constructivists often talk of this process as one of "closure," which is ultimately achieved when debate and controversy about either the design or the operation of a system is resolved (Bloor, 1976). The merits of a social constructivist approach to technological disaster are clear. Many design decisions about the structure of sociotechnical systems are arrived at in the course of debate and achieve their final form when a stakeholder imposes its solutions on other interested parties by one means or another; that is, by compromise, persuasion, or fiat.

Ultimately, the question we confront is not *why* things come apart in one specific instance but *what* in general constrains the capacity of human beings to manage and control complex and potentially harmful technologies. A sociotechnical approach to technological disaster attempts to provide an answer to this question. Moreover, if we accept, with Thomas Hughes, that technical systems are at the same time economic, social, and political constructs, then a systems approach to technological disasters must struggle with this complex web of relationships between technical, economic, social, and political factors.

The preceding analysis leads us to the following operational definition:

A *technological disaster* is a crisis that threatens the viability of a sociotechnical system, which includes machines, design engineers, human operators, rules and role behaviors, and socio-cultural factors, and which results in the massive loss of life, property damage, or environmental deterioration. By contrast, a technological failure, as we indicated in Chapter 1, is defined as the unanticipated break-down of one or more components of a technological system that may impair its functioning and that may cause minor property damage or loss of life.

CONCLUSION

Technological disasters are failures of sociotechnical systems. This fact should lead design engineers to examine not only the failures of technical design but also the impact of human, organizational, and socio-cultural factors. The hypothesis we are advancing is that those who design and operate sociotechnical systems must concern themselves not only with technology as merely a material artifact but also with the ways in which the hardware and the software are related to economic, social, and political interests of various stakeholders.

The performance, stability, and reliability of sociotechnical systems, as well as their ability to tolerate environmental disturbances, are dependent upon the nature, formation, and interaction of human and socio-cultural subsystems and suprasystems, not simply upon the technical and organizational facets of systems, as Perrow's model suggests. Each sub- and supra-system forms a link in a complex chain. Each has a role in the overall performance of the system. Disturbances in one link can affect the overall integrity of the system. In fact, many of the large-scale technological disasters, discussed in Chapter 8, have been caused by nontechnical factors, namely, the human, organizational, and socio-cultural factors (Meshkati, 1991).

Given that sociotechnical systems are socially constructed, they can be designed and redesigned to minimize the probability of technological disaster. This concept is clearly compatible with High Reliability Theory (HRT). As for Normal Accident Theory (NAT), Perrow sets forth three categories of systems: (1) systems that should be abandoned because the risks outweigh the benefits (2) systems that should be made less risky, and (3) systems that are self-correcting and could be further improved. Underlying Perrow's last two categories is an assumption that it is feasible and desirable to redesign systems to minimize risk. By focusing on the challenge of redesigning sociotechnical systems to minimize risk, we can potentially bridge the theories of HRT and NAT. In the Afterword to his new edition of *Normal Accidents* (1999), Perrow himself envisions the possibility of bridging these two theories (p. 372).

To provide such integration, Ackoff's "interactive idealized design" may be helpful (Ackoff, 1994: 79–81). A renowned systems theorist and operations researcher, Ackoff's design system must fulfill three requirements: technological feasibility, operational viability, and capability of learning from its own experience and "adapting to internal and external changes." To formulate an idealized design requires the participation of "all the systems' current stakeholders, or their representatives" (Ackoff, 1994: 80).

By accepting the primacy of system safety as a fundamental value, idealized system designers may be guided by the principle of minimizing risk. For top management, with a short-time horizon, efficiency all too often trumps safety. By focusing on net returns for the next quarter, they may, however, subject the corporation to design decisions that will prove vulnerable to technological disasters, and hence, to costly losses, such as we witnessed in the Challenger disaster.

To avoid error-inducing designs, idealized system designers build-in a multiplicity of negative feedback mechanisms. In addition, given that tight coupling and interactive complexity increase the risk of failure, idealized system designers should explore the possibility of re-designing sociotechnical systems so that they loosen the coupling and simplify system complexity— even if such system characteristics diminish efficiency in some measure.

In sum, the sociotechnical systems approach to technology, when informed by Ackoff's interactive idealized design process, contributes to our understanding of technological disaster in at least two ways. First, as the work of Hughes and his followers demonstrates, many failures of technology can only be understood when interrelated with a wide range of nontechnological and specifically social and cultural factors. Second, only a sociotechnical systems approach can do justice to the complex set of causes of technological disaster.

We now turn in the following chapter to a discussion of the root causes of technological disaster.

References

Ackoff, R.A. (1994). *The democratic corporation.* New York: Oxford University Press.

Barnes, B. (1974). *Scientific knowledge and sociological theory.* London: Rutledge and Kegan Paul.

Barnes, B. (1982). *T.S. Kuhn and social science.* London: Macmillan.

Bierly, P.E., and Spender, J.C. (1995). "Culture and high reliability organizations: The case of the nuclear submarine," *Journal of Management* 21: 639.

Bignell, V., and Fortune, J. (1984). *Understanding systems failures.* England: Manchester University Press.

Bloor, D. (1976). *Knowledge and social imagery.* London: Routledge and Kegan Paul.

Burns, T.R., and Dietz, T. (1992). "Technology, sociotechnical systems, technological development: An evolutionary perspective." In M. Dierkes and U. Hoffmann (Eds.), *New technology at the outset: Social forces in the shaping of technological innovation.* New York: Campus Verlag.

Cherns, A. (1978). "The principles of sociotechnical design." In W. Passmore and J. Sherwood (Eds.), *Sociotechnical systems: A sourcebook.* San Diego, CA: University Associates: 61–67.

Emery, F. (1959). *Characteristics of sociotechnical systems.* Document No. 527. London: Tavistock Institute.

Evan, W.M. (1993). "Organization theory and organizational effectiveness." In *Organization theory: Research and design.* New York: Macmillan: 369–389.

Hirschhorn, L. (1985). "Normal accidents," *Science* 228 (2).

Hughes, T.P. (1994). "The evolution of large technological systems." In W. Bijker, T. Hughes, and T. Pinch (Eds.), *The social construction of technological systems: New directions in the sociology and history of technology.* Cambridge, MA: MIT Press.

Khandwalla, P. (1977). *The design of organizations.* New York: Harcourt Brace Jovanovich Publishers.

Landau, M. (1969). "Redundancy, rationality, and the problem of duplication and overlap," *Public Administration Review* 29 (4): 346–358.

Latour, B., and Woolgar, S. (1986). *Laboratory life: The social construction of scientific facts.* London: Sage Press.

Meshkati, N. (1991). "Human factors in large-scale technological systems' accidents: Three Mile Island, Bhopal, Chernobyl," *Industrial Crisis Quarterly* 5: 133–154.

Miller, J.G. (1978). *Living systems.* New York: McGraw-Hill.

Pasmore, W., and Sherwood, J. (Eds.). (1978). *Sociotechnical systems: A sourcebook.* San Diego, CA: University Press Associates.

Perrow, C. (1979). *Complex organizations: A Critical Essay.* New York: Scott, Foresman and Wadsworth.

Perrow, C. (1983). "The organizational context of human factors engineering," *Administrative Science Quarterly* 28: 521–541.

Perrow, C. (1984). *Normal accidents: Living with high-risk technologies.* New York: Basic Books.

Perrow, C. (1999). *Normal accidents: Living with high-risk technologies* (2nd ed.). Princeton, NJ: Princeton University Press.

Rapoport, A. (1968). "A philosophical view." In J.H. Milsum (Ed.), *Positive feedback: A general systems approach to positive/negative feedback and mutual causality.* New York: Pergamon Press.

Roberts, K.H. (1990a). "Some characteristics of one type of high reliability organization," *Organization Science* 1: 161.

Roberts, K.H. (1990b). "Managing high reliability organizations," *California Management Review* 32: 102.

Sagan, S. (1993). *The limits of safety: Organizations, accidents, and nuclear weapons.* Princeton, NJ: Princeton University Press.

Schulman, P.R. (1993). "The negotiated order of organizational reliability," *Administration and Society* 2: 356.

Simon, H. (1957). *Models of man: Social and rational.* New York: John Wiley & Sons.

Trist, E.L. (1978). "On socio-technical systems." In W. Passmore and J. Sherwood (Eds.), *Sociotechnical systems: A sourcebook.* San Diego, CA: University Associates: 43–58.

Turner, B. (1978). *Man-made disasters.* London: Wykeham Publications.

Tversky, A., and Kahneman, D. (1974). "Judgment under uncertainty: Heuristics and biases," *Science* 185 (4): 1124–1130.

von Bertalanffy, L. (1950). "The theory of open systems in physics and biology," *Science* 111: 23–29.

Wenk, E. (1989). *Tradeoffs: Imperatives of choice in a high tech world.* Baltimore, MD: The Johns Hopkins Press: 6.

CHAPTER 5

The Root Causes of Technological Disasters

*"No one wants to learn by mistakes, but
we cannot learn enough from successes to
go beyond the state of the art."*

—Henry Petroski

What are the causes of technological disasters? It would be a great comfort if we were able to identify a single cause, for that would simplify the development of a preventive strategy. Instead, we must resign ourselves to searching for multiple causes of disaster and hence for multiple strategies for assessment, management, and prevention. Analyzing the causes of technological disaster is of theoretical as well as practical significance. Pursuing our discussion of the previous chapter on sociotechnical systems, we will now focus on two systemic dimensions in an effort to throw light on the causes of technological disaster: (1) internal vs. external systemic factors and (2) technological systems vs. social systems.

In Figure 5–1 we combine these dimensions in a 2 × 2 matrix, yielding four categories of technological disaster, which we will discuss in this chapter. These factors are highly interpenetrating, often have unclear boundaries, and exert mutual influence on each other. The close study of technological disasters reveals that there are often multiple causes of disaster. In practice it may be difficult to isolate a single cause—technical design factors, human factors, organizational systems factors, or socio-cultural factors—because many technological disasters are the result of a combination of such causes. Nevertheless, the identification of the four root causes of technological disasters provides a diagnostic tool with which to analyze such disasters.

	Internal Systemic Factors	External Systemic Factors
Technological Systems	Technical Design Factors	Human Factors Factors
Social Systems	Organizational Systems Factors	Socio-Cultural System Factors

Figure 5–1
Systemic dimensions underlying the causes of technological disasters.

The assumption underlying this matrix is that we are setting forth a mutually exclusive and jointly exhaustive classification of disasters. For analytic purposes this assumption may further our understanding of the problem, but it may pose empirically verifiable difficulties. For example, the possible failure of future technologies such as recombinant DNA and nanotechnology might be associated with causes that have yet to be anticipated.

TECHNICAL DESIGN FACTORS

Design engineers are constantly working under constraints such as limited data on the properties and reliabilities of the various materials they use. Information about the operating conditions under which the materials and pieces of equipment will be expected to function is often absent, so that estimates of the resulting stresses and strains on materials are sometimes little more than informed guesses.

It is often assumed that science and technology are fully integrated. If this were true, the design engineer would be completely confident in his or her design decisions. Yet all too often scientific theory is not available to ground technological decision making. Therein lies one of the root causes of technical design errors. Furthermore, even if all information about those components were available, the existing physical, chemical, and engineering theories may not yield an unequivocal solution to a design problem.

Examining R.R. Whyte's 1975 anthology of case studies of design and equipment failure, a first of its kind, reveals many similarities among various case studies of engineering design failure from which significant generalizations can be drawn (Whyte, 1975). Whyte's book presents a wealth of case studies concerning engineering design failures, including jet and automobile engines, industrial plant turbines and generators, boilers, heat exchangers, metal fatigue, and stress. Most of these failures are related either to uncertainties in the materials used or in the application of the design in question.

Turner's extensive review of the cases presented in Whyte's book reveals three classes of technical design failures (Turner, 1984: 19–23). The first class of design failures involves designs that extend beyond the knowledge or experience of the designer and that stretch the limits of the previous design. These types of failures are usually the result of "scaling up" existing satisfactory designs to achieve operational parameters beyond the original design. Engineers call this situation *incremental design*. An example of such a failure is when design engineers use an existing plan for a larger version of a well-tested design, introducing factors that were not anticipated in the original. Other failures are the result of just the opposite: the scaling-down of existing satisfactory designs. Engineers call such practices *streamlining* and *fine-tuning*. Starbuck and Milliken (1988) argue convincingly that 24 previous successful flights had created such confidence at NASA that they began systematically "fine-tuning" the technology and design of the space shuttle Challenger and its rockets until it "broke." As Dumas puts it, "being too ready to extrapolate well beyond previous experience is asking for trouble" (Dumas, 1999: 235). Dumas cites as an example the rapid development of commercial nuclear power plants during the 1960s and 1970s, where, within a few years of constructing the first commercial nuclear power plant, plants six times the size of the original design were built.

A second class of design failures arises when designs are forced to operate under conditions that will ultimately lead to a much wider range of unknown variations and fluctuations of stress. This is the well-studied relationship between static loads and dynamic loads. Many design failures are caused by engineers who failed to take into account variations in dynamic loads, the actual fluctuations and perturbations that will be exerted on a technological system—from a bridge to a piece of copper wiring. Perhaps the most famous case in this category is the collapse of the Tacoma Narrows Bridge in 1940. The design engineers did not take into account the excessive wind over the Narrows—a dynamic load that proved catastrophic. This case demonstrates the importance of taking wind aerodynamics into account in designing a bridge.

A third class of failures concerns uncertainties in the nature, quality, and manufacture of the materials necessary for the design to become a reality. For example, Martin and Schinzinger (1996) recommend that variations from the standard quality of a given grade of steel should be taken into account in the design process. According to these researchers, the design engineer needs to realize that a supplier's data on items like steel, glass, and other materials apply to statistical averages only (Martin and Schinzinger, 1996: 143). Individual components can vary considerably from the mean.

In order to cope with the various uncertainties about materials and components, as well as incomplete knowledge about the conditions under which the products they design actually operate, engineers have traditionally introduced *factors of safety* into their designs to cope with as many unanticipated problems as possible. Factors of safety are intended to prevent problems from arising when stresses from anticipated loads ("duty") and stresses the designed product is supposed to withstand ("capability") depart from their expected values (Nixon and Frost, 1975: 138). A product is considered safe if its capability exceeds its duty. Determining the maximum stress a material must undergo in a particular application and then designing the product to withstand, for example, double that maximum stress load would result in a factor of safety of two. In his books, *To Engineer Is Human* (1985) and *Design Paradigms: Case Histories of Error and Judgment in Engineering* (1994), the engineer Henry Petroski catalogues and discusses case after case of design failures caused by unanticipated stresses, fatigue, problems with static vs. dynamic loads, and uncertainties about materials. He then develops a theory of engineering design through the study of structural, materials, and electrical failures.

Glegg, a lecturer in electrical engineering at Cambridge University, has also tellingly addressed the engineer's problem of margin of safety:

> We all automatically insure against risks, often very remote ones. Everyone takes out a policy against fire destroying a new house, which fortunately is a rare occurrence.
>
> By contrast, the insurance of a new design against much less remote disasters is not so automatic. Temperature stresses, torsional vibration, feedback oscillations and similar destroying forces may break out without warning with devastating results. Nearly all machines have some potential disaster lurking in the background. . . . The best protection is a large margin of safety. . . . Develop as large a safety margin as you can against such things as laziness, impatience, prejudice, and arrogance and all the other human failings that inhibit constructive achievement. . . . of creative engineering design. (Glegg, 1971: 88, 93)

A fourth category of failure, which Whyte does not address, pertains to inadequacies in the proper testing and/or prototyping of technological products or processes. There are limitations to the testing process itself, for sometimes many technological systems, such as nuclear power plants, cannot be tested to destruction. Hence, uncertainties will always exist when trying to determine the maximum threshold of that particular technology. More common, of course, is the fact that prototype tests and routine quality assurance tests are sometimes carried out improperly. One notorious case involves the "fudging" of data by technical writers employed by B.F. Goodrich, which had a contract with the U.S. Air Force to design and produce a braking system for the A7-D aircraft (Vandivier, 1972). In short, one cannot uncritically trust testing procedures or those who manage and carry them out.

These categories of technical design failure are endemic to engineering design. Combinations of these factors were at the root of the collapse of the Dee Bridge, the Hyatt Regency walkway collapse, the Intel Pentium Chip crisis, and the Therac-25 computer failure.

In the Dee Bridge collapse, inadequate factors of safety, combined with extending a successful design beyond its load capacity through "fine-tuning" engineering, resulted in the collapse of the Dee Bridge at Chester, England, on May 21, 1887. The Dee Bridge, designed by Robert Stephenson and constructed in September 1846, collapsed due to torsional instability, indicating a major flaw in the new design. Five people were killed and 18 were injured as a locomotive and five carriages plunged into the river, destroying the bridge. According to Petroski, "The Dee Bridge failed because torsional instability was a failure mode safely ignored in shorter, stubbier girders whose tie rods exerted insufficient tension to induce the instability that would become prominent in longer, slenderer girders . . ." (Petroski, 1994: 95).

Petroski's analysis demonstrates that: "The critical flaw in the human design process that produced the ill-fated Dee Bridge is paradigmatic; the same fundamental flaw was rooted in the design process and environment that produced . . . a host of other doomed bridge designs" (1994: 83). Another example is the famous Tacoma Narrows Bridge, affectionately called "Galloping Gertie" because of its tendency to sway immensely when high crosswinds blew across the Narrows. Just months after its completion, the bridge came crashing down in September 1940. Countless bridges have collapsed because of design flaws. Petroski attributes these failures to what he terms the "success syndrome"; namely, the tendency for engineers to become overconfident by the past successes of a particular design in the process of extending it beyond the available fund of engineering knowledge and experience (1994: 83).

On July 17, 1981, the Hyatt Regency Hotel in Kansas City, Missouri, held a dance contest in its atrium lobby. The weight of the dancers and partygoers who populated the suspended walkways at the fourth and second floors proved to be more than the structures could withstand, causing the connections supporting the ceiling rods holding the second and fourth floor walkways to fail. Both walkways collapsed onto the crowded first-floor atrium. The ensuing investigation of the accident revealed a number of unsettling facts about the technical causes of the disaster and raised questions as to who should be held responsible for the loss of life and property.

During 1976, Crown Center Redevelopment Corporation (CCRC) began plans to design and build a Hyatt Regency Hotel in downtown Kansas City and hired GCE International (GCEI) as consulting structural engineers on the project. In 1978, Havens Steel Company was hired by CCRC to fabricate and install steel for the atrium lobby following GCEI's original design.

Hyatt Regency Walkway Collapse, July 17, 1981. Copyright AP/Wide World
Photos. Used by permission.

The hotel consisted of a 40-story guest tower and a function
block consisting of administrative offices and shops. Between these
two structures was an atrium, which included three suspended
walkways connecting the guest tower with the function block at the
second, third, and fourth floors. The fourth floor walkway hung directly over the second floor walkway, which was connected to the
fourth floor walkway with steel hanging rods. The third floor
walkway hung separately from steel rods attached to the ceiling of
the atrium. The original design developed by GCEI was to run a single support rod from the ceiling, through the fourth floor support
box beams and right through the centerline of the second floor box
beams. Finally, the steel rod would be attached at the bottom of the
second floor beam with bolts and washers. As designed, the structural load from the box beam—the amount of weight the beam is
supposed to support—would transfer to the supporting hanging
rod. A washer and bolt on the rod below the beam would support

the weight shift from the walkway beam supports to the steel hanging rods fastened to the ceiling.

In 1979, during a series of miscommunications between Havens Steel and GCEI, a design change was made in the fourth floor rod-beam connection from a single rod design to a two rod design. This change basically doubled the weight load placed on the bolt and washer set of the fourth floor connection, because instead of supporting just one box beam, as in the original design, it now had to support the weight of two box beams.

The official investigation determined that the modified design was the technical cause of the disaster. In particular, the report found that:

- The collapse initiated at the fourth-floor box-beam hanger-rod connection
- The as-constructed beam-rod connection did not meet Kansas City Building codes, nor did the original continuous rod design
- The change in rod design essentially doubled the transfer load from the beam through the nut to the steel hanging rod (Roddis, 1986: 1549–1550)

In other words, an improper design led to a structural failure that could have been avoided had the structural engineers who produced the original design examined carefully the design changes introduced by the steel fabricator.

The Missouri Board of Architects, Professional Engineers, and Land Surveyors brought civil charges against the engineers involved in the case, and a 27-week administrative law hearing ensued. The hearing board concluded that the design engineers for GCEI were fully responsible for the accident. It ruled that, in the preparation of their structural drawings, "depicting the box beam hanger-rod connection for the Hyatt atrium walkways, [GCE] . . . failed to conform to acceptable engineering practice. [This is based] upon evidence of a number of mistakes, errors, omissions and inadequacies contained on this section detail itself and of [GCE's] alleged failure to conform to the accepted custom and practice of engineering for proper communication of the engineer's design intent" (Anonymous [a], 1985).

The failure was ultimately caused by improper design changes that resulted in inattention to the impact of changes on the load forces acting on the connections while in operation. This case leads to questions concerning the responsibilities and obligations engineers have for the safe construction and implementation of their designs. After the hearing, the case was repeatedly discussed in the civil engineering community in professional journals and at academic conferences. Questions concerning the "legal costs of failure, professional liability, insurance, professional responsibility, project quality assurance, and professionalism in engineering" (Roddis, 1993: 1551) were discussed in the civil engineering community and resulted in a widespread examination of building codes and engineering practices. In particular, civil engineers learned from the disaster that "improved performance in the areas of detailing and connections, the recognition that structural detailing needs more attention in routine design and engineering education, and that structural schemes that lack redundancy demand an especially thorough design and careful review" (Roddis, 1993: 1551). These lessons should be heeded in order to avoid preventable technical design flaws in the future that could lead to disaster.

A major design flaw—more aptly designated as a technological failure rather than a technological disaster—is exemplified by Intel Corporation's Pentium computer chip. This case demonstrates that as designs become more complex, especially in the world of microprocessor and software design, tools and prototype testing become less capable of preventing and detecting design flaws (Betts, 1994: 4).

In November 1994, a mathematics professor disclosed on the Internet that Intel's new Pentium chip had a flaw that could lead to errors in some complex mathematical calculations. The flaw was discovered in the Pentium's FPU, or floating-point unit. The floating-point unit, generally implemented in computer hardware, is a method of calculation with varying numbers of decimal places. Floating-point units are used in spreadsheets, financial analysis software, graphics software, and scientific and engineering calculations. Intel had known of this flaw months earlier, but decided neither to announce the flaw to anyone nor to recall the defective chip. In retrospect, this decision was a costly public relations blunder, although, at the time, the decision was based on what top management at Intel considered to be plausible reasons. As Andrew Grove, former CEO of Intel, put it:

> We were already familiar with this problem, having encountered it several months earlier. It was due to a minor design error on the chip, which caused a rounding error in division once every nine billion times . . . [this meant] that an average spreadsheet user would run into the problem only once every 27,000 years of spreadsheet use. This is a long time, much longer than it would take for other types of problems . . . so, while we created and tested ways to correct the defect we went about our business. (Grove, 1996:12)

A week after customers began logging complaints on the Internet, Grove wisely decided to post an apology on the Internet. After a torrent of negative publicity based on these Internet exchanges, Intel offered to replace the defective chips for all customers, at a cost of $475 million to the company. As Andrew Grove eventually admitted:

> We basically defied our consumer population . . . We said 'we know what's good for them,' and we were pretty obstinate about that. In effect, we were ranking the consumers . . . we said, 'those of you that we decide merit a new chip will get a new one. The rest of you go away. You don't need one.' In retrospect, that was incredibly condescending and arrogant. (Grove, 1998: 117)

The $475 million that Intel took as a "write-off" consisted of estimates of the cost of replacement parts plus the value of the materials pulled off the line. According to Grove, the cost of replacement was the equivalent of half a year's R&D budget or five years' worth of the Pentium processor's advertising budget.

The flaw was the result of a design error that went undetected despite exhaustive design verification and testing. Such flaws are inevitable if microprocessor design engineers are pressured to dramatically increase quantity of output at the expense of quality of product. Contrary to the company's initial stance that mistakes with the FPU were limited to complex math only encountered by scientists, such calculation errors could conceivably affect business people designing currency trades, insurance contracts, and even computer-aided engineering design applications. Such flaws are often unacceptable in safety-critical applications.

Pentium's chip problem is simply one example of numerous product failures—both software as well as hardware—that have taken place in the computer industry. Peter G. Neumann (1995) has classified literally hundreds of computer-related risks causing millions of dollars of damage each year and causing potential safety hazards to scores of people. One hopes the Intel case will serve as a warning for the entire semiconductor industry (Williams, 1997).

According to Neumann, there have been a significant number of deaths due to computer-related failures. One such case involved the Therac-25 radiation device, designed to deliver doses of radiation to cancerous growths in patients in order to shrink their tumors. A flaw in the software that runs the device caused it to overradiate at least three patients, which ultimately lead to their deaths.

Between 1985 and 1987, the computer-controlled radiation linear accelerator, the Therac-25, gave massive overdoses of radiation to six patients—several of whom later died—at four different medical centers. The Therac-25 had a number of design flaws, including gaps in proper safety design, insufficient testing, and bugs in the software controlling the devices. In one case, a woman in Marietta, Georgia, received 20,000 rads (radiation absorbed dose) instead of 200 rads—100 times the intended dosage of radiation. The Therac-25 is equipped to deliver either electron beams (namely, "E-mode") or X-rays (namely, "X-mode"). In what has been described as a software programming error, the device had "scrambled" the "X-mode" and "E-mode" functions, delivering the deadly overdose (Jacky, 1990). Although original reports had attributed the failure to "operator error," it is now well recognized that the problem was caused by defects in the computer program and systems design that controlled the Therac-25 apparatus. One investigation of the Therac-25 disasters suggests that a partial cause of the problem is an organizational tendency of software development firms to not use proper software and safety engineering procedures; reliability testing procedures are notoriously absent even in the development of safety-critical systems such as medical devices (Leveson and Turner, 1993).

This case illustrates the problem of technical design factors in safety-critical systems, especially in the growing use of embedded microprocessors and microcomputers in the medical electronics industry. Researchers at the Federal Drug Administration report that almost all devices produced by the multi-billion dollar medical electronics industry now include embedded mini- or microchips. The Therac-25 case is one of the most serious computer-related failures to date (Joyce, 1987).

Henry Petroski, a profound student of engineering failures, develops a provocative assessment of engineering design successes and failures in his book *To Engineer is Human* (1985):

> [T]he colossal disasters that do occur are ultimately failures of design, but the lessons learned from those disasters can do more to advance engineering knowledge than all the successful machines and structures in the world. Indeed, failures appear to be inevitable in the wake of prolonged success, which encourages lower margins of safety. (p. xii)

Granted, many technological failures result from inadequacies in technical design. However, in looking for general principles that may elucidate the occurrence of technological disasters, we would nevertheless be unwise to restrict our attention to purely technical causes, for human, organizational, and sociocultural factors are also likely to be involved. The complexities created by such interrelationships are extremely important to examine as society continually tries to minimize the incidence of technological disasters.

HUMAN FACTORS

The field of human-factors engineering deals with various cognitive, perceptual, and workplace design problems specific to technological systems operated by human beings, especially in the ever-increasing complexities of human-machine interfaces (Bignell and Fortune, 1984). Human factor causes of technological failures and disasters have led to the development of interdisciplinary fields such as industrial psychology and aviation psychology, which develop strategies for avoiding such systemic errors (Whetzel, 1997; Wiener, 1988).

Individual error is simply insufficient to account for the failure of increasingly complex and sophisticated technologies; hence the need to understand the structure of human-machine systems

(Baron, 1990; Redmill, 1997). Of course, in some cases, the classic "pilot error" syndrome may account for some failures. "Pilot error," according to human-factors engineers, actually explains two categories of operator error. The first category of errors is referred to as *Type I*. Rochlin defines Type I errors as "overlooking, ignoring, or misunderstanding the information presented even when it occurs within the envelope of the predicted or anticipated flow of events" (1991: 114). Experts refer to the second category of error as *Type II*. Rochlin defines such errors as "accepting as true, accurate, or significant information that is misleading, incorrect, or irrelevant, or by extension, projecting into a situation 'external' beliefs or assumptions about the nature of the situation or state of the system" (Rochlin, 1991: 112; see also Landau and Stout, 1979). Both types of operator error seemed to be operating in the *Vincennes,* Three Mile Island, and Chernobyl cases. In the steamship collision involving the SS *Mendi,* we can identify Type I errors.

On the morning of February 21, 1917, the SS *Mendi* and the SS *Darro* collided 11 miles southwest of St. Catherine's Point on the Isle of Wight, causing 600 deaths out of 800 members of a native African labor battalion heading back to South Africa. The subsequent formal investigation into the high seas collision found the captain of the SS *Darro,* Henry Winchester Stump, guilty of failing to comply with articles 15 and 16 of the Regulations for Preventing Collisions at Sea, "as to sound signals and speed in a fog, and by its more serious default in failing, without reasonable cause, to send away a boat or boats to ascertain the extent of the damage to the SS *Mendi,* and to render her master, crew and passengers such assistance as was practicable and necessary, as required by section 422(1)(a) of the Merchant Shipping Act, 1894" (Clothier, 1987).

The accident at the Three Mile Island (TMI) nuclear power plant was a turning point for the nuclear power industry because it emphasized the central importance of human factors to safe plant operations. The President's Commission on the Accident at Three Mile Island stated that: "There are many examples in our report that indicate the lack of attention to the human factor in nuclear safety" (O'Hara, 1996: 46). The control room, through which the operations of the TMI plant were carried out, was lacking in many ways. The control panel was huge, including hundreds of alarm devices. Key indicator lights and switches were placed in locations where operators could not see them. The controls were described as "seriously deficient under accident conditions" (O'Hara, 1996: 46). One source of confusion during the crisis was the mismatch between what the computerized control mechanisms were communicating to the operators and their interpretation of the data. The operators interpreted the system as reporting that a crucial valve was closed when it was, in fact, left open for hours, allowing hundreds of gallons of coolant to pour out of the reactor core. Due to the myriad of human factors involved in the TMI accident, a Human Factors Evaluation Program Review Board was established in order to assist the Nuclear Regulatory Commission in conducting human-factors evaluations of the human-systems interfaces of advanced nuclear power plants (O'Hara, 1996).

As is well known from such reports, the causes of many technological disasters can be traced back to problems of cognitive processing of information, visual perception and its relation to visual displays (especially in information systems technologies), insufficiencies in human memory and retention, capacity for stress, cognitive and emotional overload, and operational burnout. These factors lead to bad judgment, inaccurate perceptions, and a host of other negative outcomes (Meshkati, 1992). It was such a set of factors that led to the shooting down of a civilian Iranian airliner by the USS *Vincennes*.

On the afternoon of July 3, 1988, during open battle with Iranian gunships, the USS *Vincennes* fired at and shot down a civilian Iranian Air jetliner, killing all 290 civilians aboard. The Hearing Board ruled that mistakes made in the interpretation of hyper-complex sociotechnical systems like the Navy's fully computerized AEGIS aircraft detection and warning system led to stress, "task-fixation," and unconscious distortion of data, all of which played a role in the *Vincennes* incident (Rochlin, 1991). As the findings reported, the AEGIS computer detection system initially misinterpreted the altitude of the Iranian Airbus, indicating it was descending, and initially identified the plane as an F-14 fighter jet. It eventually corrected itself, but it was reported that the crewmen monitoring the screens failed to detect the initial computer error.

Industrial psychologists call such events the "glass cockpit" syndrome, "a computer information overload in which the flood of technical information, faulty communications and outside stress lead to judgment errors" (Neumann, 1989: 7).

The Fogarty report on the USS *Vincennes* accident was unanimous in its recommendations:

> Since it appears that combat-induced stress on personnel may have played a significant role in the incident, it is recommended the CNO [Chief of Naval Operations] direct further study into the stress factors impacting on personnel in modern warships with highly sophisticated command, control, communications, and intelligence systems such as AEGIS. This study should also address the possibility of establishing a psychological profile for personnel who must function in this environment. (Rochlin, 1991: 107)

One can conclude that this case is an example of a human-machine mismatch. As one researcher put it, "the crew involved believed the initial system identification and altitude reading and did not double check them, nor did they change their evaluation when given new, conflicting information" (Neumann, 1989: 9).

The crucial role of human factors also surfaced in investigations into the Chernobyl nuclear disaster. During the early hours of April 26, 1986, Reactor 4 of the Chernobyl nuclear power plant exploded with two large blasts, releasing more than 140 million curies of radioactive material in a vapor cloud that eventually covered most of Eastern and Western Europe. Overall, 200,000 people were evacuated from 71 villages within an 18-mile radius of the plant.

Soviet experts carried out an exhaustive investigation of the disaster, using mathematical models and computer simulations to reconstruct the facts of the events. They presented their findings in a 430-page report, which they delivered to a meeting of the International Atomic Energy Agency (IAEA) held in Vienna, Austria, in August 1986. Their general conclusion was that the main cause of the accident was operator error. The Soviet report claimed that operator error led to "violations of the established order in the preparation of tests," "violation of the testing program itself," and "inadequate understanding on the part of the personnel of the operating processes in a nuclear reactor" (Serrill, 1986). The report also concluded that workers, operating with a sense of overconfidence, exhibited a "loss of a sense of danger" in managing the complex safety systems of the reactor (Serrill,

1986). Andronik M. Petrosyants, chairman of the Soviet Committee for the Peaceful Uses of Atomic Energy, echoed these findings when he stated that "the accident took place as a result of a whole series of gross violations of operating regulations by the workers . . . The sequence of such human actions was so unlikely . . . that the [design] engineers did not include such a scenario in [their] projects" (Serrill, 1986: 27).

Although initially skeptical of the Soviet report, nuclear scientists tended to agree with the unprecedented candor of the Soviets. The conclusions of the report concerning operator error were echoed in various subsequent International Atomic Energy Agency (IAEA) reports. "The root cause of the Chernobyl accident it is concluded is to be found in the so-called human element. . . . The lessons drawn from the Chernobyl accident are valuable for all reactor types" (IAEA, 1986: 76). Or consider the following statement: "The Chernobyl accident illustrated the critical contribution of the human factor in nuclear safety" (IAEA, 1987: 43). Finally, in the words of Valery Legasov, Soviet delegate to IAEA: "I advocate the respect for human engineering and sound man-machine interaction" (Meshkati, 1991: 148). According to the report, operators at the plant committed six major errors, all of which led to one of the worst technological disasters in history.

On April 25, 1986, a test was being conducted on Reactor 4 of the Chernobyl plant in order to determine how long the turbines would continue rotating when cut off from the steam supply. This information was needed to ascertain how much energy would be generated from the mechanical inertia of the spinning turbines and, hence, for how long this rotation could run the generators if the plant was accidentally cut off from the power grid. The test was scheduled for a routine shutdown of the plant when power output was virtually discontinued, even though the reaction process continued to operate. This was the only time the test could be conducted, so operations management was anxious to have it done since it would mean waiting a full year if it was not. The plant operators began reducing the reactor's power output levels so they could run the turbine test.

To prevent the core from being flooded with excess water that would halt the test, the operators switched off the emergency backup cooling system. They proceeded to bring the power output down to between 700 and 1,000 megawatts (thermal) and then cut off steam to one of the turbines. At this time, an order came in from a grid manager at Kiev, who asked the Chernobyl power plant to maintain the power output at normal levels to service requests for power from the local community. Therefore, the test had to be postponed until 11:00 p.m. that evening.

The *first* of six major human factors errors occurred when, violating safety rules, the operators failed to switch the emergency cooling system back on. The test resumed at 11:10 p.m. and control was transferred from a local to an automatic regulating system. Committing their *second* major error, the operators did not set the automatic system above the minimum output level of 700 MW(t). This was a serious omission because the reactor had a tendency to become unstable at low output levels. This caused the reactor output level to drop quickly to around 30 MW(t). Reducing the power output to this low level caused the nuclear fission reaction to slow down, causing poisonous xenon gas buildup in the fuel rods. Xenon gas absorbs electrons and slows down the fission process, thus decreasing overall energy output.

The operators had available a variety of manual controls with which to increase the power output levels. By withdrawing all but eight or ten of the control rods, violating safety regulations which required at least 30 control rods to be plunged into the reactor at all times, they managed to bring the power up to 200 MW(t). This resulted in their *third* major mistake. In hindsight, given the difficulty of maintaining stability in the core, the best thing the operators could have done at this point was to either bring the test to a halt, or postpone it until the xenon gas had decayed. Instead, the operators only compounded their earlier blunders with reckless attempts to force the output level to increase power. To accomplish this they switched on additional pumps in order to increase the water flow to the core. However, because the reactor was running at lower power than originally planned, too much cooling water flowed through the core, which in turn caused the steam pressure and the water level in the steam separators to drop. In order to prevent the reactor from shutting down automatically when these parameters fell below the critical point, the operators blocked signals from pressure and water sensors, thereby disabling a key part of the emergency shutdown system. This constituted their *fourth* major violation of safety procedures.

Chernobyl Nuclear Catastrophe, April 25, 1986. Copyright Getty Images, Inc. Used by permission.

The decrease in steam generation caused by the excess cooling water prompted the automatic control rods to be withdrawn completely. The operators then withdrew virtually all of the manual rods in order to maintain the power level at 200 MW(t). This further reduced the operating margin and the capacity to respond quickly in an emergency. In what constituted their *fifth* major error, operators switched off a second defense mechanism, which would shut down the reactor in the event that steam levels in the steam separator fell below normal. Nevertheless, the reactor appeared to be stabilized and, at 1:23 a.m., the steam supply to the generator was shut off, thus finally initiating the actual test. The flow of water into the reactor was decreased radically as four of the main circulation pumps and the two main pumps feeding water into the reactor were disengaged. The sudden decrease in water caused the temperature to rise, greatly increasing the amount of steam being generated. This was their *sixth* and final error (Sweet, 1989).

The reactor was operating at low power, a state in which a small increase in power causes a larger increase in steam, which in turn causes an unexpected increase in power. Boiling increased and, because of the positive void coefficient, the power started to climb sharply. At 1:23 a.m. the shift manager gave the order to hit the emergency button, which plunged the control rods into the reactor to stop the nuclear reaction. However, because the rods were almost completely withdrawn, the response time was too slow.

By this time, the situation was out of control. Intense steam generation was taking place around the fuel elements, which in turn reduced the system's heat removal ability. The power output continued to surge and the fuel started to disintegrate and fall into the cooling water. The result was a sharp increase in pressure, which ruptured the cooling channels and prompted a thermal explosion that destroyed the reactor and part of the structural components of the building. This ruptured some of the pressure tubes passing through the massive graphite moderator and circulating water around the fuel rods. Without cooling water, the fuel elements rapidly heated up and began to melt. Experts theorize that a highly combustible mixture of gases such as hydrogen and carbon monoxide was formed when steam came into contact with the hot graphite that, combined with the zirconium in the fuel rods, reacted with the superheated steam to produce even more hydrogen.

Within a few seconds, a mighty surge erupted, from 200 to 32,000 MW(t) or 100 times full power. This was followed immediately by a further surge of power to about 440 times normal level. The result was catastrophic.

Here we witness the complex interactions of the human-machine interface. The failure at Chernobyl demonstrates the complex interaction between human factors and technical design factors. In addition to the human factors discussed previously, three principal design defects of the RMBK-1000 (Russian Graphite-Moderated Reactor) are:

1. The fact that the reactor tends to gain power rather than slow down as water is lost or is turned to steam. This makes the reactor highly unstable when there is low power output.

2. Inadequate containment surrounds the reactor core.

3. The design of the system does not provide or protect against operator interference with the safety systems (Norman, 1986).

The Chernobyl case involves a combination of human factors and technical design factors that resulted in a catastrophe.

After 14 years of pondering this disaster, the Ukrainian government, in cooperation with a group of seven industrial nations, signed a formal agreement to shut down the Chernobyl power plant by the year 2000. In signing the agreement, the European Union and the United States agreed to help Ukraine devise plans to mitigate the effects of the shutdown on local populations (Shcherbak, 1996). This agreement has come to fruition. The Chernobyl plant was finally shut down on December 15, 2000 (Sciolino and Gordon, 2000: 1).

We now turn to the third root category of technological failure: organizational systems factors.

ORGANIZATIONAL SYSTEMS FACTORS

The organizational contexts in which technological systems operate add to their complexity and susceptibility to failure in at least three ways. First, organizational strategies, policies, and financial and human resources determine the level of attention devoted to managing technological systems and therefore dictate how technological systems are operated. Thus, the nature and the culture of the organization establish the level of reliability and safety at which a technology is operated. Second, the vagaries in communication and decision making in large organizations are aspects of corporate culture that are especially vulnerable to faulty group decision making and often lead to organizational failure. Many researchers have examined the complexities of corporate communication and group decision making, focusing on inadequate procedures that may eventually lead to breakdown and failure of technological systems (e.g., see Janis, 1972; Fisher, 1986; Baskin and Aronoff, 1988; Corman *et al.*, 1990). Third, organizational failures in the form of inadequate attention to safety create preconditions for failure in corporations that are involved with high-risk technologies.

Organizational systems factors were at work in the sinking of the *Sultana*. On April 27, 1865, as the Civil War was ending, the greatest marine disaster in the history of the United States occurred on the Mississippi River, a few miles above Memphis, Tennessee, when the boilers on the steamship *Sultana* exploded and burned, killing at least 1,250 passengers and crew. The *Sultana* was a packet side-wheeler: 260 feet long, 42 feet wide, displacing 719 tons of water, with a cargo capacity of 1,300 tons and licensed to carry 376 passengers. Two steam engines drove the side wheels. The steam was provided by four tubular boilers that were, at the time, an experimental design. While docked at Vicksburg, Virginia, the captain was alleged to have taken bribes to allow more than 1,800 men, former prisoners of war, to be loaded onto the boat. That fateful day, the *Sultana*'s boilers were pressured beyond capacity due to the excess weight of the passengers. The boilers suddenly burst with a terrific explosion. In the opinion of Inspector Witzig and P.B. Stillman, President of the Steamboat Inspection Board, the cause of the disaster was laxity in the enforcement of safety regulations during the war that had produced a climate of "carelessness, recklessness, and poor judgments" which made steamboat travel on the Mississippi dangerous (Salecker, 1986; Yager, 1976).

Violations of government safety regulations, compounded by pressures to conform to policies of large private corporations, are frequent causes of technological disasters. This is exemplified in the B.F. Goodrich brake scandal. In August 1969, Kermit Vandivier, a data analyst and technical writer for B.F. Goodrich, testified before Senator William Proxmire's Economy in Government Subcommittee hearing that he was part of a conspiracy to commit fraud and deception in an alleged cover-up that included lying to the U.S. Air Force and fudging technical data to make an unsafe brake system appear safe (Vandivier, 1972). It all began when B.F. Goodrich secured a contract with LTV Aerospace Technologies to supply 202 braking assemblies, which would be installed in a new attack plane, the A7D.

B.F. Goodrich had fallen out of grace with LTV, and had not received any offers from them in years. Contracts to build brake systems for military aircraft are generally lucrative, so when an offer finally came to Goodrich in June 1967, top managers decided they must secure the contract at all costs, and, bidding outrageously low, Goodrich received the contract to build the braking systems for the A7D. John Warren, a seasoned design engineer and one of Goodrich's best design engineers, was named project engineer and went to work. In a relatively brief time Warren produced a preliminary four-pad design. After several unsuccessful tests, Searle Lawson, a newcomer at Goodrich, arrived at the conclusion that the brake design was fundamentally flawed. During every trial, the brake became very hot, sometimes glowing red and radiating "incandescent particles of metal and lining." After these trial runs, the brake lining had disintegrated and turned to dust.

After extensive testing, Lawson concluded that the brake was too small. The four-disc design was inadequate, and a five-disc brake design was needed. Lawson reported his findings and concerns about the flawed design to his boss, John Warren. Warren rejected the suggestion that the brake was too small and simply requested that Lawson return to the prototype testing. Unconvinced, Lawson went to an engineering project supervisor, Robert Sink, who dismissed the claim that Warren's design could be faulty but seemed to be concerned about the data the tests were generating.

On the 13th test, Vandivier became involved when he was given the task of putting together the test data for the A7D and creating a qualification report. In reviewing the data, Vandivier discovered an irregularity. It seemed as if the recording instrument used to calibrate brake pressure had been deliberately miscalibrated to make it seem that less pressure was required to stop the aircraft than was the case.

Vandivier immediately showed the miscalibrated test logs to his lab supervisor, Ralph Gretzinger, who was aware of the deliberate data tampering, because he heard that Lawson had ordered the miscalibration on instructions from Robert Sink. Both men vowed not to have anything to do with the intentional "cooking" of the data. In a discussion between Vandivier, Gretzinger, and Lawson, Lawson said he was told, "regardless of what the brake does on the test, it's going to be qualified" (Vandivier, 1972: 214). Angered, Gretzinger went directly to Russell Line, who was responsible for engineering lab testing. Gretzinger returned, greatly saddened by his conversation with Line. He confided to Vandivier and Lawson, "I've been an engineer for a long time, and I've always believed that ethics and integrity were every bit as important as theorems and formula . . . Now this . . . Hell, I've got two sons I've got to put through school. . . ." As Vandivier put it: "He [Gretzinger] had been beaten down. He had reached the point where the decision had to be made. Defy them now while there is still time or knuckle under, sell out" (Vandivier, 1972: 215).

These men found themselves impaled on the horns of a dilemma: write the deceptive report and save their jobs at the expense of their consciences, or refuse to write the report, thus honoring their moral beliefs but risking being fired. Vandivier called Lawson and told him he would write the report as requested, directed from "upstairs" to "fix" the data.

> Lawson and I proceeded to prepare page after page of elaborate, detailed engineering curves, charts, and test logs, which purported to show what had happened during the formal qualification tests. When temperatures were too high, we deliberately chopped them down a few hundred degrees, and where they were too low, we raised them to a value that would appear reasonable to the LTV and military engineers—everything of consequence was tailored to fit the occasion. Occasionally, we would find that some test or other either hadn't been performed at all or had been conducted improperly. On those occasions, we 'conducted' the test—successfully, of course—on paper. (Vandivier, 1972)

In June 1968, the report was completed and sent to all concerned parties. Vandivier, Lawson, and Warren refused to sign the report, but, soon after, test flights commenced. Test flights led to several near crashes on landings, the brakes even welding together during one test, causing the plane to skid for over 1,500 feet. It was at this point that Vandivier's lawyer advised him to go to the authorities, and he told everything to the Federal Bureau of Investigation (FBI). The FBI tipped off the Air Force and they stopped all testing and demanded from Goodrich all raw data compiled on the prototype testing. Vandivier handed in his letter of resignation, and in the letter he stated:

> As you are aware, this report contained numerous, deliberate, and willful misrepresentations which, according to legal counsel, constitutes fraud and exposes both myself and others to criminal charges or conspiracy to defraud. The events of the past seven months have created an atmosphere of deceit and distrust in which it is impossible to work. (Vandivier, 1972)

Vandivier left the company and became a newspaper reporter for the *Daily News* in Troy, Ohio. At the Senate hearings, lawyers for B.F. Goodrich portrayed Lawson as too young and inexperienced and denigrated Vandivier for his "lack of technical training," which was confirmed by Sink, who represented B.F. Goodrich at the hearings. Within only four hours, the committee adjourned, making no recommendations and finding no one guilty. Lawson stayed on the A7D project; Line was promoted to production superintendent and Sink moved into Line's old job.

The A7D brake scandal is a particularly valuable case because it provides insight into group dynamics and the mechanisms leading to corporate deviance (see Chapter 11). In this case, top engineers committed to a course of actions and capitulating to managerial pressures endangered human lives. Given that senior engineers were prepared to falsify data, it is difficult to conceive of watertight testing procedures that are not amenable to manipulation if people are really intent on falsifying technical documents.

Another case study demonstrating the importance of whistle blowing (see also Chapter 11) is the case of the three engineers involved in the construction and testing of the Bay Area Rapid Transit (BART) system. In September 1972, the San Francisco BART system began operations. The system was hailed as the nation's first space-age, completely automated transit system. Less than a month after its first run, on October 2, a train failed to stop at one of the stations, shot off the track, and crashed into a nearby commuter parking lot. BART management downplayed the incident, as well as other malfunctions that plagued the system, as a process of "getting the bugs out." Other problems included doors suddenly opening as trains raced down the track, trains speeding through stations when programmed to stop, brake controls malfunctioning and detection of "phantom trains" causing real trains to stop. In an anonymous memo to the BART Board of Directors, three engineers alleged that the automatic train control system was unsafely designed, that testing and operator training were inadequate, that there were excessive cost overruns, and that computer software problems plagued the system (Anderson *et al.*, 1980). On March 2, 1972, the three engineers were fired.

Another prominent case of technological disaster, the Challenger explosion, is attributable to the negative effects of the hierarchical structure of corporations, the vagaries of communication in large organizations, and the impact of corporate culture on decision making. The Challenger space shuttle 51-L exploded on January 28, 1986, 73 seconds after liftoff, killing its crew of seven, including the first potential civilian in space. In addition to the tragic loss of life, $55 million of taxpayer dollars went up in smoke along with the shuttle and rocket. There is a growing amount of literature discussing the various organizational and communication factors that led to this disaster, each researcher stressing different facets of the precipitating organizational and communicative breakdowns (Hirokawa, 1988; Gouran *et al.*, 1986; Herndl, 1994; Dombrowski, 1992; Winsor, 1990; Moore, 1992; Pace, 1988; Miller, 1993; Rowland, 1986; Renz and Greg, 1988).

The Challenger case demonstrates unequivocally that flawed decision making—arising from complex interactions between engineers and managers with different objectives in the hierarchies of NASA and its contractor, Morton Thiokol—is one of the primary causes of technological disaster. In a much discussed three-way teleconference the night before the launch, engineers from Morton Thiokol sought to persuade NASA and Morton Thiokol management to abort the launch. At that point Jerry Mason, senior vice president of Morton Thiokol, turned to Robert Lund, vice president of engineering at Morton Thiokol, and asked him to "take off his engineering hat, and put on his management hat." In complying with this request, Lund rejected the no-launch recommendation of his engineers and made the decision to proceed with the launch. In addition, miscommunication, fostered by diverse corporate cultures that emphasize compliance with rules and regulations, created a great deal of stress. When one superimposes unquestioning obedience to the chain of command, we have the precondition for the occurrence of a technological disaster.

In a nine-year study of the Challenger case, Diane Vaughn presents an elaborate analysis of the corporate cultures of NASA and Morton Thiokol, in which she interprets the acceptance of risk by Morton Thiokol and NASA engineers, in spite of growing evidence of O-ring erosion problems, as a "normalization of deviance."

> The two engineering communities . . . came to the same conclusion: the design, although deviating from performance expectations, was an acceptable risk. . . . These two early conclusions—accept the risk and proceed with flight; correct rather than redesign—would become norms guiding subsequent decision-making, characterizing the work group culture in the years to come . . . . (Vaughn, 1996: 106)

In her concluding chapter, "Lessons Learned," Vaughn interprets the Challenger disaster as an example of Perrow's Normal Accident Theory: "an organizational-technical system failure that was the inevitable product of the two complex systems" (Vaughn, 1996: 415).

As the foregoing discussion of various cases has demonstrated, technical, human, and organizational factors cause many technological disasters and, consequently, have been studied extensively. Nevertheless, these three categories do not exhaust the causes of technological disaster. Crisis management theorists label these "internal" factors (Shrivastava, *et al.*, 1998). As explanatory as these three factors may be, they do not include "external" factors. Since the latter also generate technological hazards, their analysis is essential to understanding technological disasters. To our view, no theory of hazard management and technological disaster is complete without understanding *both* internal and external forces. Much of the literature does analyze external factors (economic, political, social, and cultural); however, it restricts the analysis to factors internal to organizational systems. The societal context of technological disasters must also be studied, analyzed, and understood. We now turn to these external factors.

SOCIO-CULTURAL FACTORS

Socio-cultural factors combine sociological and anthropological concepts of social structure and culture. Social structure refers to social institutions such as economic, political, legal, familial, religious, and educational organizations. Culture refers to the system of norms and values governing the behavior of members of a society. Economic, legal, and political institutions can either deter or promote the occurrence of technological disasters. Cultural factors include the values placed on safety and human life, attitudes towards risk—risk-aversive behavior vs. risk-taking behavior—as well as the individual's autonomy vs. his or her responsibility toward the group.

On December 6, 1907, 366 coal miners lost their lives in one of the worst coal mine disasters in U.S. history when a runaway cable car smashed into electric cable lines in a mine outside of Monongah, West Virginia (Jackson, 1982). In addition to the Monongah disaster, several other major mine explosions occurred in 1907. The resulting public outcry and grief demanded governmental intervention. This led directly to the establishment of the Bureau of Mines, which was the culmination of the U.S. government's growing involvement in the mining industry. The mine disasters of 1907 laid a capstone on a deteriorating situation. Between 1906 and 1908, a total of 22,840 coal miners had been killed in the United States alone; the number of fatal accidents had nearly doubled in the preceding six years. In 1907, 3,200 miners were killed, one-half in Pennsylvania. The frightening carnage was blamed on the lack of reliable information concerning the safety of explosives and the lack of enforceable mine regulations (Jackson, 1982).

On March 25, 1911, a fatal conflagration occurred in a sweatshop, The Triangle Shirtwaist Factory, which resulted, eventually, in a national movement for safer working conditions in the United States (Stein, 1962). The fire started on the eighth floor of the Asch building in lower Manhattan. One hundred and forty-six workers died, mostly young women, many of them as young as 14 years old. Like most other sweatshops, the Asch building had no sprinkler system, its doors opened inward, it had two staircases rather than the required three, and the fire escape exit led only to the roof. Some workers, having no way of opening the doors that had been locked to prevent employee theft, leapt to their deaths from seventh and eighth story windows.

The American poet laureate Robert Pinsky captures the poignancy of this tragedy in his poem "Shirt" (Pinsky, 1990).

> . . . At the Triangle Factory in nineteen-eleven.
> One hundred and forty-six died in the flames
> On the ninth floor, no hydrants, no fire escapes—
>
> The witness in a building across the street
> Who watched how a young man helped a girl to step
> Up to the windowsill, then held her out
>
> Away from the masonry wall and let her drop.
> And then another. As if he were helping them up
> To enter a streetcar, and not eternity.

A third before he dropped her put her arms
Around his neck and kissed him. Then he held
Her into space, and dropped her. Almost at once

He stepped up to the sill himself, his jacket flared
And fluttered up from his shirt as he came down,
Air filling up the legs of his gray trousers . . .

The disaster led to the creation of health and safety legislation, including factory fire codes and child labor laws, and helped shape future labor laws.

In December 1991, a San Francisco court found Dow Corning guilty of manufacturing a dubiously safe product, namely, its line of breast implants. The plaintiff, Mariann Hopkins, was awarded $7.3 million, $840,000 in compensatory damages and $6.5 million in punitive damages. This caused a wave of silicon-implant litigation, and, by 1993, the Food and Drug Administration (FDA) had received an estimated 3,000 reports of illnesses or injuries associated with silicon implants, which had been implanted in more than 1 million women. Women claiming sickness and disease from the implants filed more than 1,000 lawsuits. In 1993, a Texas state court awarded $25 million to a claimant against the Bristol-Meyers Squibb Corporation. Among the many health problems alleged to have been caused by silicon implants are: autoimmune diseases such as lupus, arthritis, scleroderma, breast cancer, abnormal tissue growth, nerve damage, inflammations, swollen joints, rashes, and fatigue.

In the Hopkins case, the jury found Dow Corning guilty of fraud after having obtained access to reams of company memos that showed numerous complaints by company researchers and management about the implants. These complaints included ruptures, leakages, infections, and autoimmune system problems. During the trial, Thomas Talcott, a former Dow Corning materials engineer who conducted silicon breast implant research testified that, "The manufacturers and surgeons have been performing experimental surgery on humans" (Hartley, 1993). In 1992 Dr. Sydney Wolfe, director of the Public Citizen Health Research Group, accused Dow Corning of actively covering up important information concerning health risks associated with silicon breast implants.

They [Dow Corning Corporation (DCC)] are reckless and they have a reckless attitude about women. . . . DCC was only thinking of themselves when they repeatedly assured women and their doctors that the implants were safe . . . [keeping] guard over hundreds of internal memos that suggested that some of Dow Corning's employees have long been dissatisfied with the scientific data on the implants. (Hartley, 1993)

These unsettling facts raise the issue of why the silicon implant devices were not more closely regulated by the FDA. A brief recounting of the role of the FDA in the history of the breast implant controversy will shed light on socio-cultural factors, such as regulatory and other institutional mechanisms that often fail in their roles to manage and control new and/or potentially risky technologies. If adequate regulation is not effective, the potential for technological disasters increases.

In 1976, Congress passed the Medical Device Amendment to the Federal Food, Drug, and Cosmetic Act of 1938, placing, for the first time, medical devices such as silicon breast implants under regulatory scrutiny. This amendment was motivated largely by the Dalkon Shield catastrophe. Among other things, the 1976 amendment directed the FDA to set up an approval process for new devices and classify existing ones into Class I, II, or III, depending on the degree of testing necessary to provide reasonable proof of the safety of the device (see Chapter 11 for a further discussion).

A preliminary review was carried out by the FDA's General and Plastic Surgery Devices Advisory Panel, a group dominated by plastic surgeons. This review placed silicon breast implants in the less restrictive Class II category, which required minimal review and did not even require testing for the product to remain on the market. No further inquiries into breast implants were made until a growing list of consumer and scientific complaints forced the FDA to look into the safety and efficacy of the implants. Finally, in early 1989, the FDA acknowledged silicon breast implants to be Class III devices and required all implant manufacturers to produce detailed information and scientific research about the implants' safety and reliability. In 1991, satisfied with the information, the FDA ruled to allow all manufacturers, including Dow Corning, to continue to manufacture, market, and sell silicon breast implants without further research.

At the same time, though, concern was growing among research physicians and consumer groups about the overall safety of the implants. A growing number of FDA scientists disputed the initial findings based on feedback from physicians, who reported numerous cases of leaking or ruptured implants. In 1992, David Kessler, FDA commissioner, announced a 45-day moratorium on the sale of silicon breast implants. Until recently, the FDA had placed a total ban on breast implants, save for legitimate circumstances of reconstructive surgery.

The Institute of Medicine (IOM) published its review of all scientific research on breast implant safety in early 2000 (Bondurant, Ernster, and Herdman, 2000). The IOM report concluded that silicon breast implants could pose enough localized complications that their safe use could be compromised. The IOM report stated that deflation, rupture, and leakage can indeed occur. Other localized complications identified were injury, pain, infection, hematoma, necrosis, and tissue atrophy, among other complications. Finally, the IOM report, the most authoritative and up-to-date of its kind, recommended that the use of silicon breast implants be limited to persons eligible for reconstruction after breast cancer surgery. One reason the report gave for limiting implants to augmentation patients is that the risks outweighed the benefits of the surgery for patients seeking breast enhancement.

Even more recent than the IOM report is a study in 2001 by researchers at the National Cancer Institute demonstrating that women with breast implants "suffer higher rates of lung and brain cancer than other plastic surgery patients" (Stolberg, 2001: A19).

The socio-cultural factors operative in this case include the ever-changing cultural standards of feminine beauty. As regards breast implants, one influential group, the American Society for Plastic and Reconstructive Surgery (ASPRS), a professional association representing 97 percent of all board-certified plastic surgeons, lobbied hard to redefine female flat-chestedness as a medical disease requiring medical treatment. In July 1982, the ASPRS filed a formal recommendation to the FDA arguing that:

> There is a substantial and enlarging body of medical opinion to the effect that these deformities [small breasts] are really a disease, which in most patients results in feelings of inadequacy, lack of self-confidence, distortion of body image and a total

lack of well-being due to a lack of self-perceived femininity. The enlargement of the underdeveloped female breast is, therefore, often very necessary to insure an improved quality of life for the patient. (Lawrence, 1993: 245)

It seems fair to remark that the ASPRS is at least partly responsible for the rush to manufacture and market silicon breast implants without the benefit of appropriate scientific research and development.

The current social pressures to conform to cultural standards of feminine beauty are summed up in a special 100th anniversary edition of *Vogue* magazine, published in 1992. The magazine states:

> . . . And in women's bodies, the fashion now is a combination of hard, muscular stomach and shapely breasts. Increasingly, women are willing to regard their bodies as photographic images, unpublishable until retouched and perfected at the hands of a surgeon. (Lawrence, 1993: 247)

The three foregoing cases—the mine disaster in Monongah, the Triangle Shirtwaist Factory fire, and the Dow Corning silicon implant injuries—underscore the impact of social structure and culture on technology as it affects our lives. Without effective regulation to protect employees and consumers, some corporate enterprises operate with a bottom-line focus to the detriment of the lives of consumers affected by their products and operations.

Socio-cultural systems factors contributing to technological disaster are important because they identify failures that have causes external to organizations, involving social, political, and cultural variables. Hence, technological disasters cannot be understood simply as design failures, operator failures, or organizational failures. In other words, the prevention and management of technological disasters cannot be achieved at the organizational level alone. The harder task is effecting changes in the institutions and the culture of a society, and strategies must be developed to promote more effective societal control of technologies.

The four root causes underlying technological disaster, together with their associated elements, are summarized in Figure 5–2.

Technical Design Factors	Human Factors
• Faulty design • Defective equipment • Contaminated or defective materials • Contaminated or defective supplies • Faulty testing procedures	• Human-machine mismatches • Operator error • Perceptual constraints • Fatigue or stress • Ignorance, hubris, or folly
Organizational Systems Factors	Socio-Cultural Factors

• Communication faulures
• Faulty group decision making
• Policy failures
• Cost pressures curtailing attention to safety
	• Cultural values and norms
	• Institutional mechanisms
	    Regulatory mechanisms
	    Educational systems

Figure 5–2
A classification of causes of technological disasters.

TERRORISM IN THE NUCLEAR-INFORMATION AGE

In reviewing our four categories of causes of technological disasters we have not attended to terrorism as a menace intrinsic to the technologies of our age. Fortunately, terrorists have not yet applied their criminal designs to potentially vulnerable technologies such as nuclear power plants. According to Dumas, however, there is little comfort in this observation.

> [T]here is nothing inherent in the nature of terrorism that makes it self-limiting. Those who are ready, even eager, to die for their cause, who stand willing to abandon every constraint of civilized behavior and moral decency against the slaughter of innocents, cannot be expected to permanently observe some artificial restriction on the amount of havoc they wreak. (Dumas, 1999: 56)

Events in recent years attest to this dismaying fact.

For many years, the United States Government has been required, by statute, to publish annual country reports on terrorism. The statutory definition of terrorism is as follows: "premeditated, politically motivated violence perpetrated against noncombatant targets by subnational groups or clandestine agents" (U.S. Code Title 22, section 2656f [d]). Pillar has cogently explicated four principal elements in this definition: (1) "premeditation, means there must be an intent and prior decision to act that would qualify as terrorism;" (2) "political motivation, excludes political violence motivated by monetary gain or personal vengeance;" (3) "that the targets are noncombatants, means that terrorists attack people who cannot defend themselves with violence in return;" and (4) "that the perpetrators are either subnational groups or clandestine agents, is another difference between terrorism and normal military operations" (Pillar, 2001: 13–14).

Some legal scholars have pointed to the need for a general definition of terrorism that would cover both state-sponsored terrorism as well as terrorism by non-state actors. The increasing occurrence of threats of highjacking, bio-chemical terrorism, and nuclear terrorism may produce a steady growth of international treaties addressing such acts. "What is needed at this juncture is the establishment of a global legal regime dealing specifically with terrorism" (Mendlovitz, 2001: 3).

Terrorist attacks have been more common in the last few decades than we would like to think. According to statistics from the U.S. Department of State, there were approximately 13,000 incidents of terrorist attacks from 1968 through 1997, or "an average of more than 430 per year for three decades" (Dumas, 1999: 55).

In the 1980s, a wave of hijackings and bombings targeted the airlines, especially foreign carriers flying without the stringent security measures imposed by the FAA on U.S. airlines. "This isn't just a matter of economics but security may be a primary reason why foreign carriers have been successfully sabotaged far more often than U.S. airlines" (Nader and Smith, 1993: 294). Table 5–1 presents a list of explosions on aircraft between 1980 and 1989. These grim statistics provide a compelling argument for developing uniform international security measures governing airlines such as those El Al airlines has implemented over the years. "Israel's El Al airline spends 8% of its revenue on security, 75% of which is subsidized by the government. American airlines, by contrast, pay only 0.2%-0.3% of their revenue toward security" (Lopez, 2001: 1).

In the 1990s, two enormously destructive terrorist bombings occurred. On February 26, 1993, Mohamed Salemeh and Ramzi Yousef, two alleged terrorists, placed a bomb in a van driven into

Table 5–1
Explosions on aircraft between 1980 and 1989.

Year	Airline	Location	Circumstances	Casualties
1980	United (USA)	Sacramento	Explosion in cargo hold while plane being unloaded	2 injured
1981	Air Malta (Malta)	Cairo	2 bombs explode as luggage being unloaded	2 killed 8 injured
1982	Pan American World Airways (USA)	In flight near Hawaii	Bomb under seat cushion exploded	1 killed 15 injured
1983	Gulf Air (Bahrain)	In flight	Bomb exploded in baggage compartment	112 killed
1984	Union Des Transport (France)	Chad	Bomb exploded in cargo hold after landing	24 injured
1985	Air India	In flight	Bomb exploded in cargo hold	329 killed
1986	TWA (USA)	In flight	Bomb exploded in cabin over Greece	4 killed 9 injured
1987	Korea Airlines	In flight	Bomb in cabin area	115 killed
1988	Pan Am 103	In flight	Bomb in cargo area	270 killed
1989	UTA	In flight	Mid-air explosion	171 killed
1989	Avianca (Columbia)	In flight	Bomb under seat	107 killed

Source: Data compiled by the President's Commission on Aviation and Security; reproduced in Ralph Nader and Wesley J. Smith (1993). *Collision Course: The Truth about Aviation Safety.* Blue Ridge Summit, PA, pp. 294–295.

the basement of the World Trade Center in New York. The eventual explosion killed six people and injured more than 1,000 (Farley, 1995). On April 19, 1995, Timothy McVeigh blew up the Alfred P. Murrah Federal Building in Oklahoma City, killing 168 people (Witkin and Roebuck, 1998; Yardley, 2000).

In 1996, an intellectual Luddite by the name of Theodore Kaczynski, otherwise known as the "Unabomber," was arrested. The author of an anti-technology "manifesto" entitled "Industrial Society and its Future" (Kaczynski, 1995), Kaczynski was determined to stop the drift of civilization to self-destruction by launching a campaign of terror. It began on May 26, 1978, when his first bomb slightly injured a Northwestern University safety officer, and ended on April 24, 1995, when a bomb he mailed killed the president of the California Forestry Association. Altogether, he mailed or delivered sixteen package bombs to scientists, academics, and others over a period of 17 years, killing 3 people and injuring 23. Kaczynski pleaded guilty to all of his crimes and was sentenced to life in prison without the possibility of parole (Chase, 2000).

When Americans think of the form a conventional terrorist attack might take, they probably think of hijackings like those that occurred during the 1980s, or bombings like the ones perpetrated by Salameh and Yousef and Timothy McVeigh, or assassinations such as those committed by Kaczynski. These are not the only forms of terrorism, however. Terrorists have begun to exploit the vulnerabilities of information technologies such as the Internet. The use of information technology to terrorize is called cyberterrorism. Cyberterrorism takes many forms, including infecting computer networks with deadly computer viruses that sabotage and delete information and erase computer files, or distributed denial-of-service attacks (DDoS), which can shut down large computer servers, disrupting financial transactions and other business functions, as well as other means to exploit computer network vulnerabilities (Cilluffo, Berkowitz, and Lanz, 1998).

Invasions of computer security by hackers and other cyber-criminals pose a costly threat (Manion and Goodrum, 2000). The "Melissa" virus of 1999, for example, caused damage estimated at $399 million; the financial toll from the "Love Bug" virus of 2000 was estimated as high as $10 billion (Lovelace, 2000). On February 8, 2000, a cyberhacker attacked Yahoo, Amazon, eBay, CNN, and Buy.com, closing them for several hours. The perpetrator, known as "mafia-boy," used a "distributed-denial-of-service" (DDoS) attack against these popular e-commerce Web sites (Anonymous [b] 2000: A1). In order to mount a DDoS attack, the perpetrator first breaks into a weakly-secured computer server, then installs special software on the server. When the attacker is ready, he or she sends a command via the installed software to all the "captured" machines connected to the server, instructing them to immediately send streams of requests to log onto the target Web site address. Such attacks, originating from hundreds of independent computers, eventually "flood" the targeted Web site with millions of simultaneous requests. This increase in fake service requests effectively blocks legitimate users from accessing the site (Denning, 1999: 235–239).

Malicious viruses and DDoS attacks are not the only forms of cyberterrorism, however. Cyberterrorists have also used their computer skills to extort huge sums of money from large financial institutions. They usually break into a bank's computer system and leave a message that will be found by the bank's computer security specialists, threatening to destroy the bank's computer files if specific sums of money are not sent to the terrorists.

According to a source in Great Britain, terrorists have extorted up to 400 million British pounds from 1993 to 1995 by making intimidating threats to various financial institutions.

> Over the three years, there were 40 reported threats made to banks in the U.S. and Britain. . . . In January of 1993 . . . a brokerage house paid out 10 million pounds after receiving a threat and one of their machines crashed . . . a blue chip bank paid to blackmailers 12.5 million pounds after receiving threats . . . another brokerage house paid out 10 million pounds. (Anonymous [c], 1997)

Another use of computer technologies that terrorists can exploit is called "information warfare," which "consists of those actions intended to protect, exploit, corrupt, deny, or destroy information or information resources in order to achieve a significant advantage, object, or victory over an adversary" (Schwartau, 1994: 12). Both state and non-state actors have utilized this potentially threatening form of warfare.

In the Gulf War, for example, the United States implanted viruses and made other computer intrusions into Iraqi air defenses. And in the war against Serbia, the U.S. military "strove to distort information Serb gunners saw on their screens, helping keep U.S. planes safe during their bombing runs" (Arquilla, 2000: 16).

Terrorists now have at their disposal powerful "weapons of mass disruption" such as computer viruses, DDoS attacks, and information warfare. Will terrorists escalate their attacks in the future by using weapons of mass destruction? Of the three types of weapons of mass destruction, nuclear weapons pose the greatest potential danger (as mentioned in Chapter 2), followed by biological weapons and chemical weapons. Biological weapons are more deadly than chemical weapons.

Nuclear weapons require a complex body of knowledge, along with sophisticated technologies, raw materials, and, of course, capital, making their acquisition beyond the capabilities of all but governments. Biological and chemical weapons, however, are relatively cheaper and easier to build, provided the raw materials can be obtained, and their potential to inflict a high number of casualties should not be underestimated. For example, former U.S. Secretary of Defense William Cohen has said that a supply of anthrax the size of a 5-pound bag of sugar, properly weaponzied, would kill half the population of Washington, D.C. (Lluma, 1999: 14). Historical experience, limited as it is, is enough to arouse substantial anxiety about biochemical terrorism. For example, the use of mustard gas during WWI was sufficiently alarming to pave the way for the 1925 Geneva Protocol, prohibiting gas and bacteriological warfare.

In the late 1980s and early 1990s there were reports of Iraqi leader Saddam Hussein using chemical weapons against Kurds, Iraqi citizens whose long quest for independence made them enemies of the Iraqi government. In March of 1988, during the Iran-Iraq war, Iraqi forces used mustard gas and nerve toxins on Kurdish towns, allegedly killing approximately 5,000 people (Anonymous [d], 1988: 2). Shiite dissidents have also been the targets of chemical attacks by Saddam Hussein (Anonymous [e], 1993: A22). Assuming these reports are valid, especially the repeated attacks on the Kurds, these attacks are not only in violation of the 1925 Geneva Protocol, but they are also in violation of the 1951 Genocide Treaty, prohibiting the deliberate destruction of a particular group of people.

A Japanese terrorist group, Aum Shinrikyo, had sought to use chemical and biological weapons between 1990 and 1993. This Japanese cult subscribed to an apocalyptic vision. Its founder, Shoko Asahara, prophesied that the United States would attack Japan with nuclear weapons and destroy 90 percent of its population. To avert this catastrophe, he asserted that 30,000 followers must be recruited to ward off the eventual Armageddon. Between 1990 and 1993 this sect attempted, on at least nine different occasions, to spread biological agents in the vicinity of Tokyo and nearby U.S. military bases. Its deadliest and most well-known attack was in March of 1995, when it released sarin in a Tokyo subway, killing 12 people and injuring more than 1,000 (Kaplan, 2000: 207–226).

A comprehensive study by the U.S. General Accounting Office (GAO) has identified an array of biological and chemical agents that might be used by terrorists. The GAO report examines the technical ease or difficulty on the part of terrorists to acquire, process, improvise, and deploy chemical and biological agents to cause at least 1,000 casualties without the assistance of a government-sponsored program (see Figures 5–3 and 5–4).

Agent	Ease of manufacture and precursor availability	Agent persistence	Lethality
Choking agents			
Chlorine (CL)	Industrial product. No precursors required.	Not persistent	Low
Phosgene (CG)	Industrial product. No precursors required.	Not persistent	Low
Nerve agents			
Tabun (GA)	No readily available manufacturing instructions, but precursors available. Relatively easy to manufacture.	Intermediate	High
Sarin (GB)	Moderately difficult to manufacture. Precursor chemical covered by Chemical Weapons Convention (CWC).	Not persistent	High
Soman (GD)	Difficult to manufacture. Precursor chemical covered by CWC.	Intermediate	High
GF	Moderately difficult to manufacture. Precursor chemical covered by CWC.	Intermediate	High
VX	Difficult to manufacture. Precursor chemicals covered by CWC.	High	Very high
Blood Agents			
Hydrogen cyanide (AC)	Industrial product. Precursor chemicals covered by CWC.	Very low	Low to moderate
Cyanogen chloride (CK)	Not easily produced. Available as commercial product.	Low	Low to moderate
Blister agents			
Sulfur mustard (HD)	Easy to synthesize. Large quantity buys of precursor chemicals without detection difficult. Precursors are covered by CWC.	Intermediate to high	Can produce incapacitation because of blistering. Can also produce death if inhaled or a toxic dose absorbed.

Figure 5–3
Characteristics of selected agents of chemical terrorism.
Source: United States General Accounting Office (September, 1999). *Combating terrorism: Need for comprehensive threat and risk assessments of chemical and biological attacks*, GAO/NSIAD-99-163.

Agent	Ease of manufacture and precursor availability	Agent persistence	Lethality
Nitrogen mustard (HN-2)	Easy to synthesize. Large quantity buys of precursor chemicals without detection difficult. Precursors are covered by CWC.	Intermediate	Can produce incapacitation because of blistering. Can also produce death if inhaled or a toxic dose absorbed.
Nitrogen mustard (HN-3)	Easy to synthesize. Large quantity buys of precursor chemicals without detection difficult but available.	High	Can produce incapacitation because of blistering. Can also produce death if inhaled or a toxic dose absorbed.
Lewisite (L, HL)	Moderately difficult to manufacture and moderately difficult to acquire precursor chemicals.	Intermediate to high	Can produce incapacitation because of blistering. Can also produce death if inhaled or a toxic dose absorbed.

Figure 5–3 (*continued*)

The chemical and biological agents listed in Figures 5–3 and 5–4 are the ones experts on terrorism consider the major threats as weapons of mass destruction because they can be relatively easy to obtain and deploy, cause a high level of mortality, spark panic, and require public health officials to be prepared for a program of counter-measures.

September 11 Attacks

The world's worst terrorist attacks occurred in New York City, Washington D.C., and Pennsylvania. On this fateful day, nineteen suicide bombers hijacked four commercial jet airliners with a total of 265 passengers and crews aboard, transforming these airliners into weapons of mass destruction. Two of the hijacked

Agent	Ease to acquire and process	Agent stability	Lethality
Bacterial agents			
Inhalation Anthrax	Difficult to obtain virulent seed stock and to successfully process and disseminate.	Spores are very stable. Resistant to sun, heat, and some disinfectants.	Very high
Plague	Very difficult to acquire seed stock and to successfully process and disseminate.	Can be long-lasting, but heat, disinfectants, and sun render harmless.	Very high
Glanders	Difficult to acquire seed stock. Moderately difficult to process.	Very stable.	Moderate to high
Tularemia	Difficult to acquire correct strain. Moderately difficult to process.	Generally unstable in environment. Resists cold but is killed by mild heat and disinfectants.	Moderate if untreated, low if treated
Brucellosis	Difficult to acquire seed stock. Moderately difficult to produce.	Very stable. Long persistence in wet soil or food.	Very low
Q Fever (rickettsial organism)	Difficult to acquire seed stock. Moderately difficult to process and weaponize.	Stable. Months on wood and in sand.	Very low if treated
Viral agents			
Hemorrhagic fevers (e.g., Ebola)	Very difficult to obtain and process. Unsafe to handle.	Relatively unstable.	Depending on strain, can be very high
Smallpox	Difficult to obtain seed to high stock. Only confirmed sources in United States and Russia. Difficult to process.	Very stable.	Moderate

Figure 5–4
Characteristics of selected agents of biological terrorism
Source: United States General Accounting Office (September, 1999). *Combating Terrorism: Need for Comprehensive Threat and Risk Assessments of Chemical and Biological Attacks.* GAO/NSIAD-99-163.

Agent	Ease to acquire and process	Agent stability	Lethality
Venezuelan Equine Encephalitis	Difficult to obtain seed stock, process, and weaponize.	Relatively unstable. Destroyed by heat and disinfectants.	Low
Toxins			
Ricin	Readily available. Moderately easy to process but requires ton quantities for mass casualties.	Stable.	Very high
Botulinum (Types A-G)	Widely available, but high toxin producers not readily available or easy to process or weaponize.	Stable. Weeks in non-moving water and food. Deteriorates in bright sun.	High without respiratory support
Staphylococcl Enterotoxin B	Difficult to acquire high yielding seed stock. Moderately difficult to process.	Very stable in dry form.	Low

Figure 5–4 (*continued*)

planes crashed into the twin towers of the World Trade Center, one airliner crashed into the Pentagon, and the fourth plunged to its destruction in a western Pennsylvania field. The terrorists committed a crime against humanity of unimaginable proportions. The fateful facts as of January 17, 2002, are as follows:

World Trade Center (New York City):

- American Airline, flight 11: 92 killed
- United Airline, flight 175: 65 killed
- Approximately 2,889 dead or missing

The Pentagon (Washington D.C.):

- American Airline, flight 77: 64 killed
- 125 dead or missing

Shanksville, Pennsylvania:

- United Airline, flight 93: 44 killed.

In addition to the 265 people who lost their lives on the hijacked airplanes, approximately 3,014 lives were destroyed in the twin towers of the World Trade Center and in the Pentagon. This catastrophe, as of the middle of September 2001, had also taken an enormous financial toll of $60 billion in direct costs to the U.S. economy, and approximately $600 billion in stock-market losses (Scherer and Paulson, 2001: 1). The events of September 11 have forced us to confront the monstrous reality of terrorism, and have led to a heightened state of awareness of the potentially destructive power of diverse terrorist weapons: information, nuclear, chemical, and biological.

For many years into the 21st century, terrorists will probably seek to convert technologies, whether high-tech or low-tech, into weapons of mass destruction. The fact that one of the alleged suicide bombers, Mohammed Atta, who crashed an airplane into the World Trade Center, is suspected of trying to buy a crop-dusting airplane in order to spread biological and chemical agents bears out this threat (Settle, 2001: 3). In order to achieve a maximum amount of chaos, terrorists will probably target cultural symbols of Western Civilization in the United States and elsewhere. To forestall and ward off their criminal designs we will have to surpass terrorists in ingenuity and creativity. To defeat terrorism in the United States, according to Ullman (2001: 17), we must devote the best minds to a new Manhattan Project. Useful as such a strategy might be, a technological approach alone may not suffice. We will also have to make a commitment to eradicate the economic and political roots of terrorism.

TERRORISM AND COUNTER-TERRORISM

How shall we apply our four-fold causal framework to the phenomenon of terrorism? Because of the complexities of terrorism, a multiple-perspective analysis is required, namely, the application of all four of our categories of causal analysis (Bowonder and Linstone, 1987).

Since terrorism involves a deliberate human act, one would be tempted to categorize it as an exclusively human factors problem. This, however, would not do justice to the complexities of the threat. In addition to involving a deliberate action of one or more human beings, terrorism also utilizes technical design factors since it exploits the vulnerabilities of technological systems. Penetrating the security system of a computer network, for example, requires that the terrorist understands the technical details of its operating system.

In the case of biochemical weapons, technical design factors are very much involved, namely, the relative ease of manufacturing such weapons compared to nuclear weapons.

Organizational systems factors are also involved in terrorist attacks; they require the coordination of tasks on the part of the perpetrators in planning, production, and implementation of their criminal acts. For example, the events of September 11 required several years of planning and coordination among the members of the terrorist network.

Finally, socio-cultural factors are also involved in information system terrorism as well as in biochemical terrorism. Hackers may define breaching a computer security system as a technical, game-like challenge and as an opportunity to express their disdain for the authority system of corporate organizations. Their underlying motivation may be rooted in a profound alienation from the economic and legal values of a society. In the case of conventional international terrorism and biochemical terrorism, the root causes may be located in the great disparities in economic resources of developed and developing countries and in the political oppression of totalitarian regimes. Economic privation and political oppression generate widespread perceptions of injustice and resentment. Such perceptions provide the fertile grounds for hatred that eventuates in expressions of terrorism.

Our four-fold causal framework is also applicable to assessing a counter-terrorism program. Counter-terrorism involves a complex, uncertain, and costly process. It is complex because it requires attention to many dimensions of threats—chemical, biological, radiological, or nuclear; it is uncertain because it is inherently difficult to know the duration of the effort required and when, if ever, it is possible to declare victory, so to speak; and it is costly because it is an open-ended struggle necessitating the recruitment of a large number of individuals and organizations. The September 11 terrorist attacks have alerted the U.S. government and the citizenry to multiple threats. These can be categorized according to the following four counter-measures:

Technical Factor Counter-Measures

Many technical design factors have been considered as counter-terrorism measures, such as installing steel-reinforced cockpit doors on airliners. Garwin, a distinguished physicist, has advanced a number of technological proposals:

- Ensure that the radar transponder, once switched to emergency mode by an airline pilot, cannot be switched back.
- To counter biological warfare, individuals, firms, governments, and other organizations should consider installing a unit to provide positive pressure High Efficiency Particulate Air filter (HEPA-filtered) makeup air to their buildings.*
- To facilitate the movement of cargo, more use should be made of the sealing of containers, ships, aircraft, or trucks at the departure point so that inspection would occur there with adequate time and space, rather than on the fly at bridges or other choke points. Electronic manifests could be sent ahead and would also accompany the vehicle. (Garwin, 2001:18)

Another technical factor counter-measure is the implementation of biometric technologies in airports, railroad stations, and anywhere else terrorists may try to carry out their criminal plans.

---

*"It takes a very small capital expenditure and a very small expenditure in power to provide a positive pressure so that normal winds will not infiltrate a building, and the anthrax spores or other microbes will be kept out. To do this the air intake to a normal building . . . should be provided with a small blower that delivers air through a High Efficiency Particulate Air filter at a rate that exceeds the leakage of air in or out of the building. Such 'makeup' air will then produce excess pressure in the building so that air flows out through any cracks or apertures, blocking any inflow of unfiltered air" (Garwin, 2001: 18).

Biometric technologies use a person's physical characteristics to identify suspected criminals or terrorists. Digitized fingerprints, voiceprints, iris and retinal images, hand geometry, and hand-written signatures are examples of physical characteristics that can identify an individual suspected of malicious intent (Pillar and Kaplan, 2001: 3).

Although there is no high-tech silver bullet to solve the problem of terrorism, if technologies such as those noted above were used, terrorists would find it more difficult to carry out their operations.

Human Factor Counter-Measures

The U.S. government has urged the citizenry to be vigilant about possible terrorist threats. Airline pilots have urged passengers to resist attempts by terrorists to hijack airliners, and air passengers have also vowed to resist any hijackers (Verhôvek, 2001:1). In the wake of the September 11 attacks, members of Congress have sponsored a bill that would increase human responses to suspected terrorists. The bill includes having federal air marshals assigned to guard passengers and aircraft, the implementation of anti-hijack training for flight crews, and the arming of properly-trained pilots to protect the cockpit controls from threatening assailants (Simon, 2001). Another idea for helping ordinary citizens to get personally involved in combating terrorism is to initiate civilian defense programs to help guard critical infrastructures such as water and electric supply systems, buses, subways and rail transport, television and radio stations, and other possible targets of terrorist attacks. Such programs were active during WWII, when many concerned citizens volunteered. Such volunteerism creates a sense of common purpose and a sense of community, which may compel citizens to educate themselves about the multiple guises of terrorism (Anonymous [f], 2001: A38).

Organizational Systems Factor Counter-Measures

The events surrounding the September 11 attacks have pointed out weaknesses in U.S. intelligence-gathering capabilities, both at the national and international levels. In response to such weaknesses, an Office of Homeland Security has been established in the United States to coordinate the efforts of about four dozen government agencies to secure the nation's borders, protect nuclear power plants, share intelligence, secure public facilities, and fight bioterrorism. One major criticism leveled against agencies such as the Federal Bureau of Investigation (FBI), the Central Intelligence Agency (CIA), and the National Security Agency (NSA)—the principal agencies whose job it is to look after national security—is their over-reliance on high-tech surveillance items such as satellites and spy planes, which can monitor actions on the ground, and electronic eavesdropping systems that can intercept and decipher fax, telephone, and e-mail communications throughout the world. Such forms of signal intelligence, or "signet," have proven to be of limited value in light of September 11. One task for the director of the Office of Homeland Security is to conduct a thorough review of "signet" technologies. Another is to focus more attention on "humint" or human intelligence capabilities, which translates into the use of operatives on the ground who can infiltrate terrorist networks and relay information back to officials that only an "insider" can know (Pincus, 2001: A11).

Socio-Cultural Factor Counter-Measures

Congress is updating a 1996 law pertaining to biological agents and toxins that require export licenses to ensure that they do not fall into terrorist hands. There is also a proposal to have optional national ID cards, which would make it "more difficult for potential terrorists to hide in open view, as many of the September 11 hijackers apparently managed to do" (Dershowitz, 2001:A23). Such an ID card would provide a tradeoff: a reduction in privacy for an increase in security.

Paradoxically, many technologies that can be used to combat terrorism can also threaten our individual privacy as well as our civil liberties. Civil libertarians claim that it is against our constitutional rights to have our biometric information analyzed and put in a database without our consent. Another example is the use of DNA to combat crime. The same DNA databases that can be used to identify and convict suspected criminals could also be used to identify genes associated with certain physical and mental health conditions that sufferers might wish to keep private. Do we want to live in a society where law enforcement personnel have the ability to monitor citizens' comings and goings by means of national ID cards, scan large crowds with face-recognition technologies in an attempt to identify known criminals, and have access to DNA databases that contain information individuals wish to keep out of the hands of employers and insurance companies? These technologies can certainly be used to combat terrorism, but they also conjure up dystopian fears of "Big Brother" and a loss of privacy and freedom. What is needed is a concentrated effort to educate the public about such technologies and a process of democratic deliberation over their possible uses. If such technologies are put to use with little or no public debate, it will create mistrust in the citizenry. In order to take advantage of surveillance technologies such as national ID cards, biometrics, or DNA databases, public debates are essential and a democratic consensus needs to be achieved—this is the only way such technologies can be made consistent with the cultural values of individual freedom and democracy that are central to our way of life.

CONCLUSION

To explain terrorism and to assess counter-terrorism programs in the nuclear-information age, we have drawn on the combined conceptual power of our four root causes of technological disasters. In discussing these causes of technological disaster, we assumed that they were applicable independent of time and place. In Chapter 6, we attempt to provide a context for technological disasters by placing them in an historical framework—the three industrial revolutions. What our investigation reveals is that the four root causes discussed in this chapter throw light on the course of technological development.

References

Anderson, Robert M., Schendel, Dan E., and Trachtman, Leone E. (1980). *Divided loyalties: Whistleblowing at BART*. Purdue, IN: Purdue Research Foundation.

Anonymous (a). (2001). "The Kansas City Hyatt Regency walkways collapse," Department of Philosophy and Department of Mechanical Engineering, Texas A&M University. Accessed from the World Wide Web at: http://lowery.tamu.edu/ethics/ethics/hyatt/ hyatt1.htm

Anonymous (b). (2000). "Fifteen year old accused in CNN case," *The Houston Chronicle,* Thursday, April 20: A1.

Anonymous (c). (1997). Accessed from the World Wide Web at: www-cs.etsu.tn.edu/gotterbarn/stdntppr/stats.htm. "Statistics on Cyber-terrorism."

Anonymous (d). (1988). "Iraq chemical attacks kill Kurds," *The Financial Times* (London), April 2: 2.

Anonymous (e). (1993). "Hundreds of Shiites reported killed by Iraqi chemicals," *The Toronto Star,* October 23: A22.

Anonymous (f). (2001). "Turn again to civil defense," *The Washington Post,* September 28: A38.

Arquilla, John. (2000). "Preparing for cyberterrorism—badly," *The New Republic,* May 1: 16.

Baron, S. (Ed.). (1990). *Quantitative modeling of human performance in complex, dynamic systems.* Washington, DC: National Academy Press.

Baskin, O., and Aronoff, C. (1988). *Interpersonal communication in organizations.* Santa Monica, CA: Goodyear Publishing Co.

Betts, Mitch. (1994). "Pentium flaw joins long list of computer math mistakes," *Computerworld,* December 19; 28 (51): 4.

Bignell, V., and Fortune, J. (1984). "Human factors paradigm." Chapter 12 in *Understanding systems failures.* Manchester, UK: Manchester University Press: 190–205.

Bondurant, S., Ernster, V., and Herdman, R. (Eds.). (2000). *Safety of silicon breast implants.* Committee on the Safety of Silicon Breast Implants. Washington, DC: Institute of Medicine.

Bowonder, B., and Linstone, H.A. (1987). "Notes on the Bhopal accident: Risk analysis and multiple perspectives," *Technology Forecasting and Social Change* 32: 183–202.

Chase, Alston. (2000). "Harvard and the making of the Unabomber," *The Atlanta Monthly* 285 (6): June: 41–65.

Cilluffo, Frank, Berkowitz, Bruce, and Lanz, Stephanie (Eds.). (1998). *Cybercrime, cyberterrorism and cyberwarfare.* Washington, DC: The Center for Strategic and International Studies.

Clothier, Norman. (1987). *Black valor: The South African native labor contingent, 1916–1918, and the sinking of the Mendi.* Pieternaritzburg, South Africa: University of Natal Press.

Corman, S., Mayer, Michael E., Bantz, Charles R., and Banks, Stephen P. (Eds.). (1990). *Foundations of organizational communication: A reader.* New York: Longman Press.

Denning, Dorothy. (1999). *Information warfare and security.* Boston, MA: Addison-Wesley.

Dershowitz, Alan M. (2001). "Why Fear National ID Cards," *The New York Times,* October 13: A23.

Dombrowski, P. (1992). "Challenger and the social contingency of meaning," *Technical Communication Quarterly* 1 (3).

Dumas, Lloyd. (1999). *Lethal arrogance: Human fallibility and dangerous technologies.* New York: St. Martin's Press.

Farley, C.J. (1995). "The man who wasn't there (world trade center bombing arrest)," *Time,* February 20: 24.

Fisher, A. (1986). *Small group decision-making: Communication and the group process.* New York: McGraw-Hill.

Garwin, Richard L. (2001). "The many threats of terror," *The New York Review of Books* Vol. XLVIII (November 1, 2001): 16–19.

Glegg, Gordon L. (1971). *The design of design.* Cambridge, England: Cambridge University Press.

Gouran, D., Hirokawa, R., Martz, A. (1986). "A critical analysis of factors related to decisional processes involved in the Challenger disaster," *Central States Speech Journal* 37 (3): 119–135.

Grove, Andrew S. (1996). *Only the paranoid survive: How to exploit the crisis points that challenge every company and career.* New York: Doubleday Books.

Grove, Andrew S. (1998). "My biggest mistake," *Inc.,* May 20 (6): 117.

Hartley, Robert. (1993). "Dow Corning and silicone breast implants: Another Dalkon Shield?" In Robert Hartley, *Business ethics: Violations of the public trust.* New York: John Wiley & Sons: 235–251.

Herndl, C., et al. (1994). "Understanding failures in organizational discourse: The accident at Three Mile Island and the shuttle Challenger disaster." In C. Bazerman (Ed.), *Textual dynamics of the professions.* Milwaukee: University of Wisconsin Press.

Hirokawa, R., Gouran, D., Martz, A. (1988). "Understanding the sources of faulty group decision-making: A lesson from the Challenger disaster," *Small Group Behavior* 19 (4): 411–433.

International Atomic Energy Agency. (1986). *Yearbook 1986.* Vienna: International Atomic Energy Agency: 76.

International Atomic Energy Agency. (1987). *Yearbook 1987.* Vienna: International Atomic Energy Agency: 43.

Jackson, Carlton. (1982). *The dreadful month.* Bowling Green, OH: Bowling Green State University Press.

Jacky, Jonathan. (1990). "Risks in medical electronics," *Communications of the ACM* 33 (12): 138.

Janis, I. (1972). *Victims of groupthink.* Boston, MA: Houghton Mifflin.

Joyce, Ed. (1987). "Software bugs: A matter of life and liability," *Datamation* 33: 88–93.

Kaczynski, T. (1995). "Excerpts from the Unabomber's manifesto," *The Washington Post,* August 2: A16.

Kaplan, D. (2000). "Aum Shinrikyo." In Jonathan Tucker (Ed.), *Toxic terror: Assessing terrorist use of chemical and biological weapons.* Cambridge, MA: MIT Press: 207–226.

Kerstetter, Jim. (2000). "How many 'love bugs' will it take?" *Business Week,* May 22: 50–56.

Landau, Martin, and Stout, Russell. (1979). "To manage is not to control: Or the folly of type II errors," *Public Administration Review,* March/April; 148–156.

Lawrence, Anne. (1993). "Dow Corning and the silicone breast implant controversy." In John Boatright (Ed.), *Cases in ethics and the conduct of business.* Englewood Cliffs, NJ: Prentice Hall: 237–263.

Leveson, Nancy, and Turner, Clark. (1993). "An investigation of the Therac-25 accidents," *Computer* 26 (7): 18–41.

Lluma, Diego. (1999). "Low probability, high consequence: Biological terrorism," *Bulletin of the Atomic Scientists* 55 (6): 14.

Lopez, Steve. (2001). "Tighter airport security is just a flight of fancy," *Los Angeles Times,* October 5: A1.

Lovelace, Herbert. (2000). "A clear and present danger," *Informationweek,* May 22: 166–173.

Manion, M., and Goodrum, A. (2000). "Terrorism of civil disobedience?: Towards a hacktivist ethic," *Computers and Society* 20, June: 14–19.

Martin, M., and Schinzinger R. (1996). *Engineering ethics* (3rd ed.). New York: McGraw-Hill.

Mendlovitz, Saul. (2001). "Crime(s) of terrorism: Developing law and legal institutions," *Newsletter of the Lawyers' Committee on Nuclear Policy* (Fall) 13 (2): 3.

Meshkati, N. (1991). "Human factors in large-scale technological systems' accidents: Three-mile island, Bhopal, and Chernobyl," *Industrial Crisis Quarterly* 5: 131–154.

Meshkati, N. (1992). "Ergonomics of large-scale technological systems," *Impact of Science on Society* 42 (165): 87–97.

Miller, C. (1993). "Framing arguments in a technical controversy: Assumptions about science and technology in the decision to launch the space shuttle Challenger," *Journal of Technical Writing and Communication* 23 (2): 99–144.

Moore, P. (1992). "When politeness is fatal: Technical communications and the Challenger accident," *Journal of Business and Technical Communication* 6 (2): 269–292.

Nader, Ralph, and Smith, Wesley J. (1993). *Collision course: The truth about aviation safety.* New York: McGraw-Hill: 294–295.

Neumann, Peter. (1989). "Risks to the public in computers and related systems," *Software Engineering Notes* 14 (1): 6–21.

Neumann, Peter. (1995). *Computer-related risks*. New York: ACM Press; Reading, MA: Addison-Wesley.

Nixon, P., and Frost, M. (1975). "Choosing a factor of safety." In R.R. Whyte (Ed.), *Engineering progress through trouble*. London: The Institution of Mechanical Engineers: 136–141.

Norman, Colin. (1986). "Chernobyl: Errors and design flaws," *Science*, September 5; 233: 1029–1033.

O'Hara, J. (1996). "Human factors evaluation of advanced nuclear power plants." In T. O'Brien (Ed.), *Handbook of human factors testing and evaluation*. Mahwah, NJ: Lawrence Erlbaum Publishers.

Pace, R. (1988). "Technical communication, group differentiation, and the decision to launch the space shuttle Challenger," *Journal of Technical Writing and Communication* 18 (3): 207–220.

Petroski, Henry. (1985). *To engineer is human: The role of failure in successful design*. New York: St. Martin's Press.

Petroski, Henry. (1994). *Design paradigms: Case histories in error and judgment in engineering*. Cambridge, UK: Cambridge University Press.

Pillar, Charles, and Kaplan, Karen. (2001). "America attacked: Technology implications; technology tools," *Los Angeles Times*, September 12: 3.

Pillar, Paul R. (2001). *Terrorism and U.S. foreign policy*. Washington, DC: Brookings Institution Press.

Pincus, Walter. (2001). "House panel suggests revamping intelligence," *The Washington Post*, October 2: A11.

Pinsky, Robert. (1990). "Shirt," *The want bone*. New York: Farrar, Straus and Giroux.

Redmill, F. (Ed.). (1997). *Human factors in safety critical systems*. Boston, MA: Butterworth-Heinemann.

Renz, M.A., and Greg, J. (1988). "Flaws in the decision-making process: Assessment of risk in the decision to launch Flight 51-L," *Central States Speech Journal* 39 (1): 67–75.

Rochlin, G. (1991). "Iran Air Flight 655 and the USS Vincennes: Complex, large-scale military systems and the failure of control." In T. Laporte (Ed.), *Social responses to large technical systems*. Dordrecht, Netherlands: Kluwer Academic Publishers: 99–125.

Roddis, W.M. Kim. (1993). "Structural failures and engineering ethics," *Journal of Structural Engineering* 119 (5): 1539–1555.

Rowland, R. (1986). "The relationship between the public and the technical spheres of argument: A case study of the Challenger disaster," *Central States Speech Journal* 37 (3): 134–146.

Salecker, Gene. (1996). *Disaster on the Mississippi: The Sultana explosion, April 25, 1865*. Annapolis, MD: Naval Institute Press.

Scherer, Ron, and Paulson, Amanda. (2001). "Costliest Disaster in U.S. history," *The Christian Science Monitor,* September 20: 1.

Schwartau, Winn. (1994). *Infowarfare: Cyberterrorism: Protecting your personal security in the information age.* New York: Thunder's Mouth Press.

Sciolino, Elaine, and Gordon, Michael. (2000). "Ukraine consents to shut Chernobyl before years end," *The New York Times,* June 6: A1.

Serrill, Michael. (1986). "Anatomy of a catastrophe: Moscow blames 'gross' human error for the Chernobyl accident," *Time,* September 1; 128: 26–30.

Settle, Michael. (2001). "Terrorist tried to buy crop-duster aeroplane," *The Herald* (Glasgow), September 26: 3.

Shcherbak, Yuri. (1996). "Ten years of the Chernobyl era," *Scientific American* 274(4): 44–54.

Shrivastava, P., Mitroff, I., Miller, C., and Miglani, R.I. (1988). "Understanding industrial crises," *Journal of Management Studies* 25 (4): 285–303.

Simon, Richard. (2001). "Aviation bill clears U.S. Senate," *Los Angeles Times,* October 12: A8.

Starbuck, W., and Milliken, F. (1988). "Challenger: Fine-tuning the odds until something breaks," *Journal of Management Studies* 25 (4): 319–340.

Stein, Leon. (1962). *The triangle fire.* Philadelphia, PA: J.B. Lippincott.

Stolberg, Sheryl. (2001). "Study links breast implants to lung and brain cancers," *The New York Times,* April 26: 36.

Sweet, William. (1989). "Chernobyl: What really happened?" *Technology Review* 92: 43–52.

Turner, B. (1984). *Man-made disasters.* London: Wykam Press.

Ullman, Harlan. (2001). "Intellect over intelligence," *The Financial Times,* October 19: 17.

Vandivier, K. (1972). "Why should my conscience bother me?" In M. David Ermann and Richard J. Lundman (Eds.), *Corporate and governmental deviance: Problems of organizational behavior in contemporary society.* New York: Oxford University Press: 205–226.

Vaughn, Diane. (1996). *The Challenger launch decision: Risky technology, culture, and deviance at NASA.* Chicago, IL: University of Chicago Press.

Verhovek, Sam Howe. (2001) "Air passengers vow to resist any hijackers," *The New York Times,* October 11: 1.

Whetzel, D. (Ed.). (1997). *Applied measurement methods in industrial psychology.* Palo Alto, CA: Davis-Black Publishers.

Whyte, R.R. (1975). *Engineering progress through trouble: Case histories drawn from the proceedings of the Institution of Mechanical Engineers*. London: Institution of Mechanical Engineers.

Wiener, E. (Ed.). (1988). *Human factors in aviation*. San Diego, CA: Academic Press.

Williams, Cindy. (1997). "Intel's Pentium chip crisis: An ethical analysis," *IEEE Transactions on Professional Communication* (March), 40 (1): 13–20.

Winsor, D. (1990). "The construction of knowledge in organizations: Asking the right questions about the Challenger," *Journal of Business and Technical Communication* 4 (2).

Witkin, Gordon, and Roebuck, Karen. (1998). "Torments that will not end: Why Terry Nichols escaped execution," *U.S. News & World Report* 124 (2), January 19: 33.

Yager, Wilson. (1976). "The Sultana disaster," *Tennessee Historical Quarterly* 35 (3): 117–132.

Yardley, Jim. (2000). "Five years after terrorist act: A memorial to the 168 victims," *The New York Times*, April 20: 16, 20.

Technological Disasters Since the Industrial Revolution

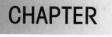

Three Industrial Revolutions and Beyond

*"We must welcome the future,
remembering that soon it will be the past
and we must respect the past, knowing that
once it was all that was humanly possible."*
—George Santayana

The management of technological risks and the prevention of technological disasters pose economic, political, and conceptual challenges. This follows because the increasing complexity of technological systems leads to increasing risk, especially when scientists and engineers are constantly pushing the limits of technological knowledge and harnessing the vast powers of nature without full knowledge of the underlying forces or the risks involved. To fully understand and manage technological disasters, one must analyze their underlying causes. An analysis of the root causes alone, however, is insufficient and will not yield an adequate theory of technological disaster. One reason is that an exclusive focus on analyzing causes does not give us a sense of the growing urgency to deal with the increasing dangers and risks of high technology. An historical analysis will reveal the cumulative effects of technology and the interrelationships between history, technological development, and the ever increasing magnitudes of technological hazard. When one turns to the history of technology, it becomes increasingly evident that with each developmental revolution in technology, the potential risks increase.

THREE TECHNOLOGICAL REVOLUTIONS

Historians and social scientists have documented three major technological revolutions that Western civilization has passed through in the past 250 years.

In his book *Technics and Civilization* (1963) the historian of technology Lewis Mumford categorizes the history of technology into three phases: the Eotechnic Phase (1000–1750), the Paleotechnic phase (1750–1900), and the Neotechnic Phase (1900–1960). For its time, Mumford's work was unprecedented in its breadth and depth. His focus on the developmental history of technics shows how each phase is characterized by the major sources of the materials and energy that dominated the period. During the Eotechnic period the technologies pertaining to wood, water, wind power, and glass basically exhausted the level of technological understanding and hence cultural development. The Paleotechnic Phase (1750–1900) ushered in what historians and economists call the *Industrial Revolution*. Dominating the age of industry were the technologies of coal, iron, and, of course, steam. A new form of energy, electricity, dominates Mumford's Neotechnic Phase (1900–1960). Technologies of electricity move from a stress on coal, associated with steam and iron technologies, to a stress on light and rare metals such as aluminum and copper. Also prominent in the Neotechnic phase are chemical synthetics, such as celluloid, and plastics, such as Bakelite.

What is important to note is that Mumford's emphasis on defining the three periods by the particular technologies associated with each is more explanatory than simply defining each period based on arbitrary historical dates. The focus on sources of power helps define a particular age. The specific kind of power and energy also significantly shapes the cultural experience of that age. Mumford's divisions mark the phases during which central technologies dominated and influenced the cultural milieu of a given age.

In his book *The Visible Hand: The Managerial Revolution in American Business* (1977), Alfred Chandler, the historian of American business enterprise, also defines periods by referring to specific technologies such as the "Revolution in Transportation" and the "Revolution in Communication." Chandler's recent survey of American, British, and German capitalism explains how managerial styles, described as Competitive (American), Personal (British), and Cooperative (German), shaped the structures of modern capitalism. These styles, in turn, reflect underlying sociopolitical realities (Chandler, 1990). Chandler's work demonstrates the symbiotic connection between technological and economic development on the one hand and socio-cultural development on the other.

McCraw, in his book *Creating Modern Capitalism* (1997), presents a similar tripartite schema for categorizing the three industrial revolutions. McCraw categorizes the three revolutions in terms of the development of new technologies. The First Industrial Revolution (1760s–1840s) is characterized by agricultural and manufacturing developments due to steam technologies and mass railroad construction. McCraw characterizes the Second Industrial Revolution (1840s–1950s) in terms of innovations in communications and mass transportation due to the telegraph, telephone, automobile, truck, and airplane. In addition, innovations in the chemical industries led to a profuse array of new consumer goods.

McCraw's Third Industrial Revolution (1950–present) is characterized by computer technologies, which, among other things, initiated an electronics revolution in manufacturing processes.

Daniel Bell, in *The Coming of Post-Industrial Society* (1973), also uses a tripartite schema to distinguish the various revolutions in technology in the past 250 years. In a subsequent article entitled "The Third Technological Revolution and its Possible Socioeconomic Consequences" (1989), Bell again distinguishes three technological revolutions, each associated with a type of industrial power. According to his schema, the First Industrial Revolution is fueled by steam power, the second is powered by technological advances in electricity and chemistry, and the third concerns information. Bell distinguishes four separate technological innovations underlying the Third Technological Revolution:

1. the change of all mechanical, electric, and electro-mechanical systems to electronic systems
2. miniaturization
3. digitalization
4. software engineering

According to Bell, society becomes "post-industrial" as the locus of power and social change shifts from industrial/mechanical processes to the various service industries and the focus on information and knowledge. This shift is facilitated through the use of computers and telecommunications technologies. Bell develops at great length the role of telecommunications and information technologies in the third industrial revolution in an article titled "The Social Framework of the Information Society" (1983).

Bell analyzes the transformation from steam and coal technologies to electricity and the various synthetics of industrial chemistry. This mirrors Mumford's analysis of technological history in the move from the Paleotechnic to the Neotechnic phase. In fact, Bell's characterization of the shift from the Second Industrial Revolution to the Third Industrial Revolution follows Marshal McLuhan's analysis in *Understanding Media: The Extensions of Man* (1966), identifying what he calls the "mechanical age" and the "electronic age."

The various distinctions made between the three industrial revolutions are summarized in Figure 6–1.

Important information is gained by comparing and contrasting the historians who stress energy and material culture as the

	First Industrial Revolution	Second Industrial Revolution	Third Industrial Revolution
Mumford	Coal, steam, and iron	Electricity, lighter metals, and synthetics	_____
Chandler	Transportation	Communications	_____
McCraw	Steam, railroads	Communications, mass transportation	Computerization
Bell	Steam	Electricity, chemistry	Information
McLuhan	Mechanical age	_____	Electronic age

Figure 6–1
Alternative conceptions of three industrial revolutions.

driving force (Mumford and Chandler, for example) with social scientists who stress information and administration as the driving force (McLuhan and Bell, for example). By focusing on media as the most significant technology of a period, as an example, McLuhan tries to show that information technology does more to modify a culture than the uses of energy and materials.

As William Kuhns has argued, these types of comparisons are particularly valuable because they enable the analyst to observe, for example, how the mechanized technologies of the Second Industrial Revolution interact with the electronic technologies of the Third Industrial Revolution (Kuhns, 1971). Understanding these relationships can, in turn, help us evaluate how a new or future technology will transform society by looking at the impact of previous technologies.

While the core of the industrial revolution is, by and large, the substitution of mechanical energy for human and animal energy, the core of the post-industrial revolution is the decline of mechanical energy and the rise of electronic energy embedded in information and cybernetic processing technologies such as computers and telecommunications. The progression is from the physical (animal/man) to the mechanical (machines) to the mental or analytical (data, information processing). This progression may help us anticipate the interrelationships between electronic technologies and the coming developments such as biotechnology, nanotechnology, and robotics.

In the following sections we assess the benefits and burdens that are the result of each revolution, focusing on the increase in technological disasters as society moves through each technological era.

THE FIRST INDUSTRIAL REVOLUTION

Coal and steam are the quintessential energy sources that fueled the First Industrial Revolution (Deane, 1965). Archeological evidence points to the burning of coal during the Bronze Age 3,000 to 4,000 years ago in Wales. The Hopi Indians made use of "outcropping" coal as early as the 12th Century A.D. The Chinese, according to Marco Polo, made use of coal in the 13th Century. Underground or shaft mining began in England as early as the

14th Century, but these shallow shaft mines were soon exhausted. By 1684, 70 mines were in operation near Bristol. These were hampered by the difficulties in draining the water in the shafts. It was not until 1710 that the water problem was eased by Thomas Newcomen's steam atmosphere engine, which supplied a cheap and reliable power source for a vertical reciprocating lift pump.

With the patenting of the steam engine by James Watt in 1769, the application of this invention diffused to coal mining, textile manufacturing, ship construction, and several other industries.

> Applied to water-pumping and to hoisting-machinery it made it possible to get cheap coal from deeper and deeper seams. Applied to the blast-furnace, it provided a blast strong enough to burn coke instead of charcoal and ensured continuous operation of expensive blast-furnace equipment wherever coal and iron ore were available, instead of being dependent on a seasonal localized water supply. Applied to industrial machinery it powered spinning and weaving factories, breweries, flour-mills and paper-mills and effectively removed an important limiting factor to the large-scale operation of a wide variety of industries. (Deane, 1965: 130)

Along with the application of coal and steam as sources of fuel in mining and steamboating, accidents began to accumulate. The statistics on mine accidents from 1705 to 1975 are presented in Table 6–1. The death toll during this period was more than 48,000. The percentage of fatalities steadily increased from the 18th Cen-

Table 6–1
Mine disasters, 1705–1975.

Years	Deaths	Percentage of Total
1705–1805	396	0.8%
1812–1850	1,433	3.0%
1851–1878	4,654	9.7%
1879–1899	6,948	14.5%
1900–1920	10,704	22.4%
1921–1950	16,994	35.6%
1951–1965	4,945	10.4%
1966–1975	1,699	3.6%
Total	47,773	100.0%

Source: Data compiled from Nash, Jay Robert (1976), *Darkest hours*. Chicago, IL: Nelson Hall: 710–720.

tury until the middle of the 20th Century, when it began to decline, possibly because of advances in the technology of coal mining.

Steam navigation created new hazards. "Steamboat accidents fell into five main classes: explosions and other accidents caused by the escape of steam, snagging, collisions, fires, and a miscellaneous class which came to be listed in government reports as 'wrecked and foundered' and included accidents due to storms, grounding, and striking rocks and other obstructions" (Hunter, 1969: 271).

Table 6–2 presents accident data from the beginning of steamboat operations until 1851. As informative as this table is, it omits data on the loss of life. Between 1830 and 1840, 104 steamboat accidents resulted in the loss of 1,018 lives (Hunter, 1969: 272–273). In the early decades of the 19th Century, boiler explosions on steamboats navigating U.S. rivers became common occurrences, resulting in a rising number of fatalities.

Lloyd's *Steamboat Directory* (1856) presents grim narratives of recurrent steamboat disasters. One example of such a disaster, the explosion of the steamboat *Ben Franklin* at Mobile, Alabama, in 1836 will suffice:

> Scarcely had she disengaged herself from the wharf, when the explosion took place, producing a concussion that seemed to shake the whole city to its foundations. The entire population of Mobile, alarmed by the terrific detonation, was drawn to the spot to witness a spectacle that must have harrowed every soul

Table 6–2

Steamboat accidents on the western rivers, 1811–1851.

Cause of Accident	Total Number of Accidents	Percentage of Total Accidents	Average Loss (Dollars)	Total Property Loss (Dollars)	Percent of Total Loss
Collision	44	4.5	8,635	379,933	4.5
Fire	166	17.0	10,948	1,817,428	21.0
Explosion	209	21.0	13,302	2,780,118	32.0
Snags, other obstructions, etc.	576	57.5	6,391	3,681,297	42.5

Source: Hunter, Louis C. (1969). *Steamboats on the western rivers.* New York, NY: Octagon Books, a division of Hippcrene.

with astonishment and horror. This fine boat, which had on that very morning floated so gallantly on the bosom of the lake, was now a shattered wreck, while numbers of her passengers and crew were lying on the decks, either motionless and mutilated corpses, or agonized sufferers panting and struggling in the grasp of death. Many others had been hurled overboard at the moment of the explosion, and such were the numbers of drowning people who called for assistance, that the crowd of sympathizing spectators were distracted and irresolute, not knowing where or how to begin the work of rescue. Many—how many, it is impossible to say—perished in the turbid waters before any human succor could reach them.

Apart from the loss of life, which at that time was unexampled, the destruction produced by this accident was very extensive. The boiler-deck, the boilers, the chimneys, and other parts of the machinery, besides much of the lading, were blown overboard and scattered into fragments over the wharf and the surface of the river.

The cause of the accident is believed to have been a deficiency of water in the boiler. The boat was injured to that degree that repairs were out of the question, and she was never afterwards brought into service. (Lloyd, 1856: 74–75)

The frequency of such disasters prompted the Franklin Institute of Philadelphia to appoint a committee "to examine the causes of the explosions of the boilers used on board of steamboats, and to devise the most effectual means of preventing the accidents, or of diminishing the extent of their injurious effects" (Sinclair, 1966: 4). The committee's report inquired into five different causes of steamboat explosion:

I. Explosions from undue pressure within a boiler, the pressure being gradually increased.

II. Explosions produced by the presence of unduly heated metal within a steam-boiler.

III. Explosions arising from defects in the construction of a boiler or its appendages.

IV. Explosions resulting from the carelessness or ignorance of those entrusted with the management of the steam-engine.

V. An examination of the particular cases of collapse of a boiler, or its flues, by rarefaction within. (Sinclair, 1966: 6)

Especially noteworthy are two of the committee's findings: "defects in the construction of a boiler" and "carelessness or ignorance of those entrusted with the management of the steam-engine" (Sinclair 1966: 39).

The Franklin Institute report on steam-boiler explosions was presented to Congress in March 1836. It included detailed recommendations, for example, requiring inspections of all boilers every six months, prohibiting the licensing of ships using boilers with an unsafe design, imposing penalties when explosions result from improper maintenance, incompetence, or negligence of the master or engineer. The recommendations were incorporated into a law passed by Congress on July 7, 1838 (Burke, 1972: 105–106).

This law, however, did not have the intended effect of preventing boiler explosions. Provisions of the law pertaining, for example, to licensing of engineers and inspections of boilers were not enforced and few steamboat owners were ever prosecuted under the manslaughter section of the law. The inefficacy of this law was further reflected in the rising death toll: 277 from explosions in 1850, and 407 in 1851 (Burke, 1972: 109).

Thus, in 1852 Congress passed a more stringent law over the opposition of those who viewed such legislation as a threat to private property rights. In the course of eight years, this follow-up law had the desired effect of reducing the loss of life due to boiler explosions on steamboats by 65 percent (Burke, 1972: 112).

THE SECOND INDUSTRIAL REVOLUTION

Just two years before Edison's incandescent electric light created a sensation at the 1881 International Electrical Exhibition in Paris, Ernst Werner von Siemans was astonishing the public with his "electric traction" streetcar, the first electric-powered train at the Berlin Exhibition of 1879 (Lilley, 1966: 122). The supply of electrical currents to long-distance trains, however, posed numerous safety hazards. Therefore, locomotives of trains traversing the country continued to be powered by coal and steam. Not until the Second World War did national railway systems begin to use traction power from electricity. This was made possible by the development of the power station (Cardwell, 1994: 353). The development of mass

Table 6-3
Railroad wrecks, 1833–1975.

Years	Deaths	Percentage
1833–1861	836	3.4%
1864–1884	1,853	7.5%
1886–1895	972	3.9%
1896–1907	1,629	6.6%
1908–1918	3,513	14.3%
1919–1928	1,804	7.3%
1929–1939	1,598	6.5%
1940–1949	3,241	13.2%
1950–1959	4,750	19.3%
1960–1975	4,423	18.0%
Total	24,619	100.0%

Source: Nash, Jay Robert (1976). *Darkest hours*. Chicago, IL: Nelson Hall: 736–743.

transport networks based upon electric railways dominated the landscape and the city in developed countries. They have also taken their toll in human lives. Table 6–3 reports the number of deaths attributed to railroad crashes from 1833 to 1975.

As Table 6–3 indicates, the increasing use of electrically powered trains after World War II brought about a marked increase in the percentage of railroad wrecks and fatalities.

At the beginning of the 20th Century, electrified mass transit seemed to be the technological marvel of the future. With its coupling of speed and power, the railway set the pattern for most subsequent reactions to new technologies (Fichman, 1993). However, railroads were soon surpassed by another manifestation of the Second Industrial Revolution, the automobile. Even more so than the railroad, the automobile provided an outlet for mankind's deep-seated desire for rapid mobility and infatuation with speed and power (Fichman, 1993).

The development of the automobile was made possible by the invention of the internal combustion engine. In fact, one of the crucial shifts from the First Industrial Revolution to the Second Industrial Revolution was the change of transport systems, from those dominated by the steam engine to transport systems dominated by the internal combustion engine—the prototype of the modern automobile, ship, aircraft, and rocket engines. The first successful gas engine was developed by Nicolaus Otto in 1876. The "Silent Otto" was about twice as efficient as the best steam engine and was an instant commercial success (Derry and Williams, 1961: 602). But, as long as internal combustion engines depended on coal for fuel, such engines were limited. For example, Otto's engines were stationary. It was not until the introduction of oil fuels that the development of a genuinely mobile internal combustion engine was made possible. It was left to the work of Gottlieb Daimler and Karl Benz to develop the first successful engine "on wheels." These two German engineers, unknown to each other, successfully adapted a petrol engine for road vehicles. Gottlieb Daimler produced the first motorcycle and Karl Benz the first motorcar, both in 1885 (Cardwell, 1994: 343–346).

Henry Ford demonstrated that the complete process of automobile manufacture and assembly could be achieved within one carefully integrated operation, with specially designed machine tools and a moving assembly belt. His first mass-produced cars came off his production lines at Dearborn, Michigan, in 1903. In 1909, he began mass-producing his famous Model T. By 1927 some 15 million units had been built, by which time the car had acquired a world market (Derry and Williams, 1961: 395).

Since then, the automobile has become an all-pervasive force in 20th Century society, transforming life and countryside. However, ever since their inception and mass production, automobiles have faced a long and problematic relation to safety and sound engineering. As automotive technologies were diffused throughout modern society with their undoubted advantages, cars and trucks have had increasingly cumulative harmful effects on human lives in terms of crashes and accidents and on the natural environment in terms of substantial environmental pollution. In the United States each year, more than 40,000 people are killed and more than 3 million people are injured in automobile accidents (U.S. Census Bureau, *Statistical Abstract of the U.S.*, 2000: 635). At the global level, each year cars and trucks kill more than 150,000 people, maim 500,000, and injure 10 million others (Fichman, 1993: 155).

The staggering statistics in Table 6–4 attest to the unanticipated and devastating consequences of a major technological innovation in the 20th Century.

Table 6–4
Motor vehicle traffic fatalities and injuries in the United States.

Years	Fatalities	Percentage of Total Fatalities	Years	Injuries	Percentage of Total Injuries
1899–1920	76,353	2.49%	1980–1983[1]	12,885,000	24.67%
1921–1940	546,429	17.80%	1988–1991[2]	13,028,000	24.95%
1941–1960	656,115	21.37%	1992–1995	12,950,000	24.80%
1961–1980	967,055	31.49%	1996–1999	13,359,000	25.58%
1981–1999	824,399	26.85%			
Total Fatalities	3,070,351	100.00%	Total Injuries	52,222,000	100.00%

Source: U.S. Department of Transportation, National Highway Traffic Safety Administration.
[1] Data from 1980–1983 are from National Accident Sampling System (NASS), which is a random sample of police-reported crashes.
[2] Data from 1988–1999 are from NASS General Estimate System (GEM), which is a random sample of police-reported crashes.

In a wide-ranging assessment of the impact of the invention of the automobile on American society, Flink, in his book *The Car Culture* (1975), traces a multiplicity of negative, unanticipated effects: the invention of mass production, the rise of giant automobile corporations paving the way to oligopolistic market practices, their increasing domination of a network of automobile dealers, the construction of a network of highways instead of developing public transit systems, the misuse of limited energy resources and the growing dependence on foreign gasoline and oil imports.

The other great technological innovation made possible by the internal combustion engine is the airplane. The first successful, controlled flight by the Wright brothers was carried out in December 1903. The development of the airplane was advanced during the First World War, when military applications acted as an impetus to the airplane manufacturing industry (Lilley, 1966: 130–131). Alcock and Brown's Atlantic crossing in 1919 helped convince the world of the usefulness and reliability of the airplane. In the coming decades, regular air services were firmly established, eventually totaling millions of miles traveled each year. Before and during the Second World War, a growing aerospace industry was on the rise in the developed countries, which began producing a variety of high-powered, metal-framed machines as

Table 6–5
Air crashes, 1913–2000.

Years	Deaths	Percentage
1913–1947[a]	3,572	12.0%
1948–1957[a]	4,322	14.6%
1958–1967[a]	9,425	31.9%
1968–1975[a]	9,901	33.6%
1982–1989[b]	1,436	4.9%
1990–2000[b]	867	3.0%
Total	29,523	100.0%

Sources: Data compiled from (a) Nash, Jay Robert, 1976. *Darkest hours*. Chicago Il: Nelson: 631–641, and (b) National Traffic Safety Board, 2000.

fighters, bombers, and transport aircraft. The conversion of wartime bombers to passenger aircraft soon demonstrated the potential for civilian airlines. By the 1940s, civilian airlines became firmly established in society. The introduction of the jet engine in civilian aviation during the late 1950s made possible the realization of transoceanic flight and played a central role in the establishment of routine air transport. Again, along with social benefits came social risks. Table 6–5 shows the number of deaths caused by airplane crashes from 1913 to 2000.

With the marked increase in civilian aviation from 1958 to the present came a corresponding increase in air crashes and fatalities—as Table 6–5 indicates—up until the year 1975. The striking decline thereafter may be attributed to substantial advances in aviation technology.

THE THIRD INDUSTRIAL REVOLUTION

The development of nuclear energy, which ushered in a new technological age, was accompanied by another technological achievement, no less profound in its influence—the invention of the computer. These twin developments have brought about the Third Industrial Revolution, which is generating a complex of economic, political, social, and psychological effects more powerful than those produced in the First and Second Industrial Revolutions.

The impacts of the First Industrial Revolution were spread over two centuries. The Second Industrial Revolution featured a much more rapid diffusion of technology and techniques over four or more decades. The impacts of the Third Industrial Revolution, however, are overwhelmingly rapid. The time available to adapt to these transformations is being dramatically compressed; the pace of change threatens our individual and institutional capacities to make the appropriate adaptations.

Now that we are in the midst of the Third Industrial Revolution, we are witnessing new and more disastrous breakdowns associated with the proliferation of more complex technological innovations—nuclear power, the microchip, fiber optics, superconductors, bioengineering, and space exploration. Many of us are aware of the costs, burdens, and failures of the space industries. Failures of nuclear technologies have been unprecedented, and nuclear war and nuclear terrorism are continuing threats, notwithstanding the end of the Cold War. By far the greatest effort in the nuclear industry has gone into the production and upgrading of nuclear weaponry, though the various nuclear test ban agreements raise hopes for an eventual reduction of nuclear arsenals and of the threat of nuclear war.

Risks of Nuclear Power Plants

The Nuclear Age was ushered in on July 16, 1945, before a tense group of renowned scientists and military men gathered in a New Mexico desert where the first atomic bomb was detonated at a remote section of Alamogordo Air Base, 120 miles southeast of Albuquerque. The first applications of this astonishing invention, of course, were the bombings of Hiroshima and Nagasaki on August 6 and 9, 1945.

The world's first nuclear power station went into operation near Moscow on June 27, 1954. It was a small experimental plant, producing only 5,000 KW energy, but it served to demonstrate that the production of energy from nuclear fission was feasible. Britain's Calder Hall was the first full-scale nuclear power plant. It went into operation in May 1956 and delivered power to an electrical grid five months later. The reactors at Calder Hall reached an output capacity of 37,500 KW. Both the Soviet and British plants were designed primarily to produce plutonium for military purposes, the electricity production and sales being a minor byproduct, which helped to reduce the cost of producing the plutonium.

The American program of civilian power stations also started as an offshoot of wartime projects. The prototype of the first pressurized water reactor made its debut in 1955 when it powered the first nuclear submarine, the USS *Nautilus*, which made her maiden voyage in January of that year. As a result of the success of the *Nautilus* nuclear reactor, the U.S. Atomic Energy Commission decided to embark on an experimental electric power plant at Shippingport, Pennsylvania, which had an output capacity of 60,000 KW, almost double that of Calder Hall. As of December 1998, 437 nuclear reactors were operating around the world, with approximately 109 operating in the United States, producing more than 345 billion KW of electricity (Dumas, 1999: 19–21).

All reactors need special elements for control. Such control is achieved either by varying the parameters of the coolant circuit or through the use of special absorbing assemblies called *control rods*. Most reactors are equipped with three types of rods, each serving different functions: (1) safety rods for starting up and shutting down the reactor, (2) regulating rods for adjusting the reactor's power rate, and (3) shim rods for compensating for changes in reactivity as fuel is depleted by the fission process. The control rods stop the fission process by absorbing the neutrons that maintain the nuclear chain reaction. The most important function of the safety rods is to shut down the reactor, either when such a shutdown is scheduled or in case of a real or suspected emergency. Sometimes the reactor must be shut down immediately. This is accomplished by rapidly inserting the safety rods into the core, a practice called *scramming*.

A scram is a sudden, violent shutdown of a nuclear reactor. Since scrams, or emergency plant shutdowns, pose an actual or potential threat to the health or safety of the public, the Nuclear Regulatory Commission requires that all scrams be reported.

Table 6–6 lists various nuclear power plants in the United States that reported scrams in 1993 through 1995.

As Table 6–6 indicates, the reactors are ranked in terms of average number of scrams per three-year period. Since approximately 109 nuclear power plants are in operation, the table is not a comprehensive list of reported scrams. Also, the NRC's Performance

Table 6–6
Number of automatic scrams reported at various U.S. nuclear power plants, 1993–1995.

Reactor	1993	1994	1995	Average # of scrams 1993–1995
Cook-2	2	3	4	3.0
Calvert Cliffs-2	1	4	2	2.3
Comanche Peak-1	3	2	2	2.3
Grand Gulf-1	1	1	5	2.3
Nine Mile Point-1	2	4	1	2.3
Salem-1	3	4	0	2.3
Surry-2	5	0	2	2.3
Browns Ferry-2	0	3	3	2.0
Catawba-2	1	3	2	2.0
Dresden-3	2	1	3	2.0
Fermi-2	4	0	2	2.0
Hope Creek-1	1	5	0	2.0
Limerick-2	2	1	3	2.0
Quad Cities-2	4	1	1	2.0
Washington Nuclear-2	3	0	3	2.0
Comanche Peak-2	2	1	2	1.7
Farley-2	1	2	2	1.7
Hatch-1	3	2	0	1.7
LaSalle-1	1	3	1	1.7
Millstone-2	5	0	0	1.7
Oconee-2	2	2	1	1.7
Oconee-3	1	3	1	1.7
Totals	49	43	40	

Source: Adapted from Riccio, J. and Brooks, L. (1996). *Nuclear lemons: An assessment of America's worst commercial nuclear power plants*. Washington, DC: Public Citizen, Critical Mass Energy Project: 21.

Indicator Program monitors *automatic* scrams and does not include *manual* scrams. As indicated in the table, automatic scrams have remained relatively constant over the past few years. However, as a recent NRC report indicates, manual scrams have almost doubled from 1994 to 1995 (Riccio and Brooks, 1996). Nevertheless, Table 6–6 does provide an indication of the potential threat of catastrophic failure of commercial nuclear power plants. Reactor accidents must now and for many years to come, be acknowledged as posing potential threats to the lives of people living in the vicinity of the 437 nuclear power plants throughout the world.

In order to obtain a more vivid sense of the negative consequences of nuclear power plant failures, a brief description of significant nuclear power plant failures are as follows:

- On September 29, 1957, an explosion at the Soviet power plant at Kyshtym, in the Ural Mountains, released more than 20 million curies of radioactive materials. The fallout of the explosion took the lives of 8,015 people over 32 years (Perera, 1992). Moreover, the surrounding area has been and will be abandoned for decades.

- On October 10, 1957, a fire broke out at Windscale Nuclear Power plant in the northwest of England, sending huge amounts of radioactive isotopes into the environment. The fallout caused 260 cases of cancer, with 13 fatalities reported. Statistical analysis has attributed more than 1,000 deaths to the Windscale fire (May [a], 1989).

- On March 22, 1975, insulation wire caught fire, spreading to one of two reactors at the Browns Ferry nuclear power plant in Decatur, Alabama. Both reactors were shut down for 18 months, resulting in a total cost of $100 million to $200 million, including damage to the plant as well as costs associated with the loss due to the shutdown (May [b], 1989).

- On March 28, 1979, a partial meltdown of the Three Mile Island nuclear plant occurred at Middletown, Pennsylvania. Contaminated coolant water escaped into a building adjacent to the reactor core, releasing radioactive gases. As many as 200,000 people fled the region. A study performed by Dr. Ernest Sternglass, professor of radiation physics at the University of Pittsburgh, showed that iodine radiation release caused the death of at least 430 infants (Lutins, 2000).

- On April 26, 1986, two explosions occurred at the Chernobyl nuclear power plant, about 70 miles north of Kiev, capital of the Ukraine. It sent an estimated 50 million curies of radioactive material into the atmosphere, which eventually spread to Western Europe. Thirty-one people died immediately. By the early 1990s, experts estimate that 6,000 to 10,000 deaths can be attributed to cancers from radioactive fallout from the Unit 4 RMBK at Chernobyl (Medvedev, 1991: 97).

Before the Soviet Union collapsed in 1991, a Soviet nuclear industry economist estimated that:

> By the year 2000 the Chernobyl accident may cost the country 170 billion to 215 billion rubles in lost electricity production, contaminated farmland and other economic consequences. . . . At the official exchange rate in Moscow, it accounts for $283 billion to $358 billion. . . . The total bill suggests that the Soviet Union may have been better off if it had never begun building nuclear reactors in the first place. Since the Soviets opened their first nuclear power reactor in 1954, Mr. Koryakin estimates, the net economic contribution of the Soviet nuclear industry has been 10 billion to 50 billion rubles. The sum is a measure of how much money the country saved by using cheaper, nuclear-generated electricity than more costly coal-burning plants. The Chernobyl accident costs exceed that sum several times. (Hudson, 1990: A8)

Although risks associated with nuclear reactor failures, as indicated in the previous examples, are a cause for concern, risks associated with commercial nuclear power also extend to concern for radiation exposure to workers, radioactivity release during normal operational procedures, the thorny problem of nuclear waste management, and the massive costs of decommissioning nuclear plants after their mandatory 40-year life span. A recent report of the shutdown of the Chernobyl nuclear power plant on December 15, 2000, indicated that the decommissioning process will extend over the next 60 years and cost billions of dollars (Sciolino and Gordon, 2000: A1).

One of the greatest debates concerning commercial nuclear power focuses on the issue of occupational health and safety. Thousands of nuclear power plant workers are exposed to potentially harmful ionizing radiation each year. Although the various governmental agencies such as the Nuclear Regulatory Agency (NRC) and the Occupational Safety and Health Administration (OSHA) set maximum standards for workers exposure to ionizing radiation—limits which employers are required to monitor quite closely—workers are all too often subjected to radiation levels above maximum dose levels.

Radioactivity is also released outside the biological shield of a nuclear reactor if the reactor coolant contaminates underground water. Also, a certain amount of solid material is contaminated—mops, paper towels, glassware, etc.—which is simply buried in designated ground sites or dumped at sea in sealed drums. In contrast to these low-level wastes, the spent fuel of reactors is highly radioactive. The temporary storage and permanent disposal of radioactive wastes present the most challenging problems associated with the operation of nuclear reactors (McCuen, 1990). Moreover, the operation of nuclear waste management systems extends far beyond the components of the nuclear fuel cycle; the biosphere will need to be protected from these wastes for centuries to come.

Notwithstanding the objections from environmentalists in the United States and Europe about the dangers of nuclear power, new concerns about energy supplies, soaring natural gas prices, shortages of electricity, and concerns about global warming have stimulated a revival of interest in nuclear power (Francis, 2001: 3; Rhodes, 1993). The fact that nuclear power plants do not produce any carbon dioxide, which is the principal greenhouse gas, is an argument for its renaissance, according to its proponents.

The nuclear industry, both in the western world and in Russia, has learned a great deal about how to run existing reactors more safely and how to design safer reactors since the Three Mile Island and Chernobyl disasters. While this quest for greater safety has led to evolutionary improvements in the design for both light and heavy water reactors, the current front-runner is the modular high temperature gas reactor, which incorporates passive safety features.[1]

Computer-Related Risks

Whereas it took years to develop commercial uses for nuclear power, and its diffusion has been slowed by fears of the negative impacts, computer technologies developed quickly and are already affecting almost every aspect of our lives. It was the English mathematician Charles Babbage who in 1833 first conceptualized the automation of mathematical processes. Babbage's "analytical engine" was never made, however, and no further practical steps toward the electronic computer were taken until the Second World War (Lilley, 1966: 232–233). In 1937, Howard H. Aiken of Harvard University constructed a machine based on many of Babbage's ideas, incorporating a new invention—electromagnetic relay circuits—into the design. The U.S. Navy first used Aiken's machine, the Automatic Sequence-Controlled Calculator, in April 1944 (Cardwell, 1994: 468).

Meanwhile, in 1946 the U.S. Army Ordinance Department began sponsoring work on an electronic machine called the "Electronic Numerical Integrator and Calculator" (ENIAC) at the University of Pennsylvania. Originally designed to calculate the trajectories of artillery shells, it could do 5,000 additions in a second and calculate in a half a minute a ballistic path that previously took 20 hours to calculate (Lilley, 1966: 233). The first commercial computer, the Universal Automatic Computer (UNIVAC 1) was commissioned by General Electric in 1949. It took less than one ten-thousandth of a second for an addition and under a hundredth of a second for a multiplication.

From then on, the story has been one of rapid innovations due to advances in hardware and software engineering, the substitution of transistors for vacuum tubes, and the development of microelectronics. The first breakthrough was the development of the solid-state transistor at Bell Labs by Bardeen, Brattain, and Shockley. Soon after, mainframe computers became available for businesses and offices and, by the early 1960s, more than 15,000 mainframe computers were in operation worldwide (Fichman, 1993).

In 1992, modern supercomputers could make 500 million to 1 billion calculations in a second (Buchanan, 1992). This million-fold advantage in performance and speed, as compared to energy and information supplied by technologies of the Second Industrial Revolution, is indeed impressive. For example, a jet airplane allows a human being to travel 300 times faster than by foot (Buchanan, 1992). In the emerging Fourth Industrial Revolution, the memory capacity of microprocessing chips appears to be governed by Moore's Law, namely that it doubles every 18 months. If this law holds, we can look forward to the phenomenal growth of supercomputers (Dragan, 2001).

Silicon is the major raw material of the microelectronics revolution in much the same way iron was during the First Industrial

Revolution. Fortunately, silicon is the second most abundant element in the Earth's crust and hence is relatively inexpensive. In the early 1970s, Intel incorporated the entire processing unit of a computer on a silicon chip. By 1980, large-scale integrated circuits (VSLIs) were in use with millions of components fabricated and interconnected onto a single silicon chip about one-tenth the size of a postage stamp (Ceruzzi, 2000: 202).

Notwithstanding these spectacular technological advances, technological failures and disasters have occurred in a wide array of information systems technologies, including telecommunications, computerized networks, and CAD/CAM technologies. Reports of faulty missile systems and helicopter accidents pile up on the desks of bureaucrats. Financial and banking errors, some small but some with global proportions, have occurred due to the unreliability of information systems. The use of computers in medical technologies has led to massive radiation overexposures to a number of cancer patients, sometimes with fatal results. Many other information systems breakdowns abound, and considering the increasing reliance of modern society on vast computerized networks, the increasing number of technological failures and disasters compels us to search for an understanding of the potential for technological disasters associated with computer technologies. The Y2K problem, discussed in Chapter 3, is a testament to this fact.

Table 6–7 presents some data on the types of risk encountered in a variety of computer applications. The numbers in the table refer to individual incidents. Thus, the numbers in the column headed "Cases with Deaths" refer to the number of cases involving deaths, and not the number of deaths. MacKenzie has determined that, from 1985 to 1992, there have been 1,100 deaths attributed to computer-related failures (MacKenzie, 1998: 192). Although clearly an incomplete record, this table alerts us to the potential failures and disasters of information technology.

In order to appreciate the magnitude of failures associated with computer technologies, we present brief descriptions of the following significant computer-related failures:

- In 1990, a programming error in the call-handling computer systems at AT&T shut down long-distance telephone networks for nine hours. Some 74 million

Table 6–7
Computer-related failures*.

Problem Area	Cases with Deaths	Cases with Risks to Lives	Cases with Resource Losses	Totals
Communication reliability	3	28	30	61
Space	1	25	23	49
Defense	9	41	18	68
Commercial aviation	17	49	9	75
Public transport	12	36	15	63
Control systems/robotics	12	13	13	38
Medical	17	13	4	34
Electric power	2	23	2	27
Financial sectors	0	1	134	135
Law enforcement/legal	1	23	40	64
Security/privacy	2	10	72	84
Telephone fraud	0	0	22	22
Totals	76	262	382	720

Source: Adapted from Neumann, Peter G. (1995). *Computer-related risks.* New York, NY: ACM Press: 309. *As of August 1993.

long-distance and 800-number calls could not be completed, bringing phone-dependent businesses—such as car, hotel, and airline reservation systems and credit-card approval systems—to a standstill (Anonymous, 1995).

• In 1991, during the Gulf War, the failure of certain Patriot missiles to track and destroy incoming Iraqi Scud missiles were attributed to software flaws, specifically the accumulation of inaccuracies in the internal time-keeping mechanisms of the computer-controlled tracking system. During such a Scud attack, 28 American soldiers were killed when a missile slammed into their barracks in Dhahran, Saudi Arabia (Littlewood and Stringini, 1992: 433).

• In 1993, an $80 million satellite called Clementine was hopelessly lost in space after a software error caused its thruster rockets to fire continuously, consuming all its fuel before its asteroid-rendezvous mission was completed (Anonymous, 1995).

- Bugs in a computerized baggage-handling system delayed the opening of the new Denver airport, which eventually opened in 1995. The system would drive automated baggage carts into walls or deposit bags at the wrong destination. The costs to fix the system were approximately $80 million (Anonymous, 1995).

- In 1995, a computer error in a contest sponsored by Pepsi Cola in the Philippines caused 800,000 winning numbers to be generated instead of the intended 18. The face values of the winnings would have been $33 billion! Pepsi paid nearly $10 million to customers with winning numbers to maintain "goodwill," but still faced hundreds of lawsuits and criminal complaints (Baase, 1997: 121).

In addition to these dramatic computer-related failures, the Therac 25 case and the USS *Vincennes* case, discussed in Chapter 5, are examples of failures of computer technologies that have had lethal consequences.

Another failure of computer-related technology, which had monetary but fortunately no human costs, was the "Love Bug" computer virus. On May 4, 2000, the "I Love You" virus was sent as an e-mail attachment under the subject heading "I Love You." The attachment contained a Visual Basic program that was activated when the victim opened it. The program then took over the victim's PC, searching for MP3 and other files on the hard drive and replacing them with copies of itself. In addition, the virus also sent copies of itself to all individuals in the victim's Microsoft Outlook address book—unlike the "Melissa" virus in 1999, which sent itself to only the first 50 names. Like the Melissa virus, which created an estimated $80 million in damages, the Love Bug virus plagued Microsoft Outlook users and used the Internet to propagate itself, causing $8 billion to $10 billion in damages. However, unlike the Melissa virus, which took days to wreak havoc, the Love Bug virus spread around the world in a few hours, eventually affecting tens of millions of computers (Marlin, 2000).

How are such serious breaches of computer security possible? How are the safety and protection of the computerized networks that undergird our critical infrastructure to be assured? Critics have charged Microsoft with unwisely adding Web-scripting features into newer versions of its Office and Windows products. A *script* is a mini-program that allows users to make software products more versatile and to extend the capabilities of their software applications. "These tools make it easier for tech-savvy users to automate routine tasks, but virus writers who craft small, malicious codes and spread them through e-mail can also exploit such script tools" (Lovelace, 2000: 166).

Such virus attacks as the Melissa and Love Bug viruses point to systems' vulnerabilities when they are connected to computer networks and the Internet. For all its notoriety, the Love Bug attack is just one of dozens of security vulnerabilities discovered daily by computer security experts at large corporations. Tens of thousands of reported computer viruses and worms are in circulation, each with varying degrees of power to bring down computer systems and networks. It is no longer enough to be vigilant about PCs. "The focus for improved security must move to a higher level: to software design, to the corporate servers that speed mail in and out of offices, and to the Web's gateways, the internet service providers (ISPs)" (Kerstetter, 2000: 50). It is the responsibility of software manufacturers to develop more secure, safe, and reliable software.

A FOURTH INDUSTRIAL REVOLUTION?

The inventions of the first three industrial revolutions—mechanized factories, automobiles, planes, nuclear power, computers, etc.—clearly created great social benefits. However, these technological advances did not come without a price. Industrialization demanded coal and metals stripped from the earth, electrical power from dammed rivers and nuclear power plants, oil, natural gas, and other natural resources. These technologies have created social benefits but they are accompanied by social costs, including toxic wastes and a polluted environment.

Today, technological forecasters predict that other major technological revolutions, such as recombinant DNA technologies and nanotechnology, will greatly affect us in the 21st Century. The emerging fields of genetic engineering, nanotechnology, robotics, and artificial intelligence (the creation of intelligent, autonomous machines) are creating the basis of a Fourth Industrial

Revolution, and it promises to produce a spectacular array of new products without the poisonous and destructive side effects of traditional technologies.

Proponents of biotechnology argue that microbes will devour massive landfills and toxic waste dumps and in the process generate valuable by-products such as methane gas (Sylvester and Klotz, 1995). Genetic engineering of agricultural products promises to create an unprecedented abundance of foodstuffs, more efficient crops, and disease-resistant plants. In addition, the invention of genetically engineered drugs and hormones may lead to the prevention and cure of a variety of human illnesses (Kielstein, 1996). The promises of biotechnology, however, may not be without risk.

Geneticists have linked the emergence of pathogenic bacteria and antibiotic resistance to the transfer of genetic material between organisms and species, which experts refer to as *horizontal gene transfer.* Laboratory studies show that genetically modified (GM) genes increase the variety of DNA that is available for bacteria to transfer or pass on traits, including harmful ones, from one organism to the next, or from one species to the next (Ho, 2001: 45). This happens either through infection by viruses, through pieces of DNA taken up into cells from the environment, or by mating between unrelated species. Experts have identified horizontal gene transfer to be the cause of antibiotic-resistant strains responsible for cholera outbreaks in India in 1992, the streptococcus epidemic in Tayside, England, in 1993, and the E. coli outbreak in Scotland in 1995 (Ho, 1997). Two strains of E. coli, which were isolated in a transplant ward outside Cambridge, England in 1993, were found to be resistant to 21 out of 22 common antibiotics (Ho, 1997). In fact, diseases such as tuberculosis, cholera, malaria, diphtheria, and meningitis, largely thought to be under control, are coming back worldwide; many are thought to be new strains resistant to antibiotics. Finally, the World Health Organization reports that at least 30 new diseases, including AIDS, Ebola, and hepatitis C, have emerged during the past 25 years and are linked to the negative effects of horizontal gene transfer, the main technique used in genetic engineering (Ho, 1997).

As far as animal uses are concerned, one of the perils of genetic engineering includes the creation of monstrosities in animal experimentation (Wacks, 1996). The unprecedented power of bioengineering to alter the very genetic makeup of the human species—the power to alter and create life itself—may very well be the greatest philosophical challenge the human race has ever faced.

There is no more outspoken critic of biotechnology and its dangers than Jeremy Rifkin. In the Introduction to his book, *The Biotech Century* (1998), he writes:

> It was more than twenty years ago that I co-authored, with Ted Howard, a book entitled *Who Should Play God?*. In that book we wrote of the promises and perils of a fledgling new technology that few people had heard of called genetic engineering. While we discussed the many benefits that would result from the new science, we also warned of the dangers that might accompany the new technology revolution. . . . We also expressed concern over the increasing commercialization of the Earth's gene pool at the hands of the pharmaceutical, chemical, and biotech firms, and raised questions about the potentially devastating long-term impacts of releasing genetically engineered organisms into the environment. (p. ix)

As Rifkin reports, scientists, policy makers, and the media dismissed the warnings as "alarmist" and "far-fetched." Another early warning about the dysfunctions of biotechnology was presented by Amitai Etzioni in his book *The Genetic Fix* (1973). The irony is, of course, that many of the predictions expressed by Rifkin and Etzioni in the 1970s have been realized in the 21st Century.

Joseph Rotblat, a distinguished physicist who has battled for years against nuclear weapons proliferation, had this to say about the dangers of biotechnology: "My worry is that other advances in science may result in other means of mass destruction, maybe more readily available even than nuclear weapons. Genetic engineering is quite a possible area, because of these dreadful developments that are taking place there" (Ho, 1997).

Nanotechnology, like biotechnology, promises to offer us everything the previous revolutions afforded, but without the detrimental side effects of manipulation of "bulk" matter. Nanotechnology is hailed as a principled break from all previous technologies. Eric Drexler, the MIT prophet of nanotechnology, sees two types of technology: *bulk technology,* the "ancient style of technology that led from flint chips to silicon chips" that handles atoms and molecules in bulk, and *molecular technology,* "the handling of individual atoms and molecules with control and precision" (Drexler, 1986).

Nanotechnology promises to lead to the "completion" of the work of artificial intelligence (AI), by which thinking machines become a reality, "assemblers" become self-replicating, self-regulating, and self-motivating "cell-repair," and other types of nanomachines are created. Nanotechnology promises to allow us to build anything, molecule by molecule, atomic layer by atomic layer, functioning in spaces billionths of a meter in size.

Revolutions in materials engineering—namely, nanotechnology with its central idea of self-replicating assemblers—promise the inexpensive manufacturing of products. Such assemblers will be capable of moving atoms around, constructing anything they are programmed to create, including themselves. As one reviewer put it, "the design or 'programming' of an artifact will constitute ninety-nine percent of its real cost. If products use solar energy as fuel, dirt for raw materials, and can double their bulk through self-replication in a few days, what won't be cheap?" (Crandall, 1992: 80). The fusion of AI and robotics will lead researchers to the possible creation of fully autonomous cyborgs and robots. Researchers claim that intelligent machines will eventually displace humans from factories and farms, providing all of the basic human needs to all people.

However, serious questions about nanotechnology, raised by Drexler himself, force us to ponder the potential negative side effects of this technology:

> What will happen to the global order when assemblers and automated engineering eliminate the need for most international trade? What will we do when replicating assemblers can make almost anything without human labor? What will we do when AI systems can think faster than humans? What will happen if machines and robots surpass us, given their superiority and seeming immortality? (Drexler, 1986: 16)

So much for the dream of visionaries. At present, robotic technology is generally limited to applications in industrial manufacturing and production. As of the year 2000, approximately 742,000 such robots were in use worldwide (United Nations Economic Commission for Europe, 2000). However, given that the number of industrial robots continues to double every three or four years, we should expect that number to be near a million in the first decades of the 21st Century. The main motivation for developing robots is, of course, for the enjoyment and profit of *homo ludens,* who has always dreamed of freeing himself from the toils of labor and having obedient slaves at his command. Proponents even predict that, in the very near future, robots will find applications well beyond the factory floor, performing any and all forms of monotonous labor, competing with humans in all the manual occupations.

Eventually, with the development of artificial intelligence and expert systems, robots will compete with humans for almost any human professional activity. Robots are already being used in the medical and biological sciences, where experiments have been carried out to develop robot surgeons. One prophet of the new technology, Hans Moravec, has this to say:

> By 2010 we will see mobile robots as big as people but with the cognitive abilities similar in many ways to those of a lizard. The machines will be capable of carrying out simple chores, such as vacuuming, dusting, delivering packages and taking out the garbage. By 2040, I believe, we will finally achieve the original goal of robotics and the mainstay of science fiction: a freely moving machine with the intellectual capabilities of a human being. (Moravec, 1999)

Questions remain, however, as to the viability of the new technologies. Will the "slaves" remain obedient? Will they eventually come to realize their "superiorities" to their human programmers and revolt, eventually enslaving the enslavers? In other words, if the new intelligent actors become evolutionarily superior to us humans, by what moral imperative should humans—by that time an inferior species—not accept a servile role on earth?

To thwart any dangers to humans at the hands of their robot creations, Isaac Asimov created his now famous "laws of robotics" to ensure that automatons remain subservient and obedient machines. His three laws are as follows:

1. A robot may not injure a human being or, through inaction, allow a human being to come to harm.

2. A robot must obey the orders given it by human beings, except where such orders conflict with the First Law.

3. A robot must protect its own existence, as long as such protection does not conflict with the First or Second Law (Asimov, 1950).

According to Roger Clarke, however, Asimov's laws of robotics are not simply a successful literary device. They also provide a framework for exploring the practicality of developing and applying the laws for controlling robotic behavior in the 21st Century. "Asimov projected the robotic laws into future scenarios, and, in the process, discovered issues that will probably become real-world situations" (Clarke, 1993). Asimov's stories also explore how and why these laws may be inadequate to control robots unleashed in the world. Clarke provides many examples, but one telling case in point is that the "laws" do not consider areas of judgment that involve deadlock, or "robo-block." In such cases it would not be evident what is "human," especially given the potential ambiguity or vagueness of word meanings. Nor is it evident what would happen if two humans give conflicting commands to the robot. As Asimov put it:

> What was troubling the robot was what roboticists called an equipotential of contradiction on the second level. Obedience was the Second Law and the robot was suffering from two roughly equal and contradictory orders. Robot-block was what the general populations called it, or more frequently 'roblock' for short . . . [or] mental freeze-out. . . . No matter how subtle and intricate a brain might be, there is always some way of setting up a contradiction. This is a fundamental truth of mathematics. (Quoted in Clarke, 1994: 59)

These episodes of robo-lock may result in the robot attempting to override its program to "solve" the dilemma and in the process moving beyond the control of its human operators. In short, Asimov recognized that his laws, even if programmed into the robot's "brain," might not ensure subservience and obedience. These defects in the robot code of conduct led Asimov to eventually revise his laws. As Thompson puts it, "Even were we to stipulate that observing the Three Laws was *necessary* for moral sensibility, this would certainly not be *sufficient*" (1999: 412). The conclusion may be that no single set of laws can reliably constrain the behavior of intelligent machines. Clarke puts it this way,

> Robot design would have to incorporate a high-order controller (a 'conscience?') that would cause a robot to detect any potential for non-compliance with the laws and report the problem—or

immobilize itself. The implementation of such a meta-law ('a robot may not act unless its actions are subject to the laws of robotics') might well strain both the technology and the underlying science . . . this difficulty highlights the simple fact that robotic behavior cannot be entirely automated. (Clarke, 1994: 57)

Asimov's own predilections are instructive. In a later work he writes: "Even if we solve all the problems that face us today, partly by human sanity and partly by means of technology itself, there is no guarantee that we may not, in the future, be threatened with catastrophe through continuing success of technology" (Asimov, 1979: 345). In his book *God and Golem,* Norbert Weiner, father of cybernetics, also anticipated the possible failure to control robots, as in the medieval legend about the Golem (Weiner, 1964).

Prophets of biotechnology, nanotechnology, and intelligent robots such as Drexler and Moravec may be technological visionaries, but they have yet to develop an adequate methodology for technology assessment and technological forecasting. As Joseph Coates, a long-time student of technology assessment puts it:

The most common error [in technological forecasting] is that of the enthusiast who is so optimistic about a new development that he or she neglects the social, economic, and political constraints, and anticipates the arrival of a new technology far sooner than it can occur. . . . This is complemented by a second error: overlooking the side effects of the new technology. (Coates, 1998)

This seems to be the message from a recent outspoken critic of genetic engineering, nanotechnology, and robotics, Bill Joy, chief scientist of Sun Microsystems and a pioneer in the computer industry. According to Joy, the central threat posed by genetic engineering, nanotechnology, and robotics, or what he calls "GNR technologies," is their ability to self-replicate at a super-rapid rate (Joy, 2000). Only a few decades away, self-replicating machines have the potential to produce catastrophic effects on human communities and the natural environment. The combination of wild self-replication of living or inanimate things and unanticipated self-mutation could produce a biosphere-destroying accident. This could happen either by accident, or more nefariously, it can be initiated through conscious, malevolent intent. Joy gives the example of a criminal mind creating a "designer pathogen," something that could be possible in the next 20 to 30 years, according to researchers in the field. Such a disease would not be entirely biological or mechanical, but a combination of the two. Specialists in this field refer to the dangers of rapid self-replication as the "gray goo" problem. Drexler describes the problem:

> Dangerous replicators could easily be too tough, small, and rapidly spreading to stop—at least if we made no preparation. We have enough trouble controlling viruses and fruit flies. Among the cognoscenti of nanotechnology, this threat has been known as the 'gray goo' problem. Though masses of uncontrolled replicators need not be gray or gooey, the term 'gray goo' emphasizes that replicators able to obliterate life might be less inspiring than a single species of crabgrass. They might be 'superior' in an evolutionary sense, but this need not make them valuable. . . . The gray goo threat makes one thing perfectly clear: we cannot afford certain kinds of accidents with replicating assemblers. (Drexler, 1986: 173)

The dangers of rapid replication and proliferation are not limited to nanotechnology. As Ho puts it:

> The large-scale release of transgenic organisms is . . . worse than nuclear weapons or radioactive wastes, as genes can replicate indefinitely, spread and recombine. There may yet be time enough to stop the industry's dreams of turning into nightmares . . . before the critical genetic 'melt-down' is reached. (Ho, 1997)

Joy is also aware of the analogy of the dangers associated with the harnessing of nuclear fission. But, unlike nuclear, biological, and chemical weapons, which require massive material and huge capital investments as well as esoteric expertise, Joy draws our attention to the fact that the new technologies will be within the reach of individuals or small groups. "They will not require large facilities or rare raw materials. Knowledge alone will enable the use of them" (Joy, 2000: 242). This could easily lead to what Joy calls "knowledge-enabled mass destruction." According to Joy, we simply cannot afford to put that kind of power in the hand of any rogue individual or group of rogue individuals or terrorists. To quote Murphy's Law, "if anything can go wrong, it will."

Joy's conclusion is that scientists must reexamine their quests for unfettered technological progress. He warns that technologies such as genetic engineering, nanotechnology, and robotics could spiral out of control if scientists and engineers are not bound by an ethic of responsibility, which would include adequate attention to better technology assessment and technological forecasting.

Are such developments as biotechnology, nanotechnology, and intelligent machines the inevitable result of technological "progress," evolving due to some internal logic that makes their future reality an historical necessity? Some of the technological "visionaries" answer with an unqualified "yes." One such visionary is Raymond Kurzweil, author of *In the Age of Spiritual Machines* (1999):

> Once we have intelligent systems in a nonbiological medium, they're going to have their own ideas, their own agendas. They'll evolve off in completely unpredictable directions. Instead of being derived only from human civilization, new concepts will also be derived from their electronic civilization. But I see this as part of evolution—a continuation of the natural progression. (p. 26)

Because potential failures associated with these technologies will be unparalleled in human history, such futurist scenarios raise serious questions as to the nature of technological development and our ability to manage and control high technology in the emerging Fourth Industrial Revolution.

CONCLUSION

The technological optimism of researchers such as Drexler, Moravec, and Kurzweil is countered by the pessimistic assessments of emergent technologies—genetic engineering, robotics, and nanotechnology—of Rifkin, Ho, and Joy. Perhaps a middle ground between the pessimist and optimist can be found by focusing on the debate between technological determinism and social constructivism.

The crucial question here is: are the multiplicity of technological developments and achievements the result of human volition, judgment, and deliberation, or are they instead the result of a technological imperative or determinism beyond the control of human beings? The determinist focuses on technological imperatives and the negation of human free will, while the social constructivist asserts the positive role of free will and, hence, deliberate choice and responsibility concerning decisions about technology.

Technological determinists such as Kurzweil maintain that technology is the prime mover; that technology is the necessary force through which past events or states of the world determine future states. For the determinist, technology is the medium through which physical laws, identified by science, shape the course of human events. According to technological determinism, given a specific state of technology, "the subsequent development of society would be the same, no matter what people thought or desired" (Staudenmaier, 1985).

In sum, technological determinism means: "machinery and allied subhuman powers somehow function as the independent agencies in history" (Staudenmaier, 1985).

Robert Heilbronner expresses the sentiments of the determinist when he states:

> I believe there is such a sequence—that the steam-mill follows the hand-mill not by chance but because it is the next stage in a technical conquest of nature that follows one and only one grand avenue of advance. (1967: 336)

Or, take the following statement:

> If we restrict ourselves to the functional relationships directly connected with the process of production itself, I think we can indeed state that the technology of society imposes a determinate pattern of social relations on that society. (Heilbronner, 1967: 340)

While the determinist sees the development of technology as a result of inevitable and immutable historical laws that are beyond the power of human beings to control, social constructivists read the history of technological development as the history of human judgments, decisions, and actions in the design, development, and deployment of technological artifacts. As Langdon Winner puts it:

> The notion that people have lost any of their ability to make choices or exercise control over the course of technological change is unthinkable—for behind the massive process of transformation one always finds a realm of human motives and conscious decisions in which actors at various levels determine which kinds of apparatus, technique, and organization are going to be developed and applied. (1977: 53)

According to the social constructivist, there is simply no useful line that can be drawn between the technical and the social factors when it comes to understanding either the invention of technological artifacts such as the telephone, for example, or the development of large-scale technological systems such as railroads and electric power delivery systems (Bijker, Hughes, and

Pinch, 1994). Both sets of factors, the technical and the social "interpenetrate" and are often indistinguishable from one another. The metaphor adopted by this approach is the "seamless web," which means there is no meaningful way to distinguish between technical, economic, political, social, and cultural aspects of technological development (Hughes, 1986).

Hence, the discoveries of scientists and engineers can no longer be interpreted as inevitable, or simple "givens"—unquestioned axioms of the ineluctable logic of a causal technological determinism. "Because they are invented and developed by system builders and their associates, the components of technological systems are socially constructed artifacts" (Hughes, 1994: 52). In other words, emerging technological developments, and the failures that are likely to follow them, are not so much the result of a preordained determinism that develops independent of human agency as they are the result of actions subject to conscious human control.

The objective of a social constructivist approach, specifically applied to the investigation of technological disasters that have occurred throughout the three industrial revolutions, and that are anticipated in the emerging Fourth Industrial Revolution, is to investigate the interpenetration of the technical, human, organizational, political, economic, and historical factors behind such failures. This approach is an important corrective to a determinist approach, in which negative impacts and technological disasters must be accepted as the *inevitable* price to pay for technological progress.

References

Anonymous. (1995). "Notorious bugs," *Byte*, September: 127–129.

Asimov, Isaac. (1950). *I, robot*. London: Grafton Books.

Asimov, Isaac. (1979). *A choice of catastrophe: The disasters that threaten our world*. New York: Simon and Schuster.

Baase, Sara. (1997). *A gift of fire: Social, legal, and ethical issues in computing*. Upper Saddle River, NJ: Prentice Hall.

Bell, Daniel. (1973). *The coming of post-industrial society: A venture in social forecasting*. New York: Basic Books.

Bell, Daniel. (1983). "The social framework of the information society." Reprinted in Tom Forester (Ed.), *The microelectronics revolution: The complete guide to the new technology and its impact on society*. Cambridge, MA: The MIT Press.

Bell, Daniel. (1989). "The third technological revolution and its possible socioeconomic consequences," *Dissent* (Spring): 164–177.

Bijker, Wiebe, Hughes, Thomas, and Pinch, Trevor. (1994). *The social construction of technological systems: New directions in the sociology and history of technology*. Cambridge, MA: The MIT Press.

Buchanan, R.A. (1992). *The power of the machine: The impact of technology from 1700 to the present*. London: Viking Press.

Burke, John C. (1972). "Bursting boilers and the federal power." In Melvin Kranzberg and William H. Davenport (Eds.), *Technology and culture: An anthology*. New York: Shocken Books.

Cardwell, Donald. (1994). *The Norton history of technology*. New York: W.W. Norton and Company: 355.

Ceruzzi, Paul. (2000). *A history of modern computing*. Cambridge, MA: The MIT Press.

Chandler, Alfred. (1977). *The visible hand: The managerial revolution in American business*. Cambridge, MA: Harvard University Press.

Chandler, Alfred. (1990). *Scale and scope: The dynamics of industrial capitalism*. Cambridge, MA: Harvard University Press.

Clarke, Roger. (1993). "Asimov's laws of robotics: Implications for information technology, Part I," *Computer* 26 (12): 53–61.

Clarke, Roger. (1994). "Asimov's laws of robotics: Implications for information technology, Part II," *Computer* 27 (1): 57–65.

Coates, Joseph. (1998). "The next twenty-five years of technology: Opportunities and risks." In Reil Miller, Wolfgang Michalski, and Barry Stephens (Eds.), *21st century technologies: Promises and perils of a dynamic future*. Paris: Organization for Economic Cooperation and Development.

Crandall, B.C. (1992). "We are nanotechnology," *Whole Earth Review* 77: 79–83.

Deane, Phyllis. (1965). *The first industrial revolution*. Cambridge, MA: Cambridge University Press.

Derry, T.K., and Williams, Trevor I. (1961). *A short history of technology*. Oxford, England: Oxford University Press.

Dragan, Richard. (2001). "The meaning of Moore's Law—few technology predictions have held up as well, but don't misunderstand Moore's Law," *PC Magazine,* June 5: 60.

Drexler, K. Eric. (1986). *Engines of creation: The coming era of nanotechnology*. New York: Anchor Press.

Dumas, Lloyd. (1999). *Lethal arrogance: Human fallibility and dangerous technologies*. New York: St. Martin's Press.

Etzioni, Amitai. (1973). *The genetic fix*. New York: Macmillan Publishing.

Fichman, Martin. (1993). *Science, technology, and society: A historical perspective*. Dubuque, IA: Kendall Hunt Publishing Co.

Flink, James. (1975). *The car culture.* Cambridge, MA: The MIT Press.

Francis, David. (2001). "Nuclear power, long dormant, undergoes a nascent revival," *The Christian Science Monitor,* April 27: 3.

Heilbronner, Robert. (1967). "Do machines make history?" Reprinted in Merritt Roe Smith and Leo Marx (Eds.) (1996). *Does technology drive history? The dilemma of technological determinism.* Cambridge, MA: The MIT Press.

Ho, Mae-Wan. (1997). "The unholy alliance: Dangers of genetic engineering in biotechnology," *The Ecologist* 27 (4): 152–159.

Ho, Mae-Wan. (2001). "Horizonal gene transfer happens," *Synthesis/Regeneration* (Winter): 45–48.

Hudson, Richard. (1990). "Cost of Chernobyl nuclear disaster soars in new study," *The Wall Street Journal,* March 29.

Hughes, Thomas P. (1986). "The seamless web: Technology, science, etcetera, etcetera," *Social Studies of Science* 16: 281–292.

Hughes, Thomas P. (1994). "The evolution of large technological systems." In Wiebe Bijker, Thomas Hughes, and Trevor Pinch (Eds.), *The social construction of technological systems: New directions in the sociology and history of technology.* Cambridge, MA: The MIT Press: 51–82.

Hunter, Louis C. (1969). *Steamboats on the western rivers.* New York: Octagon Books.

Joy, Bill. (2000). "Why the future doesn't need us," *Wired,* April: 238–262.

Kerstetter, Jim. (2000). "How many 'love bugs' will it take?" *Business Week,* May 22: 50–56.

Kielstein, Rita. (1996). "Clinical and ethical challenges of genetic markers for severe human hereditary disorders." In Gerhold Becker (Ed.), *Changing nature's course: The ethical challenge of biotechnology.* Hong Kong: Hong Kong University Press.

Kuhns, William. (1971). *The post-industrial prophets: Interpretations of technology.* New York: Weybright and Talley.

Kurzweil, Raymond. (1999). "Man and machine become one: An interview with Raymond C. Kurzweil," *Business Week,* September 6 (3645): 26D.

Lilley, S. (1966). *Men, machines, and history: The story of tools and machines in relation to social progress.* New York: International Publishers.

Littlewood, B., and Stringini, L. (1995). "The risks of software." In Deborah Johnson and Helen Nissenbaum (Eds.), *Computers, ethics and social values.* Englewood Cliffs, NJ: Prentice Hall: 432–438.

Lloyd, James T. (1856). *Lloyd's steamboat directory.* Cincinnati, OH: James T. Lloyd & Co.

Lovelace, Herbert. (2000). "A clear and present danger," *Informationweek,* May 22: 166–173.

Lutins, Allen. (2000). "U.S. nuclear accidents." Accessed June 12, 2000, from the World Wide Web at: www.hempwine.com/alleycat/mukes.html

MacKenzie, Donald. (1998). "Computer-related accidental deaths." In *Knowing machines: Essays on technical change*. Cambridge, MA: The MIT Press: 185–215.

Marlin, S. (2000). "Tough love," *Bank Systems and Technology* 37 (7): 10–12.

May, John. (1989a). "Windscale." In *The Greenpeace book of the nuclear age: The hidden history and the hidden cost*. New York: Pantheon Books: 113–118.

May, John. (1989b). "Brown's Ferry." In *The Greenpeace book of the nuclear age: The hidden history and the hidden cost*. New York: Pantheon Books: 140–147.

McCraw, Thomas. (1997). *Creating modern capitalism: How entrepreneurs, companies, and countries triumphed in three industrial revolutions*. Cambridge, MA: Harvard University Press.

McCuen, Gary (Ed.). (1990). *Nuclear waste: The biggest cleanup in history*. New York: GEM/McCuen Publications, Inc.

McLuhan, Marshal. (1966). *Understanding media: The extensions of man*. New York: McGraw-Hill.

Medvedev, G. (1991). *The truth about Chernobyl*. New York: Basic Books.

Moravec, Hans. (1999). "Rise of the robots," *Scientific American*, December.

Mumford, Lewis. (1963). *Technics and civilization*. New York: Harcourt, Brace and World.

Nash, Jay. (1976). *Darkest hours*. Chicago: Nelson Hall.

National Traffic Safety Board. (2000). Accessed from the World Wide Web at: www.ntsb.gov/aviation/Table3.htm

Neumann, Peter G. (1995). *Computer-related risks*. New York: ACM Press.

NUREG, 0090. (2000). Report to Congress on abnormal occurrences, fiscal year, 1999. Washington, DC: February 9, 2000. Office of Nuclear Regulatory Research.

Perera, Judith. (1992). "Soviet plutonium plant killed thousands," *New Scientist*, June 20; 134 (1826): 10.

Rhodes, Richard. (1993). *Nuclear renewal: Common sense about energy*. New York: Viking Books.

Riccio, J., and Brooks, L. (1996). *Nuclear lemons: An assessment of America's worst commercial nuclear power plants*. Washington, DC: Public Citizen, Critical Mass Energy Project.

Rifkin, Jeremy. (1998). *The biotech century: Harnessing the gene and remaking the world*. New York: Jeremy P. Archer/Putnam: ix.

Sciolino, Elaine, and Gordon, Michael. (2000). "Ukraine consents to shut Chernobyl before year's end," *The New York Times,* June 6: A1.

Sinclair, Bruce. (1966). *Early research at the Franklin Institute: The investigation into the causes of steam boiler explosions, 1830–1837.* Philadelphia, PA: Franklin Institute.

Staudenmaier, John. (1985). *Technology's storytellers: Reweaving the human fabric.* Cambridge, MA: The MIT Press.

Sylvester, Edward, and Klotz, Lynn. (1995). *The gene age: Genetic engineering and the next industrial revolution.* New York: Charles Scribner's Sons.

Thompson, Henry. (1999). "Computational systems, responsibility, and moral sensibility," *Technology in Society* 21: 409–415.

United Nations Economic Commission for Europe. 2000. "The boom in robot investment continues—900,000 robots by 2003." Press Release ECE/STAT/00/10, October 17, Geneva, Switzerland. Accessed from the World Wide Web on October 31, 2001 at: www.unece.org/press/ 00stat10e.htm

Wacks, Ray. (1996). "Sacrificed for science: Are animal experiments morally defensible?" In Gerhold Becker (Ed.), *Changing nature's course: The ethical challenge of biotechnology.* Hong Kong: Hong Kong University Press: 115–123.

Weiner, Norbert. (1964). *God and Golem.* Cambridge, MA: The MIT Press.

Winner, Langdon. (1977). *Autonomous technology: Technics-out-of-control as a theme in political thought.* Cambridge, MA: The MIT Press.

Endnote

1. Passive safety means safety features that do not require the activation of components such as water pumps and valves in order to prevent the nuclear fuel from melting in the case of a loss-of-coolant accident. In a modular high temperature gas reactor, the neutrons are slowed down, namely, moderated by a solid—graphite—and the fuel is cooled by a gas—helium—rather than having both functions performed by light and heavy water. If the power of such a reactor is restricted to a level of about 100 Megawatts, instead of the standard 1,000 Megawatt capacity of current light water reactors, it is possible to realize a design in which the fuel will not melt even if the helium coolant is completely lost and never restored. In other words, the safety of the reactor does not depend on the proper functioning of systems to reintroduce coolant back into the reactor core within a short period of time.

A Matrix of Technological Disasters

*"Human history becomes more and more
a race between education and
catastrophe."*
—H.G. Wells

In Chapter 5, we developed and illustrated a fourfold classification of the causes of technological disasters: technical design factors, human factors, organizational systems factors, and socio-cultural factors. In Chapter 6, we undertook an historical analysis of the three industrial revolutions. These conceptual frameworks will now be combined into a 3 × 4 matrix in an effort to provide a more detailed explanation of technological disasters (see Figure 7–1).

Our 3 × 4 matrix combines a structural set of causes, factored through a temporal schema of technological history—the three industrial revolutions. The 12 cells of the matrix consist of exemplary case studies of technological disasters (see Figure 8–2). In reviewing our case studies of technological disaster, it became evident that technology is becoming increasingly complex as we move from the first to the third industrial revolution; in addition, increasing technological complexity increases the probability of technological disaster.

The rationale for our 3 × 4 matrix can be explained by identifying four underlying assumptions.

1. Each of the four causes of disaster applies to each of the three industrial revolutions.
2. The causal factors exhibit a differential frequency as we move from the first to the third industrial revolution.

Four Root Causes of Technological Disaster

	Technical Design Factors	Human Factors	Organizational Systems Factors	Socio-Cultural Factors
First Industrial Revolution				
Second Industrial Revolution				
Third Industrial Revolution				

(left axis label: Three Industrial Revolutions)

Figure 7–1
The 3 × 4 matrix of technological disasters.

3. There is an increasing level of complexity of technology as we move from one industrial revolution to another.

4. There is a concomitant increase of risk and danger of disaster over time.

With the aid of these assumptions we formulated three testable hypotheses.

The first hypothesis is that the First Industrial Revolution produced a predominance of technical design disasters. One explanation of this hypothesis is that the first revolution witnessed a disjunction between scientific theory and technological innovations—knowledge mastered versus innovations ventured. The second hypothesis is that as society advances through the Second Industrial Revolution, one witnesses an increasing complexity of the causes of disaster; namely, one begins to see a growing number of human factors and organizational systems as causes of disaster. This is the result of conflicts arising between corporate interests and the interests of the general public. The third hypothesis is that as society advances into the Third Industrial Revolution, one begins to witness the growing prevalence not only of human factors and organizational factors, but also of socio-cultural factors as catalysts of technological disaster. Such factors emerged as causes because even though the gap between the theoretical underpinnings of scientific knowledge and the "art" of technological innovation has become narrower and narrower, the complexity of technology has increased to such an extent that we often cannot fully comprehend, let alone anticipate, all of the risks and pitfalls of technology.

TESTING THREE HYPOTHESES ABOUT THE HISTORY OF TECHNOLOGICAL DISASTERS

A systematic empirical test of the three hypotheses is difficult because it would require examining a large sample of cases of technological disaster spanning the last 250 years. Fortunately, Schlager presents 103 cases in his compendium, *When Technology Fails* (1994), which allows us to undertake a provisional test of our hypotheses. However, since Schlager's collection of cases does not include any technological disasters of the First Industrial Revolution, an additional source was consulted. Smith's *Environmental Hazards* (1992) provides a listing of 20 cases of disasters of the First Industrial Revolution; these cases were used as our data set for the First Industrial Revolution. Thus, we have identified 123 cases of technological disaster that occurred in the course of the three industrial revolutions. This is, to be sure, a small sample, which limits the generalizability of our findings.

In preparing the data for a test of our three hypotheses, we performed a content analysis of the 123 cases by constructing a set of abstracts for each case of technological disaster. These abstracts were presented to four raters—all college professors—who were asked to classify independently each case according to our fourfold classification of causes of disaster: technical design factors, human factors, organizational systems factors, and socio-cultural factors.

A prior methodological question is this: how reliable are the nominal ratings of the 123 cases by our four raters? Upon analysis, we found that there was a fairly high degree of agreement among the raters who classified the 123 cases in the four categories of causes of disaster.[1] This level of interrater agreement

Table 7–1
Interrelationship between the three industrial revolutions and the four causes of technological disasters.

	Technical Design Factors	Human Factors	Organizational Systems Factors	Socio-Cultural Factors	Total
Cases in the First Industrial Revolution	70% 14	20% 4	5% 1	5% 1	20
Cases in the Second Industrial Revolution	62.32% 43	11.59% 8	21.74% 15	4.35% 3	20
Cases in the Third Industrial Revolution	35.29% 12	20.59% 7	23.53% 8	20.59% 7	34

provides some support for the validity of our conceptualization of the four causes of failure.

We then proceeded to classify the 123 cases of technological disaster into the 3 × 4 matrix shown in Figure 7–1; the number of cases, as classified by the raters, for each of the four causal factors, appears in each cell of the matrix. Having done this, we were then able to test our three hypotheses. Examining the data presented in Table 7–1, we see that all three of our hypotheses are *partially* confirmed, considering that we are dealing with relatively small sample sizes.

The first hypothesis, that the First Industrial Revolution saw a predominance of technical disasters, is confirmed by the finding that 70 percent of cases were classified as technical. The second hypothesis is also *partially* confirmed because we observe a rise in the percentage of cases pertaining to organizational systems factors (from 5 percent to 24 percent), as we move from the First to the Third Industrial Revolution. The third hypothesis is also *partially* confirmed; witness the rise in the percentage of sociocultural cases from only 5 percent in the First Industrial Revolution to 21 percent in the Third Industrial Revolution.

Another noteworthy finding in Table 7–1 is that our four causes of technological disaster and the three industrial revolutions are interdependent.[2] This finding supports our intuitive assumption that gave rise to our 3 × 4 matrix, namely, that the four types of causes of technological disaster are interrelated with the three industrial revolutions. The matrix thus provides a framework for analyzing technological disasters.

CONCLUSION

Our findings highlight the bearing of technological complexity on the incidence of technological disaster. The partial confirmation of our hypotheses relating the three industrial revolutions to the four causes of technological disaster are suggestive of potential future developments. As we move beyond the current Third Industrial Revolution to the emerging Fourth Industrial Revolution, technological complexity will further increase, as reflected in the ongoing innovations in genetic engineering, nanotechnology, and robotics. If these developments come to pass, ushering in ever higher levels of technological complexity, our findings in this chapter lead us to anticipate a corresponding increase in the frequency of technological disasters.

References

Fleiss, J.L. (1981). *Statistical methods for rates and proportions*. New York: John Wiley & Sons.

Landis, J.R., and Koch, G.G. (1977). "The measurement of observer agreement for categorical data," *Biometrics* 33 (1): 159–174.

Schlager, N. (1994). *When technology fails*. Chicago: Gale Publications.

Smith, K. (1992). *Environmental hazards*. New York: Routledge.

Endnotes

1. To obtain a measure of the degree of agreement among the four raters, a "Kappa statistic" was calculated (Landis and Koch, 1977; Fleiss, 1981). The overall mean Kappa statistic was 0.3399 ($z = 6.996$, $p < 0.0001$) with a 95 percent confidence interval of (0.2447, 0.4351). Even though the Kappa statistic is statistically significant, the level of reproducibility, according to Landis and Koch's "benchmarks" for evaluating Kappa statistics, is only fair.

2. To test whether the two sets of factors, the four causes of technological disaster and the three industrial revolutions, are interrelated, a chi-square test was performed. The test yields a statistically significant χ^2 value with a probability of 0.02 ($\chi^2 = 15.05$, df = 6, $p = 0.02$). Since some of the frequencies are small—and the expected frequencies are less than five—the Pearson chi-squared test of interrelationship is not suggested in the statistics literature. To avoid this problem we considered a 3 × 3 table formed by combining the data pertaining to organizational systems and socio-cultural factors. The results were substantially the same ($\chi^2 = 10.76$, d.f. = 4, $p = .03$).

IV

Analysis of Case Studies of Technological Disasters

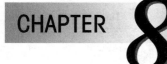
Twelve Exemplary Case Studies of Technological Disasters

"Every case study . . . is potentially a paradigm for understanding how human error and false reasoning can thwart the best laid plans."

—Henry Petroski

*I*n this chapter we present 12 outstanding accounts of techno-logical disaster classified in a 3 × 4 matrix shown in Figure 8–2. Each case study was selected, among many potential candidates, as an exemplar that would illustrate the two underlying dimen-sions of the matrix: the causes of disaster and the three industrial revolutions. A supplementary set of 12 significant technological disasters, discussed in Chapter 5, is included in Figure 8–3. We therefore provide two case studies for each of the 12 cells in the 3 × 4 matrix.

We, of course, do not claim that these 24 case studies are rep-resentative of all cases of technological disaster, since, to the best of our knowledge, there exists no single source that provides an exhaustive compilation of such disasters.

Rather than present the 12 exemplary case studies in this chapter in a straight chronological order, we have chosen to con-sider them in the time frame of the three industrial revolutions—discussed in Chapter 6—subdivided by the four causes of failure analyzed in Chapter 5.

It is instructive to consider the terminology used by the authors of the 12 case studies to characterize the failures (see Figure 8–1).

Case Study	Terminology
USS *Princeton* Explosion	tragedy
Titanic Sinking	tragedy, disaster
Aisgill Train Wreck	disaster
Johnstown Flood	disaster, calamity
DC-10 Crash	disaster
Tenerife Runway Collision	disaster
Santa Barbara Oil Spill	accident
Love Canal Toxic Waste Contamination	disaster
Apollo I Fire	accident, disaster
Three Mile Island	accident
Challenger Disaster	disaster
Bhopal Poison Gas Release	accident

Figure 8–1
Terminology used by authors of the 12 case studies to characterize technological disasters.

Clearly, the terms used do not always convey the magnitude of disaster with regard to the loss of life and property. This is obviously the case with the Bhopal Poison Gas Release, which should certainly be called a disaster or a calamity, given that 14,000 people were killed and tens of thousands were injured. In fact, the characterization of such devastating technological disasters as "accidents" inadvertently places them in the same category as totally unexpected or totally unintentional misfortunes. The implication of falsely categorizing such technological disasters in this manner is that it:

> . . . skews our perceptions in a certain direction: We see it not only as an unintended and unanticipated event, but also as an unavoidable event—something that could not have reasonably been prevented. It also implies that what happened was an act of fate—the result of impersonal forces—and to the extent that human actions were involved, they would not be regarded as "causing" the unexpected event. The net effect of this cultural classification . . . is to obscure the role of human actions in the production of . . . disaster[s] . . . (Poveda, 1994: 21–22)

Reference

Poveda, Tony. (1994). *Rethinking white-collar crime*. Westport, CT: Praeger.

Four Root Causes of Technological Disaster

	Technical Design Factors	Human Factors	Organizational Systems Factors	Socio-Cultural Factors
First Industrial Revolution	USS *Princeton* Explosion	*Titanic* Sinking	Aisgill Train Wreck	Johnstown Flood
Second Industrial Revolution	DC-10 Crash	Tenerife Runway Collision	Santa Barbara Oil Spill	Love Canal Toxic Waste Contamination
Third Industrial Revolution	Apollo I Fire	Three Mile Island	Challenger Disaster	Bhopal Poison Gas Release

(Left margin label: Three Industrial Revolutions)

Figure 8–2
The 3 × 4 matrix of case studies of technological disasters, as presented in Chapter 8.

Four Root Causes of Technological Disaster

	Technical Design Factors	Human Factors	Organizational Systems Factors	Socio-Cultural Factors
First Industrial Revolution	Dee Bridge Collapse	SS *Mendi* Collision	*Sultana* Steamboat Sinking	Monongah Mine Disaster
Second Industrial Revolution	Hyatt Regency Walkway Collapse	USS *Vincennes* (Iranian Airbus Shootdown)	B.F. Goodrich Brake Scandal	Triangle Shirtwaist Factory Fire
Third Industrial Revolution	Therac-25 Radiation Device Malfunction	Chernobyl Disaster	BART Whistleblowing Case	Dow Corning Breast Implants

(Left margin label: Three Industrial Revolutions)

Figure 8–3
Twelve case studies of technological disasters discussed in Chapter 5, classified by the 3 × 4 matrix.

CASE STUDY 1

USS Princeton Explosion[1]

Lee M. Pearson

Introductory Note by W.M.E. and M.M.: *On February 28, 1844, a gun exploded aboard the U.S. Navy's new steam frigate, the USS* Princeton, *causing a considerable loss of life. Fragments struck a group of illustrious visitors, killing five men of national prominence, two of whom were members of the U.S. Cabinet. During testing of the gun, small cracks developed near its breech. The officer in charge of commissioning the guns, Robert F. Stockton, ordered the construction of reinforcing bands that were to be added to the gun. A basic design feature, welding the reinforcing bands, was, however, overlooked. The causes of the explosion were traced to the design decision to use wrought-iron for the twelve-inch barrel guns instead of the traditional cast-iron used in previous naval ordnance, as well as the decision not to use standard welding techniques for the reinforcement bands.*

This technological disaster of the First Industrial Revolution involves technical design factors.

On February 28, 1844, a gun exploded aboard the navy's new steam frigate, the U.S.S. *Princeton,* causing a multiple tragedy. Fragments struck a group of illustrious visitors, killing five men of national prominence, two of them members of the Cabinet—Abel P. Upshur, Secretary of State, and Thomas W. Gilmer, Secretary of the Navy. But for President Tyler's interest in Julia Gardiner, his wife-to-be and the daughter of one of the men killed, he might have been among the fatalities. . . .

The *Princeton,* constructed in 1841–43, was a full-rigged sailing vessel with steam propulsion and screw propeller. She could be called epoch-making, for many of her features anticipated later practice. Her engines, novel in design, were coupled directly to the shaft and were mounted below the water line. Designed to burn anthracite coal, she was fitted with forced-draft boilers, feed-water heaters, and a telescopic smokestack—to minimize smoke and chance of detection.

Her ordnance . . . was anything but a success; however, her main battery—a limited number of guns, each of maximum size, pivoted to fire in various directions, and fitted with a compressor brake recoil system—had a strong conceptual resemblance to the batteries of later capital ships. Similarly, her range finder and self-actuating gunlocks, which fired the guns at a preselected elevation, showed appreciation of the problems of fire control as well as ingenuity in attempting to find a practical solution. . . .

The West Point Foundry of Cold Springs, New York, provided a cast-iron gun, and its president, William Young, was involved in the original decision to design a twelve-inch wrought-iron gun. One of the guns was made by the Mersey Iron Works of Liverpool, England. Despite the number of firms involved in the undertaking, the *Princeton* was the brainchild of John Ericsson and Captain Robert F. Stockton, U.S. Navy, and was constructed under their supervision. . . .

As their discussion turned . . . to guns, William Young, President of the West Point Foundry, joined Stockton, Ogden, and Ericsson. As Ogden later explained, they knew the strength of wrought iron from the experience with forged steamer shafts and reasoned that such forgings could be made into heavy guns. Later reports and testimony indicated several advantages such a gun was supposed to have over one of cast iron. A larger gun could be constructed of wrought iron for a given weight. In addition, larger guns supposedly could be built of wrought iron since, beyond a certain point, increasing the thickness of cast iron did not increase its strength.

Other advantages claimed were less erosion of the barrel, a more regular trajectory, and more destruction upon impact.

Stockton therefore determined that a wrought-iron gun of twelve-inch bore was feasible and arranged with Ogden to have such a gun constructed. As Ericsson later described the negotiations: "Captain Robert F. Stockton . . . consulted me regarding the possibility of constructing naval ordnance of wrought iron. Being an advocate of that material, I readily met the wishes of Captain Stockton, and at once prepared drawings of a gun of 12-inch caliber. The Mersey Iron-Works near Liverpool being willing to enter into a contract with Captain Stockton, received forthwith an order from the enterprising and spirited officer to build the gun at his expense." Ogden shipped this gun to America in 1841. . . .

To return to the gun, upon its arrival in America it was sent to Philadelphia, where trunnions were installed. In 1842 it was shipped to Sandy Hook and, while lying in the sand, was proved with one charge of 35 pounds of powder and a 212-pound shot. When a small crack developed near its breech, Stockton employed the Phoenix Foundry to construct reinforcing bands or hoops. These were fabricated at Hamersley Forge and shrunk on the gun at Sandy Hook.

As a companion piece, Stockton had obtained a twelve-inch cast-iron gun from the West Point Foundry. This gun was intended for the *Princeton,* and a gun carriage and friction gear were obtained for it. Stockton fired both guns at Sandy Hook for comparative purposes. The cast-iron gun broke after about eighteen rounds, but the wrought-iron gun was fired about fifty times without apparent change.

Stockton immediately requested permission from Crane to order a wrought-iron gun to replace the broken cast-iron piece and submitted a report of his 1842 tests at Sandy Hook which, he claimed, proved that the wrought-iron gun was superior to cast-iron guns. His evidence was its longer life, its greater accuracy, and its greater destructiveness compared to service weapons. A few rounds from the twelve-inch gun had destroyed a target representing a seventy-four-gun ship, and he had sent one round through a four-and-one-half-inch thick wrought-iron target representing the Army of the Stevens Battery—the first iron-clad ordered by the navy. He said little about the gun carriage other than that recoil did not exceed three feet. Alexander Wadsworth, who inspected the wrought-iron gun after Stockton had completed his tests, gave some support to Stockton's claims; neither the bore nor the crack in the breech had enlarged.

Crane forwarded Stockton's report to Lieutenant Colonel Talcott of Army Ordnance. Talcott agreed that large projectiles were more destructive than smaller ones and that their trajectories were predictable. He maintained, however, that the greater accuracy, if any, of the wrought-iron gun resulted from the fact that its projectiles had been wrapped in felt, which cut down the windage. He believed that twelve-inch guns were too heavy for use aboard ship and specifically questioned the use of welding in fabricating guns. Prophetically, he pointed out that there was "uncertainty of . . . [such guns] being homogeneous and actually welded throughout as they must of necessity be composed of many pieces, which may be welded on the exterior while many fissures exist in the interior of the mass."

In light of Talcott's derogatory opinion, Stockton, in the spring, returned to Sandy Hook to further prove his claims. Although no report of these tests have been found, he fired the English gun another seventy to one hundred times. His associates later maintained there was no discernible change in either the crack or the bore.

The above provided the basis for Stockton's decision to construct a second wrought-iron gun. As Lieutenant Hunt later testified: "a gun . . . made of American iron [would be] sufficiently strong to stand any number of pounds of powder." Ogden, who had some knowledge of metallurgy, noted that the English gun withstood repeated firings, even though it "had lost all tenacity" and "depended alone on bands 3 1/2 inches thick." He believed that the new gun "with these bands *welded* upon two inches greater diameter . . . would be perfectly safe under any trial." Ogden glossed over a basic design change, welding (rather than shrinking) the strengthening bands. None of the other recapitulations of events even mentioned that such a design change was made. Thus our knowledge of this crucial change is very limited. . . .

While this gun was under construction the *Princeton* was completed, and on October 17, 1843, having taken her crew the previous day, she proceeded down the Delaware to New York. When returning, she outdistanced the steamer "Great Western" in a race in which both ships used sails and steam. Against flood tide, the *Princeton* had required one hour and thirty-one minutes to go from Castle Garden to Sandy Hook Point, more than twenty-one miles. Thus she had a speed of more than fourteen miles per hour (above twelve knots), and Stockton boasted that was the "fastest sea-going steamer in the world. . . . "

By this time the American gun was completed. When the navy agent sought to get funds to pay for it, the secretary directed that Stockton should first prove it. Thus, while at New York, Stockton had the gun loaded aboard the station ship, "Anchor Hoy," which the *Princeton* then took in tow. They headed down the bay, "came too [*sic*] outside the Narrows and fired the "Peace Maker" on board the *Anchor Hoy.*" This curiously rash expression of confidence began with 14 pounds of powder and in five firings worked up to 45 pounds of powder and a 212-pound shot. Stockton's official report jibed at the Navy Department's meticulous insistence upon the formalities of proof-firing: "P.S. The men who made it deserve their money. It is worth all the guns on board of any frigate." The next morning was a gala day as the big gun was hoisted aboard the *Princeton.* The watch officer recorded that they "christened it 'Peace Maker' with six cheers" and then, with fitting liberality, "spliced the Main Brace."

The armament was complete when a committee from the American Institute visited the *Princeton.* This group included the shipbuilder, an engineer and a chemist, an author of improvements in steel manufacture, and an artist and designer, all of whom had some military or naval background. Other members of the committee, from their names, were scions of old New York or New Jersey families who may have been friends of the prominent Stockton. The committee came on board at 11:00 a.m. and departed at 2:30 p.m. to write a glowing report of the ship and its guns. This report is interesting for both its factual description and for its superlative praise. The guns were "the most formidable ordnance ever mounted," and the "Peacemaker" was "beyond comparison the most extraordinary forged work ever executed. . . . "

Upon arriving at Washington, Stockton demonstrated his ship to various government officials, always firing the "Peacemaker." When he took President Tyler on a cruise, he fired the big gun three times. Tyler, much impressed, requested Congress to authorize construction of several ships similar to the *Princeton* but larger and better fitted to carry heavy armament. Stockton, no doubt attempting to override opposition to Tyler's request from the chairman of the House Naval Affairs Committee, held another gay excursion on February 28. President Tyler set the tone by bringing his fiancée, Julia Gardiner. The festive group of 150 ladies and 200 gentlemen included foreign ministers and government officials. Among them were Abel P. Upshur, still secretary of state, and Thomas W. Gilmer, who thirteen days earlier had succeeded Henshaw as secretary of the navy. As part of the merrymaking, and to emphasize the soundness of the new warship,

Stockton fired the "Peacemaker" twice. The third time, it burst. Fragments struck and killed Gilmer; Upshur; Beverly Kennon, Chief of the Bureau of Construction, Equipment and Repair; Virgil Maxcy of Maryland; Colonel David Gardiner, father of the president's fiancée; and a servant of the President. . . .

Lieutenant R.E. Thompson reported in the ship's log that the gun broke off at the trunnion band and the breech and split in two. One piece of the breech passed overboard while the second fell "in the Larboard gangway." His matter-of-fact entry identified the dead and wounded, reported the departures of the guests, the later departure of the President, and the removal of injured seamen. . . .

The next morning President Tyler sent a report of the tragedy to Congress giving his opinion that it was "one of those tragedies which, are invariably incident to the temporal affairs of mankind." Seeking to save Stockton's reputation and his recommended naval construction program, Tyler added that it in no means detracted "from the value of the improvements contemplated in the construction of the *Princeton,* or from the merits of her brave and distinguished commander and projector." Despite the President's urging, the House Naval Affairs Committee promptly disapproved constructing more ships similar to the *Princeton,* wryly noting that "the success thus far of our war steamers has not been so perfectly complete as to call for immediate action. . . . "

At the instigation of Stockton, the Committee on Science and Arts of the Franklin Institute investigated the explosion. This committee limited its inquiry to material and workmanship. By avoiding legal complications and staying clear of motivation, it did not exacerbate any arguments. At the same time, it gathered facts regarding the gun's fabrication and competently analyzed the large fragment.

The forging had been made by laying up a fagot from thirty iron bars, each of which was four inches square and about eight and one-half feet long. These were welded together and rounded into a shaft of twenty to twenty-one inches in diameter, using a seven-and-one-half-ton hammer. As Talcott had predicted, the welding had indeed been inadequate, for the form of the original bars could still be detected in the fragment, while scales of iron oxide nearly penetrated the body.

Other deficiencies were more serious than poor welding in the forging proper. As Ogden had testified, bands were welded on the American gun, whereas those on the English gun had been shrunk on. In terms of modern metallurgical practice, this use of welded bands was a gross error in design. Shrunk-on bands would have served as crack arrestors, whereas welded bands permitted any cracks to enlarge.

Other weaknesses resulted from the method of fabrication. Iron segments, usually large enough to reach one-third of the way around, were welded on the forging in two strata, and the breech was thus built up to about thirty-six inches in diameter. Making the forging from iron bars and welding on the strengthening band required forty-five "turns," or a day's work, during which the gun was held at welding temperatures. . . .

Even before the Franklin Institute had completed its analysis, the ordnance officers in Washington were conjecturing that prolonged heating had weakened the iron. Talcott, in a report on wrought-iron guns that he prepared following the tragedy, included a generalized statement to this effect. Ward, in his private correspondence, was more definite but stopped short of certainty. The Franklin Institute was positive and recommended that no more guns of this design should be obtained. In all fairness, however, it must be pointed out that even twenty years later there was some disagreement as to whether prolonged heating and inadequate working did in fact weaken wrought iron. . . .

A number of conclusions, some of them pertinent to contemporary naval research and development procedures, emerge from the study of the "Peacemaker" episode:

1. Most obvious, the proof-testing of the "Peacemaker" was inadequate to determine the gun's life and strength, even though the evidence given at the Court of Inquiry indicated that this proof-testing, despite irregularities, was in keeping with existing navy practice.

2. Only slightly less apparent, the tragedy might have been avoided if Stockton had been able to examine objectively the product of his labors and had exercised normal prudence in exposing his guests.

3. Stockton's consultants and associates were equally enthusiastic and uncritical in their support of his endeavors. Thus a wide consensus was built up in favor of the endeavor, enabling him to ignore such sound criticism as was made, notably that of Talcott.

4. Stockton's ability to proceed with his demonstrations, despite doubts of the ordnance officers, stemmed from his political support. His efforts to strengthen this support transformed an accident into a national tragedy.

5. At the same time, this tragedy militated against the acceptance of sound technical innovation. There is no way of knowing whether the explosion retarded the acceptance of propeller-driven ships, but it upset Stockton's and Tyler's plans to modernize the navy.

6. The above five points all revolve around an aspect of government administration which is now recognized as crucial. Critical scientific and technological issues often have to be decided by political authority. In the event of conflicts in professional opinion or of uncertainty as to who is qualified to render sound professional advice, decisions may be made without adequate consideration of technological factors. The background to the tragedy

aboard the *Princeton* points up the fact that this problem, although perhaps only recently recognized, has been of long duration. It suggests the need for continuing study of the institutions and procedures whereby political authorities have decided issues involving scientific and technological judgments.

Endnote

1. Pearson, Lee M. "The 'Princeton' and the 'Peacemaker'; A Study in Nineteenth-Century Naval Research and Development Procedures." Technology and Culture 7:2 (1966), 37–44. © Society for the History of Technology. Reprinted by permission of The Johns Hopkins University Press.

CASE STUDY 2

Titanic Sinking[1]

Sally Van Duyne

Introductory Note by W.M.E. and M.M.: *On April 14, 1912, the state-of-the-art ocean liner* Titanic *sank off the coast of Newfoundland, taking over 1,500 passengers and crew with it to the bottom of the sea. The causes of such a disaster are many, but the major cause was attributed to irresponsible captaincy by Captain E.J. Smith.*

Although there have been bigger technological failures, some claiming more lives, few have lingered so persistently in the public consciousness. One reason is that it is a classic tale of man vs. nature, complete with its own element of Greek tragedy, where fatal pride or hubris reflected an unquestioned faith in technology and progress, resulted in disaster.

This technological disaster of the First Industrial Revolution was caused by human factors.

On the night of April 14, 1912, the White Star Line's *Titanic,* on her maiden voyage from Southampton to New York, struck an iceberg off Newfoundland. It sank just two hours and forty minutes after the collision at 11:40 p.m., with a loss of over fifteen hundred lives. The *Titanic,* the largest and most luxurious ship afloat, was also believed to be the safest—"practically unsinkable." In 1902 the White Star Line had been swallowed up by the monster trust International Mercantile Marine, which made the American financier J. Pierpont Morgan the principal owner of the *Titanic.* The venture was a marriage of American money and British technology. However, the *Titanic* was thought of as being British: it was registered as a British ship, was manned by British officers, and would revert to the British Navy in time of war . . . *Titanic* was built in Belfast alongside its sister ship the *Olympic,* which the *Titanic* exceeded in gross tonnage but not in length. The *Titanic* was launched on May 31, 1911.

The *Titanic* was eight hundred and eighty-two feet in length, ninety-two feet in width, and weighed 46,328 gross tons, its nine steel decks rose as high as an eleven-story building. The stability afforded by its vast size was deemed to be one of many safeguards against its foundering. It also embodied a greater proportionate mass of steel in its structure than had been used in previous ships. It was built with a double bottom, both skins of which were heavier and thicker than they had been in ships built before. The outer skin of the bottom was a full inch thick. The huge hull was divided by fifteen transverse bulkheads, or upright partitions, extending the width of the ship, into sixteen "watertight" compartments, any two of which

Titanic Sinking, April 14, 1912. Copyright Corbus. Used by permission.

might flood without affecting the safety of the ship. The six compartments that contained boilers had their own pumping equipment. The doors between watertight compartments could be closed all at once by a switch on the bridge, or individually by crewmen. The bulkheads rose from the double bottom to five decks above forward and aft, and to four decks above amidships (the middle).

The *Titanic* was not only the biggest ship afloat but the last word in comfort and elegance. It had the first shipboard swimming pool, a Turkish bath with a gilded cooling room, a gymnasium, and a squash court; a loading crane and compartment for automobiles; and a hospital with a modern operating room. The dining rooms, staterooms, and common rooms were furnished in various styles and periods; there was a Parisian cafe, for instance. The first-class cabins were especially opulent, some with coal-burning fireplaces in the sitting rooms and full-size four-poster beds in the bedrooms. There were few detractors of this venture, probably because the *Titanic* embodied the spirit of *overconfidence* of the times. . . .

The evening of April 14, the fifth day of the *Titanic's* maiden voyage, the sea was exceptionally calm and the sky was starry but moonless. The *Titanic* had received a number of warnings of ice in the region, and although Captain E. J. Smith is known to have seen at least four of these, he did not alter his speed of about twenty-one knots. There are several probable reasons for this. That visibility seemed to be good and that the ship was believed to be practically unsinkable must have contributed to his *overconfidence*. Also, ice fields were a hazard that he dealt with all the time, and robust captaincy was a matter of pride among sea captains, especially those with the White Star Line. Finally, he was probably trying to make good speed on the maiden voyage.

Yet April is one of the worst months for icebergs, the seaward tips of glaciers that break off, most of them coming from the west coast of Greenland. About a thousand of these reach the shipping lanes each year. On the night in question, the pair of lookouts—who were working without binoculars, which were supposed to be standard equipment on the White Star Line—did not see the iceberg until it was only a quarter mile away. The bow was swung swiftly to port (left), but it was too late. The underwater shelf of the ice tore through the plating on the starboard (right) bow, perhaps aft to amidships. Six watertight compartments were thought to have been breached. The bow started to sink, and later more compartments filled with water, which soon sloshed over the tops of the transverse bulkheads

The forty-six thousand-ton *Titanic* was traveling at twenty-one knots. So insignificant was the energy absorbed in the collision in proportion to the total energy of the vessel that the impact was scarcely felt. The gravity of the situation was comprehended only gradually, and many passengers had to be awakened. The collision occurred at 11:40 p.m., and the order to ready the lifeboats was given at 12:20 a.m.

There was great confusion in the loading of the lifeboats. There were only sixteen lifeboats and four emergency rafts—enough for roughly half of the passengers. Although the *Titanic* had complied with British Board of Trade rules regarding lifeboats, these were based on the tonnage of the ship rather than on the passenger capacity. Also, a reduction in the number of boats was allowed for ships deemed to have satisfactory watertight subdivision, as was the *Titanic*.

Since there had been no boat drill, the crew and passengers did not know to which lifeboats they should report. Furthermore, the officers in charge of loading the boats were afraid that if they were fully loaded, either the boats would buckle as they were being lowered or the davits (cranes) holding the boats over the side would break (neither of which would have happened, as boats and davits had been tested). Thus they sent the boats down only partly loaded, with instructions to come alongside the cargo ports to pick up more passengers. However, the cargo ports were never opened, so many boats went away only partly filled. There was room enough in the boats for 1,178 persons; since about seven hundred and eleven were saved in boats, about four hundred and sixty-seven needlessly lost their lives in the loading mishap.

As the *Titanic's* bow sank lower and lower, those remaining on the ship climbed to the stern, from which some ended up jumping into the 28 degrees Fahrenheit water in their lifebelts. As those in lifeboats watched spellbound, the immense stern reared up almost to the perpendicular and remained still for a few moments before sliding slowly, at 2:20 a.m., into the sea. Many survivors reported hearing sounds like thunder, or thick detonations, or a kind of "death rattle" before the stern went down. Although some passengers asserted that the great ship broke in two, the explanation generally accepted to explain the thunderous sound was that as the stern rose, the boilers crashed down through the bulkheads. From what we know now about the position of the bow and the stern on the ocean floor—they are facing in opposite directions and are about 1,970 feet apart it seems likely that the ship did break in two at or near the surface. As the bow and the stern rose, the pressure on the keel probably increased until it snapped.

The other startling discovery made by the 1984 Franco-American expedition and the 1986 American expedition led by Dr. Robert D. Ballard of the Woods Hole Oceanographic Institute concerned the gash in the hull. These expeditions located and photographed the wreck in thirteen thousand feet of water a few miles from where it had been thought to be. Dr. Ballard observed many buckled plates below the waterline, but no gash. It is possible, however, that the gash remains hidden deeper down where the bow sank with great force into the sediment.

It is likely that mishaps in wireless communication contributed to further unnecessary loss of life on the *Titanic.* On the *Californian,* a ship stopped for the night in ice fields not more than twenty miles away, the wireless operator had stopped working only fifteen or twenty minutes before the operator from the *Titanic* tried to get through with a distress call. At the time, wireless operators were employees of the Marconi company and did not follow around-the-clock shipboard watches. The next closest ship, which the *Titanic* did succeed in reaching, was the *Carpathia,* about fifty-eight miles away. The *Carpathia* picked up the first lifeboat at 4:10 a.m. If the *Titanic* had had lifeboats for everyone aboard, there might have been no loss of life.

In 1993, a team of architects and engineers released a report in which they argued that the tragedy was caused not so much by the collision with the iceberg as by the structural weakness of the ship's steel plates. Low-grade steel such as that used on the *Titanic* is subject to brittle fracture—breaking rather than bending in cold temperatures. If a better grade of steel had been used, the ship might have withstood the collision or, at the very least, sunk more slowly, thus allowing more passengers to be saved. Interestingly, the team also suggested that the roar heard by the passengers as the ship sank may have been the steel plates fracturing, not the boilers crashing through the bulkheads. . . .

Endnote

1. From "Titanic" by Sally Van Duyne. Taken from *When Technology Fails.* Edited by Neil Schlager. Copyright © 1994 Gale Research. Reprinted by permission of The Gale Group.

CASE STUDY 3

Aisgill Train Wreck[1]

David Howell

Introductory Note by W.M.E. and M.M.: *On September 2, 1913, outside the English Midlands town of Aisgill, two coal-powered trains collided, resulting in the loss of 16 lives. The train wreck was blamed on the inadequate number of personnel operating the Settle to Carlisle railway. Drastic cost-cutting measures by management led to widespread layoffs and strikes, which resulted in reduced numbers of workers and operators.*

The case points out the potential dangers when managerial pressures to increase profits create serious lapses in safety standards, a theme more common to technological failures than is generally acknowledged.

This technological disaster of the First Industrial Revolution was caused by organizational systems factors.

. . . The Aisgill disaster became a *cause célèbre* in the labor and socialist press; its consequences included the threat of a national railway strike and hurried consultations between senior government ministers. The tragic affair illuminates wider themes: the pressures on Edwardian railway management and divisions within the newly formed National Union of Railwaymen. In turn, consideration of these questions perhaps offers some insight into the complexities of the pre-war labour unrest.

Appreciation of the wider debates requires an examination of the events leading up to the disaster. Soon after midnight on 12 September 1913 two footplate crews signed on at the Midland Railway's locomotive sheds in Carlisle; prior to working express trains to the south. Driver William Nicholson and Fireman James Metcalf were booked to take engine No. 993 on the 1:35 a.m. departure from Carlisle; driver Samuel Caudle and Fireman George Follows would follow fifteen minutes later. Each crew would work the train as far south as Leeds.

All the men were highway experienced. Nicholson was forty-nine years old and had been a driver for twelve; he had driven expresses for the previous two years. Caudle was ten years older, his driving experience went back for twenty-nine years and plus service with the Midland Railway for about forty. He had received seven awards for vigilance.

At night he saw the coal tipped into his locomotive's tender. The quality of the fuel provided for the Midland's Carlisle depot had recently become a matter of concern and Caudle's response was critical:

> It struck me that the coal was worse than usual as regards its size; it was very small . . . Engine No. 446 is a good steamer

and it struck me when I saw the coal that if she would steam with this kind of coal she would do well with anything.

The 1:35 a.m. train was assembled at Carlisle Station from portions originating at Glasgow and Stranraer. No. 993 was one of the most powerful passenger engines owned by the Midland, but over the steep gradients of the Settle and Carlisle, its maximum load unassisted was laid down as 230 tons. When Nicholson found that his train had 13 tons excess weight, he asked for a pilot engine, but was told that none was available. In fact, the judgment made by the platform Inspector was that more time would be lost waiting for a pilot than in climbing to Aisgill with a train overloaded by 13 tons.

No. 993 with Nicholson and Metcalf on the footplate left three minutes late. Initially they coped well, over the first thirty miles to Applebee only one more minute was lost, but then the poor quality of the coal began to have its effect. Metcalf was emphatic:

> He had more trouble with the coal that night than ever before. The coal was a great deal more dusty and he had to pour water on it. This dogged the coal into cakes that lay dead on the fire generating no heat.

Time began to be lost as steam pressure fell; most critically the falling pressure eroded the vacuum in the train pipe and the brakes began to leak. The train came to a standstill at about 2:57 a.m. about half a mile short of Aisgill summit.

The 1:35 a.m. should have been secure; it was protected in the rear by the signals at Mallerstang about three miles to the north. The Mallerstang signalman, George Sutherland, had returned these to danger after Nicholson's train had passed. He was not permitted to allow any following train to go forward until he had been notified that its predecessor had passed the next signalbox at Aisgill. The next southbound train was that driven by Caudle. This had left Carlisle five minutes late at 1:54 a.m. No. 446 was a smaller engine than No. 993 but in this case there was no question of a pilot. Although 446 was limited to 180 tons over this steeply-graded line, the train composed of sections from Aberdeen, Inverness and Edinburgh was 23 tons under the maximum. . . .

Since leaving Carlisle, water had been supplied through the use of the left-hand injector, but the poor coal and the inferior steaming had reduced its efficiency. Follows tried to activate the auxiliary right hand injector:

> I could not get it to work. Sometimes the steam does not catch behind the water properly, for the first time or two and you have to go through the operation again. We had just come out of the tunnel and I was engaged in working at the right-hand injector when the driver came in off the outside of

the engine to the footplate. He took the injector in hand and I started firing. . . .

Caudle's concern with the supply of water to the boiler was based in part on the disciplinary consequences of an injector failure:

> If he had burned his firebox, he might as in the old days have been discharged or severely punished, though he had not heard of this being done lately.

The crisis preoccupied Caudle and Follows on their approach to Mallerstang and to the last set of signals before the stalled train. Sutherland in the Mallerstang box had three signals to protect the line to Aisgill in sequence from north to south: a distant signal which a driver could pass at caution; a home signal which could be lowered to green in the event of the line ahead being blocked only when a driver had brought a train to a halt or had almost done so; and finally a starting signal which must not be lowered until the line was clear to Aisgill. When it became apparent to Sutherland that the first train was taking an unusually long time to reach Aisgill, he phoned his colleague there, but was told that the position of the 1:35 was not known. Clearly something was amiss and on the approach of Caudle's train Sutherland initially kept his signals at danger. However his first sight of the train suggested that it was slowing down and therefore he lowered his middle signal to green. He apparently intended to minimize any delay by allowing Caudle to draw up to the starting signal. This was a normal practice but it rapidly became clear to Sutherland that the train was not obeying the signals. He threw the lowered signal back to danger and waved a red hand lamp, but No. 446 and its six coaches continued up the bank. . . .

[Caudle] . . . admitted that the injector crisis had meant that he had failed to take his customary second look at the Mallerstang distant and he had seen neither of the other signals—one of which had shown a green light for a brief period. Instead he realized on resolving the injector problem that he had just passed the final Mallerstang signal. Three miles away stood the stalled train.

The Midland Railway's regulations did not rely only on the security of the signals for the protection of a train that had come unexpectedly to a standstill between signal boxes. Rule 217(a) prescribed that such a train be *immediately* protected in the rear by detonators as a second safeguard against a following train running through signals. On this occasion the precaution was not immediately taken and it is easy to see how inflexible adherence to such a rule could produce what seemed to be unwarranted delays. Some ambiguity surrounds the failure to comply with the rule. The 1:35 train had two guards. Following the unexpected stop the front guard, Donnelly, asked Nicholson what had happened. The driver said that they were short of steam, but they'd only be a minute and he should get back to his van.

Donnelly conveyed only part of this information to the rear guard, Whitley. He suggested the stop would be brief but not that they were short of steam:

> I did not tell Guard Whitley what the driver told me, that the engine . . . was stopped for want of steam, because he must have known it himself already. The train had been traveling so slowly for such a distance.

Whitley later claimed ignorance on this point and insisted that had he been aware of the cause of the delay, he would have protected the train immediately. Certainly there remains in all the testimony a degree of doubt about how rapidly these exchanges were carried out. Subsequently it was estimated that the train stood for six or seven minutes before the collision. Only when the train crew heard and then saw Caudle's engine climbing the bank did Whitley run back with lamp and detonators. He went less than 200 yards; no detonators were put down. At the last minute Follows saw the tail lamps of the preceding train; his initial judgment that they were Aisgill signal lights was quickly abandoned as he saw Whitley's lamp.

Caudle made an emergency brake application; but there was insufficient distance and his engine ploughed into the rear of the first train. The last vehicle was a van and was completely demolished, the next one was a third-class carriage and all the fatalities occurred amongst its occupants. Fourteen died on the spot and two more later in hospitals at Leeds and Carlisle. Each train was gas lit; the last three vehicles of the stalled train were completely burned out, although witnesses disagreed how far the fire was a consequence of ignited gas and how far of spillages of hot coals.

Public concern about the disaster was perhaps intensified by the fact that its location was only two miles away from the scene of another fatal collision early on Christmas Eve 1910. When an official Midland Railway statement suggested, contrary to the claims of some witnesses, that all the deceased had been dead or unconscious before the fire reached them, an accusatory postcard arrived at the company's Derby headquarters:

> WHAT A LYING STATEMENT TO MAKE. WHAT HAVE YOU DONE SINCE THE LAST DISASTER. NEARLY NIL. FROM A PASSENGER (taken from PRO-RAIL, 491-791, a collection of press clippings, telegrams, etc., relating to the Hawes Junction and Aisgill disasters)

Both disasters could be linked to wider features of company policy; in the aftermath of the 1910 accident, the connections were barely hinted at. Less than three years later they became the subject of angry debate.

An understanding of these exchanges necessitates some analysis of the Midland Railway's response to the problems facing Edwardian railways. Some generalizations about this predicament are possible. Companies faced a rise in the proportion of costs to receipts; they were unable to make a direct response because legislation placed severe obstacles in the way of raising charges. Particularly after the Conciliation Agreement of 1907, the companies faced a further constraint in the growing self-confidence and combativity of many railway trade unionists. Some companies responded by initiating schemes of "scientific" management, whilst some leading figures in the railway world were increasingly skeptical about the alleged benefits of competition. Instead they claimed that a more efficient system would result from amalgamations or traffic-sharing agreements.

The result of the reform was a huge drop in overtime and a reduction in freight train mileage that assisted the economy drive. The more rational structuring of freight services helped to improve passenger train punctuality and this was also facilitated by the introduction of more scientific timetabling techniques. This drive for rationalization involved a system of loadings for each class of engine over each stretch of line. Excess weight required the provision of a pilot to keep to scheduled times. This systematization lay behind Nicholson's unsuccessful request for a pilot prior to the Aisgill tragedy.

This and other controversies lay in the future when Sir Ernest Page commented on the initial impact of the reforms to a meeting of Midland shareholders in February 1910:

> The Midland Railway is better operated now than I believe it has ever been since its construction. It has been brought about by refusing to agree that old methods had been used for years by men of undoubted experience, that therefore they must not be questioned, and when those old methods would not stand the test of investigation, they have been ruthlessly thrown aside and better ones put in their place.

The drive for efficiency had already produced a rift in the Derby hierarchy. A further reorganization of the management structure had led in 1909 to the resignation of the Midland's Locomotive Engineer, Richard Deeley. He objected to the proposal, subsequently carried out, to remove the running side of the Locomotive Department from his jurisdiction and place it under Paget's control. Beyond this demarcation dispute there lay a lengthy battle over the Midland's policy on locomotive construction. Deeley had favored a strategy of larger, more powerful engines, an option vetoed by the Board to avoid an expensive programme of bridge reconstruction. As a result, the Midland's locomotives remained small compared with those of its principal competitors. This policy was strengthened with the appointment of Deeley's successor, the amenable Henry Fowler. The operational consequences of this policy were a service of light, frequent trains and the widespread use of pilots.

The drive for economy also left its first somber impact on the Settle and Carlisle. The Christmas Eve 1910 disaster had as its immediate cause, an error by a busy signalman nine hours into a ten-hour shift. Most expresses in the busy Christmas season required pilots up to Aisgill. In order to save mileage the pilots were detached at the summit and ran three miles south to Hawes Junction where they could use a turntable and then return to their home stations. Early that morning the Hawes Junction signalman had to deal with no fewer than nine such engines whilst attending to many other tasks. He forgot about two of the locomotives standing on the mainline and ready to return to Carlisle. They were run into by a London to Glasgow express. The policy of cost cutting did not leave its mark on this tragedy solely through the small-engine policy. Fifteen months earlier, the Midland Traffic Committee had approved a proposal to replace the two existing signal boxes at Hawes Junction by one new installation. The remodeling had been carried out in the summer of 1910 at a cost of £770 but with an annual saving of £164 in signalmen's wages. At the subsequent inquiry, little was made of the wider issues: a man with a previously unblemished record accepted responsibility for twelve deaths. . . .

Whatever the nuances of style and opinion amongst Midland management over the trade union question, there was abundant evidence that many employees were becoming estranged from the company: the drive for economy involved regradings with loss of wages; the bureaucracy of the control system could be authoritarian; and many railwaymen felt that the company abused the conciliation system. The cautious Derbyshire Miners' MP W. E. Harvey referred to "seething discontent with the management." The Railway Servants VP Walter Hudson acknowledged the desirability of more effective control of traffic but claimed that on the Midland it was administered "with a ruthless hand and reckless as to how much men suffer in consequence. . . ."

The recovery of the experiences of Caudle, Nicholson and the other Midland men has relevance for the continuing debate about the tensions within Edwardian society; arguably it also strikes a contemporary chord with those tragedies of 1987 and 1988—the grim images of the wreck of the Herald of Free Enterprise, the King's Cross fire, the Clapham Junction disaster. Once again the question is posed as to whether managerial pressures for profitability reduce safety standards; once again the harsh choices rest with workers on the spot.

This links back to the core of the Aisgill disaster. Beyond the consultations of urbane politicians and measured public servants, the austere verdicts of courts, there stand the rank and file railwaymen. They encountered daily the pressures of their jobs and at a moment of crisis responded creatively and effectively to assist a colleague. . . .

Endnote

1. From "Railway Safety and Labor Unrest: The Aisgill Railway Disaster of 1913" by David Howell. Taken from *On the Move: Essays in Labor Transport History Presented to Philip Bagwell.* Edited by Chris Wrigley and John Shepard. Rio Grande, Ohio: The Hambledon Press, 1991. Copyright © 1991 Hambledon Press. All rights reserved. Reproduced by permission.

CASE STUDY 4

Johnstown Flood[1]

David McCullough

Introductory Note by W.M.E. and M.M: *More than 2,000 people perished on February 16, 1889, when the South Fork Reservoir dam broke, sending thousands of gallons of water rushing down the valley, totally destroying the town of Johnstown, Pennsylvania. The Johnstown flood still stands as one of the worst technological disasters in U.S. history. Criminally negligent construction standards, reflecting callous attitudes of the wealthy fishing club members toward subordinate social classes living downstream, was identified as the major cause of the disaster.*

Tycoons from such families as the Carnegies and Mellons had the decrepit dam rebuilt in order to create a 70-foot-deep lake in the midst of their summer retreat. Instead of hiring an engineer, the rich elite assigned the job to a railroad contractor who merely bolstered the breakwater with rocks, tree stumps, and dirt. Adding to the tragedy is the fact that only one club member helped with the cleanup, and the others gave a mere $20,000 to help defray $10 million in damages.

This technological disaster of the First Industrial Revolution was the result of socio-cultural factors.

. . . Feelings were running very strong against the club at South Fork. Monday after dark an angry crowd of men had gone up to the dam looking for any club members who might have been still hanging about. When they failed to find anyone, they broke into several of the cottages. Windows were smashed and a lot of furniture was destroyed. Then, apparently, they had gone over to the Unger farm to look up the Colonel [Unger]. The reporters later called it a lynch mob and said they were bent on killing Unger. Whether or not it would have come to that, there is no way of knowing, for Unger by that time was on his way to Pittsburgh. There was a good deal of grumbling among the men as they milled about outside Unger's house; threats were shouted; then the men went struggling off through the night, back down the hollow. The clubmen who had been at the lake had gone off on horseback, heading for Altoona, almost immediately after the dam broke Friday afternoon, though one of them, it seems, stuck around long enough to settle his debts with some of the local people.

He had no intention of ever coming back again, he told them, which they in turn repeated for the benefit of the newspapermen. They also emphasized that the Pittsburgh people had not made things any better for themselves by pulling out so rapidly at a time when, as anyone could see, there was such a crying need for able-bodied men in the valley. Had they stayed on to help, it was said, then people might have felt somewhat differently toward them. This way there was only contempt.

But it was when they began describing how the dam had been re-built by Ruff and his workers that their real bitterness came through, that all the old, deep-seated resentment against the rich, city men be-gan surfacing. Farmers recalled how they had sold Ruff hay to patch the leaks. A South Fork coal operator who insisted that his name be withheld, but who was almost certainly George Stineman, South Fork's leading citizen, told how, years earlier, he had gone to John-stown on more than one occasion to complain about the dam's structural weaknesses. Reporters heard that the dam had been "the bogie of the district" and how it had been the custom to frighten dis-obedient children by telling them that the dam would break. The clubmen were described as rude and imperious in their dealings with the citizens of the valley. Reporters were told of the times neighbor-hood children had been chased from the grounds; and much was made of the hated fish guards across the spillway. Old feuds, per-sonal grudges, memories of insults long forgotten until then, were trotted out one after the other for the benefit of the press.

Someone even went so far as to claim that several of the Italian work-men employed by the club had been out on the dam at the time it failed and had been swept to their death, thus implying that the Pitts-burgh men had heartlessly (or stupidly) ordered them out there while they themselves had hung back on the hillsides.

One local man by the name of Burnett, who conducted a reporter on an inspection of the dam, told the reporter that if people were to hear that he was from Pittsburgh, they might jump to the conclusion that he was connected with the club and pull him from the carriage and beat him to death. "That is the feeling that predominates here," Bur-nett said, "and, we all believe, justly."

The plain fact was that no one who was interviewed had anything good to say about the South Fork Fishing and Hunting Club, its mem-bers, or its dam. And when a coroner's jury from Greensburg, in West-moreland County, showed up soon after the reporters, the local people willingly repeated the same things all over again.

The jurymen had come to investigate the cause of death of the 121 bodies that had been recovered at Nineveh, which was just across the line in Westmoreland County. They poked about the ruins of the dam, talked to people, made notes, and went home. The formal in-vestigation, with witnesses testifying under oath, was to be held on Wednesday, the 5th.

In the meantime, Mr. H. W. Brinkerhoff of *Engineering and Building Record,* a professional journal published in New York, arrived in South Fork to take a look at the dam and was soon joined by A. M. Welling-ton and F. B. Burt, editors of *Engineering News.* Most of the reporters remained cautious about passing judgment on the dam, waiting to see what the experts had to say. But on June 5 the headline on the front page of the *New York Sun* read:

CAUSE OF THE CALAMITY

The Pittsburgh Fishing Club Chiefly Responsible

The Waste Gates Closed When the Club Took Possession

The indictment which followed, based on a *Sun* reporter's "personal investigation," could not have been much more bluntly worded.

> . . . There was no massive masonry, nor any tremendous ex-hibition of engineering skill in designing the structure or put-ting it up. There was no masonry at all in fact, nor any engineering worthy of the name. The dam was simply a gi-gantic heap of earth dumped across the course of a moun-tain stream between two low hills. . . .

In Johnstown on the same day, General Hastings told a *World* corre-spondent that in his view, "It was a piece of carelessness, I might say criminal negligence." In Greensburg the Westmoreland coroner's jury began listening to one witness after another testify to the shoddy way the dam had been rebuilt and the fear it had engendered, though two key witnesses had apparently had second thoughts about speak-ing their minds quite so publicly and refused to appear until forced to do so by the sheriff.

Two days later, on the 7th, a verdict was issued: ". . . death by vio-lence due to the flood caused by the breaking of the dam of the South Fork Reservoir . . ." It seemed a comparatively mild statement, considering the talk there had been and coming as it did on the same day as Hastings' pronouncement. But on the preceding day, another coroner's inquest, this one conducted by Cambria County, had ren-dered a decision that spelled out the cause of the disaster, and fixed the blame, in no uncertain terms.

The Cambria jurors had also visited the dam and listened to dozens of witnesses. But their inquest was held to determine the death of just one flood victim, a Mrs. Ellen Hite. Their verdict was "death by drowning" and that the drowning was "caused by the breaking of the South Fork dam."

But then the following statement was added:

> We further find, from the testimony and what we saw on the ground that there was not sufficient water ware, nor was the dam constructed sufficiently strong nor of the proper material to withstand the overflow; and hence we find the owners of

said dam were culpable in not making it as secure as it should have been, especially in view of the fact that a population of many thousands were in the valley below; and we hold that the owners are responsible for the fearful loss of life and property resulting from the breaking of the dam."

Now the story broke wide open. "THE CLUB IS GUILTY" ran the *World's* headline on June 7. "Neglect Caused the Break . . . Shall the Officers of the Fishing Club Answer for the Terrible Results?"

The Cincinnati *Enquirer* said that in Johnstown, as more facts became known, the excitement was reaching a "fever heat" and that "it would not do for any of the club members to visit the Conemaugh Valley just now." The Chicago *Herald* said there was "no question whatever" as to the fact that criminal negligence was involved.

Although it would be another week before the engineering journals would publish their reports on the dam, the gist of their editors' conclusions had by now leaked to the press. On Sunday, the 9th, *The New York Times* headline ran: "An Engineering Crime; The Dam of Inferior Construction According to the Experts."

Actually, the engineering journals never worded it quite that way. The full report which appeared in the issue of *Engineering News* dated June 15 said that the original dam had been "thoroughly well built," but that contrary to a number of previously published descriptions, it had not been constructed with a solid masonry core." (From this some newspapers would conclude that the "death-dealer" was nothing but a "mud-pile.") The repairs made by Benjamin Ruff, however, had been carried out "with slight care," according to the report. Most important of all, there had been "no careful ramming in watered layers, as in the first dam." But Ruff's work was not the real issue, according to the editors. "Negligence in the mere execution of the earthwork, however, if it existed, is of minor importance, since there is no doubt that it was not a primary cause of the disaster; at worst, it merely aggravated it."

. . . But the point the editors of the report seemed most determined to hammer home was that there was no truth to any claims being made that the dam had been rebuilt by qualified engineers. "In fact, our information is positive, direct, and unimpeachable that at no time during the process of rebuilding the dam was ANY ENGINEER WHATEVER, young or old, good or bad, known or unknown, engaged or consulted as to the work—a fact which will be hailed by engineers everywhere with great satisfaction, as relieving them as a body from a heavy burden of suspicion and reproach."

Moreover, contrary to some statements made in Pittsburgh since the disaster, they had found no evidence that the dam had ever been "inspected" periodically, occasionally, or even once, by anyone "who, by any stretch of charity, could be regarded as an expert."

Johnstown Flood, February 16, 1889. Copyright Browns Brothers. Used by permission.

In other words, the job had been botched by amateurs. That they had been very rich and powerful amateurs was not considered relevant by the engineering journals, but so far as the newspapers were concerned that was to be the very heart of the matter. It was great wealth which now stood condemned, not technology. The club had been condemned by the coroners' juries, General Hastings, and by the engineering experts . . .

The South Fork Fishing and Hunting Club was now described as "the most exclusive resort in America," and its members were referred to as millionaires, aristocrats, ex nabobs. According to the Cincinnati *Enquirer* not even vast wealth was enough to gain admission, unless it was hereditary. "Millionaires who did not satisfy every member of the club might cry in vain for admission," the *Enquirer* wrote. "No amount of money could secure permission to stop overnight at the club's hotel . . ." The paper said that no one could visit the club without a permit, and called it "holy ground consecrated to pleasure by capital," but added that no one would want to go there now, "except to gaze a moment at *the Desolate Monument to the Selfishness of Man . . .*"

And a man by the name of Isaac Reed wrote a widely quoted poem that opened with the lines:

Many thousand human lives
Butchered husbands, slaughtered wives,
Mangled daughters, bleeding sons,
Hosts of martyred little ones,
(Worse than Herod's awful crime)
Sent to heaven before their time;
Lovers burnt and sweethearts drowned,
Darlings lost but never found!
All the horrors that hell could wish,
Such was the price that was paid for—fish!

. . . [If] the men of the South Fork Fishing and Hunting Club, as well as the men of responsibility in Johnstown, had in retrospect looked dispassionately to themselves, and not to their stars, to find the fault, they would have seen that they had been party to two crucial mistakes.

In the first place, they had tampered drastically with the natural order of things and had done so badly. They had ravaged much of the mountain country's protective timber, which caused dangerous flash runoff following mountain storms; they obstructed and diminished the capacity of the rivers; and they had bungled the repair and maintenance of the dam. Perhaps worst of all they had failed—out of indifference mostly—to comprehend the possible consequences of what they were doing, and particularly what those consequences might be should nature happen to behave in anything but the normal fashion, which, of course, was exactly what was to be expected of nature. As one New England newspaper wrote: "The lesson of the Conemaugh Valley flood is that the catastrophes of Nature have to be regarded in the structures of man as well as its ordinary laws . . . "

Endnote

1. Reprinted with permission of Simon & Schuster from *The Johnstown Flood* by David G. McCullough. Copyright © 1968, and renewed © 1996, by David G. McCullough.

CASE STUDY 5

DC-10 Crash[1]

Barbara Himes and Tom L. Beauchamp

Introductory Note by W.M.E. and M.M: *On March 3, 1974, the rear cargo door of a Turkish Airlines DC-10 blew open minutes after take-off from Paris, killing all 346 persons aboard. At the time it was the worst crash in the history of commercial aviation. The cause of the crash was attributed to basic design flaws, including the latch and lock mechanism of the rear cargo door. The tragedy of the disaster was made all the worse when it was discovered that senior engineers were aware of the design flaws in the rear cargo doors and even attempted to warn management of the imminent dangers if nothing was done to repair the problems. Most noteworthy in the case was the seeming negligence on the part of both managers and regulators to act on the problem, given an identical incident some months before when the rear cargo doors blew open on a DC-10 over Windsor, Ontario. Fortunately, no one was hurt in that case.*

This technological disaster of the Second Industrial Revolution was caused by technical design factors.

. . . In August 1968, McDonnell Douglas awarded Convair a contract to build the DC-10 fuselage and doors. The lower cargo doors became the subject of immediate discussion. These doors were to be outward-hinging, tension-latch doors, with latches driven by hydraulic cylinders—a design already adequately tested by DC-8 and DC-9 models. In addition, each cargo door was designed to be linked to hydraulically actuated flight controls and was to have a manual locking system designed so that the handle or latch lever could not be stowed away unless the door was properly closed and latched. McDonnell Douglas, however, decided to separate the cargo door actuation system from the hydraulically actuated primary flight controls.

This involved using electric actuators to close the cargo doors rather than the hydraulic actuators originally called for. Fewer moving parts in the electric actuators presumably made for easier maintenance, and each door would weigh 28 pounds less.

However, the Convair engineers had considered the hydraulic actuators critical to safety. They were not satisfied with these changes, and they remained dissatisfied after further modifications were introduced. As Convair engineers viewed the situation, the critical difference between the two actuator systems involved the way each would respond to the buildup of forces caused by increasing pressure. If a hydraulic latch was not secured properly, the latches would smoothly slide open when only a small amount of pressure had built up in the cabin. Although the doors would be ripped off their hinges, this would occur at a low altitude, so that the shock from decompression would be small enough to land the plane safely. By contrast, if an electric latch failed to catch, it would not gently slide open due to increasing pressure. Rather, it would be abruptly and violently forced open, most likely at a higher altitude where rapid decompression would dangerously impair the structure of the plane.

Convair's Director of Product Engineering, F. D. "Dan" Applegate, was adamant that a hydraulic system was more satisfactory. However, McDonnell Douglas did not yield to Convair's reservations about the DC-10 cargo door design.

Once a decision had been made to use an electrical system, it was necessary to devise a new and foolproof backup system of checking and locking. In the summer of 1969 McDonnell Douglas asked Convair to draft a Failure Mode and Effects Analysis, or FMEA, for the cargo door system. A FMEA's purpose is to assess the likelihood and consequences of a failure in the system. In August 1969, Convair engineers found nine possible failure sequences that could result in destruction of the craft, with loss of human lives. A major problem focused on the warning and locking-pin systems. The door could close and latch, but without being safely locked. The warning indicator lights were prone to failure, in which case a door malfunction could go undetected. The FMEA also concluded that the door design was potentially dangerous and lacked a reliable failsafe locking system. It could open in flight, presenting considerable danger to passengers (Eddy, Potter, & Page, 1976; Curd & May, 1984; French, 1982).

The Federal Aviation Administration (FAA) requires that it be given an FMEA covering all systems critical to safety, but no mention was made of this hazard to the FAA prior to "certification" of the DC-10 model. McDonnell Douglas maintains that no such report was filed because this cargo door design was not implemented until all defects expressed in the FMEA were removed. The FMEA *submitted,* they contend, was the final FMEA, and did not discuss past defects because they had been removed. As lead manufacturer, McDonnell Douglas made itself entirely responsible for the certification of the aircraft and, in seeking the certification, was expressing its position that all defects had been removed. Convair, by contrast, was not formally responsible because its contract with McDonnell Douglas forbade Convair from reporting directly to the FAA. During a model test run in May 1970, the DC-10 blew its forward lower cargo door, and the plane's cabin floor collapsed. Because the vital electric and hydraulic subsystems of the plane are located under the cabin floor (unlike in the 747, where they are above the ceiling), this collapse was doubly incapacitating (Sewell, 1982: 18).

A spokesperson at McDonnell Douglas placed the blame for this particular malfunction on the "human failure" of a mechanic who had incorrectly sealed the door. Although no serious design problems were contemplated, there were some ensuing modifications in design for the door, purportedly to provide better checks on the locking pins. As modified, the cargo door design was properly certified and authorities at McDonnell Douglas believed it safe. Five DC-10s were flight tested for over 1,500 hours prior to certification of the craft.

Certification processes are carried out in the name of the FAA, but the actual work is often performed by the manufacturers. As a regulatory agency, the FAA is charged with overseeing commercial products and regulating them in the public interest. However, the FAA is often not in an independent position. The FAA appoints designated engineering representatives (DERs) to make inspections at company plants. These are company employees chosen for their experience and integrity who have the dual obligations of loyalty to the company that pays them as design engineers and of faithful performance of inspections to see that the company has complied with federal airworthiness regulations. The manufacturers are in this respect policing themselves, and it is generally acknowledged that conflicts of interest arise in this dual-obligation system (Eddy, Potter, & Page, 1976).

During the months surrounding November 1970, a number of internal memos were written at both McDonnell Douglas and Convair that cited old and new design problems with the cargo door. New structural proposals were made, but none were implemented. McDonnell Douglas and Convair quarreled about cost accounting and about pinning fault for remaining design flaws. The FAA finally certified the DC-10 on July 29, 1971, and by late 1971 the plane had received praise for its performance at virtually all levels. Under rigorous conditions its performance ratings were excellent. The company vigorously promoted the new aircraft.

But on June 12, 1972, an aft bulk cargo door of a DC-10 in flight from Los Angeles to New York separated from the body of the aircraft at about 11,750 feet over Windsor, Ontario. Rapid cabin decompression occurred as a result, causing structural damage to the cabin floor immediately above the cargo compartment. Nine passengers and two stewardesses were injured. A National Transportation Safety Board (NTSB) investigation found that the probable cause of the malfunction was the latching mechanism in the cargo door and recommended changes in the locking system. The NTSB's specific recommendations were the following: (1) Require a modification to the DC-10 cargo door locking system to make it physically impossible to position the external locking handle and vent door to their normal locked positions unless the locking pins are fully engaged; and (2) require the installation of relief vents between the cabin and aft cargo compartment to minimize the pressure loading on the cabin flooring in the event of sudden depressurization of the compartment (National Traffic Safety Board, 1973).

The administrator of the FAA, John Shaffer, could have issued an airworthiness directive that required immediate repairs. He elected not to issue the directive, choosing instead a "gentleman's agreement" with McDonnell Douglas that allowed the company to make the necessary modifications and recommend new procedures to affected airlines. All actions by the company were to be voluntary. Fifteen days subsequent to the blowout over Windsor (June 27, 1972), Dan Applegate wrote a stern memo to his superior at Convair that expressed his doubts about the entire project and offered some reflections on "future accident liability." The following excerpts from the memo reveal Applegate's anguish and concerns:

The potential for long-term Convair liability on the DC-10 has caused me increasing concern for several reasons:

> 1. The fundamental safety of the cargo door latching system has been progressively degraded since the program began in 1968.
>
> 2. The airplane demonstrated an inherent susceptibility to catastrophic failure when exposed to explosive decompression of the cargo compartment in 1970 ground tests.
>
> 3. Douglas has taken an increasingly "hard-line" with regards to the relative division of design responsibility between Douglas and Convair during change cost negotiations.
>
> 4. The growing "consumerism" environment indicates increasing Convair exposure to accident liability claims in the years ahead. . . .

I can only say that our contract with Douglas provided that Douglas would furnish all desired criteria and loads (which in fact they did) and that we would design to satisfy these design criteria and loads (which in fact we did). There is nothing in our experience history which would have led us to expect that the DC-10 cabin floor would be inherently susceptible to catastrophic failure when exposed to explosive decompression of the cargo compartment, and I must presume that there is nothing in Douglas's experience history which would have led them to expect that the airplane would have this inherent characteristic or they would have provided for this in their loads and criteria which they furnished to us.

My only criticism of Douglas in this regard is that once this inherent weakness was demonstrated by the July 1970 test failure, they did not take immediate steps to correct it. It seems to me inevitable that, in the twenty years ahead of us, DC-10 cargo doors will come open and I would expect this to usually result in the loss of the airplane. [Emphasis added.] This fundamental failure mode has been discussed in the past and is being discussed again in the bowels of both the Douglas and Convair organizations. It appears however that Douglas is waiting and hoping for government direction or regulations in the hope of passing costs on to us or their customers.

If you can judge from Douglas' position during ongoing contract change negotiations they may feel that any liability incurred in the meantime for loss of life, property and equipment may be legally passed on to us. It is recommended that overtures be made at the highest management level to persuade Douglas to immediately make a decision to incorporate changes in the DC-10, which will correct the fundamental cabin floor catastrophic failure mode. Correction will take a good bit of time; hopefully there is time before the National Transportation Safety Board (NTSB) or the FAA ground the airplane which would have disastrous effects upon sales and production both near and long term. This corrective action becomes more expensive than the cost of damages resulting from the loss of one planeload of people.

F. D. Applegate
Director of Product
Engineering

(Eddy, Potter, & Page, 1976, 183–5)

If this memo had reached outside authorities, Applegate conceivably might have been able to prevent the occurrence of events that (to some extent) he correctly foresaw. However, this memo was never sent either to McDonnell Douglas or to the FAA. Applegate received a reply to his memo from his immediate supervisor, J. B. Hurt. By now it was clear to both Applegate and Hurt that such major safety questions would not be addressed further at McDonnell Douglas. Hurt's reply to Applegate pointed out that if further questions were now raised, Convair, not McDonnell Douglas, would most likely have to bear the costs of necessary modifications. Higher management at Convair subsequently agreed with Hurt. Without taking other routes to express his grave misgivings about the DC-10, Applegate filed away his memo.

In July 1972, Ship 29 of the DC-10 line was inspected by three different inspectors at the Long Beach plant of McDonnell Douglas. All three certified that the ship had been successfully altered to meet FAA specifications. Two years later, Ship 29 was owned by Turkish Airlines. This ship crashed near Paris in 1974, killing all 335 passengers and 11 crewmembers—the worst single-plane disaster in aviation history. Experts agreed that the immediate cause of the crash was a blowout of the rear cargo door, at approximately twelve minutes after lift-off. Decompression of the cargo bay caused a collapse of the cabin floor, thereby severing control cables. It was alleged by Sanford Douglas, President of McDonnell Douglas, that the Turkish airline involved in the crash had attempted to "rework" the door rigging or latching mechanism, was working with an inadequately trained ground crew, and failed to follow specified procedures for proper latching. The Turkish airline denied the charges. Recovery of a flight recorder indicated that there was no explosion, fire, or evident sabotage, and that the cargo door blew because it was not securely sealed. . . .

References

Curd, Martin, and May, Larry. (1984). *Professional responsibility for harmful actions.* Dubuque, IA: Kendall/Hunt Publishing Co.

Eddy, Paul, Potter, Elaine, and Page, Bruce. (1976). *Destination disaster: From the tri-motor to the DC-10.* New York: Quadrangle Books, New York Times Book Co.

French, Peter. (1982). "What is Hamlet to McDonnell-Douglas or McDonnell-Douglas to Hamlet: DC-10," *Business and Professional Ethics Journal* 1 (1): 5–6.

National Transportation Safety Board (NTSB). (1973). Aircraft Accident Report no. NTSB-AAR-73-2, February 28.

Sewell, Homer. (1982). "Commentary," *Business and Professional Ethics Journal* 1: 17–19.

Endnote

1. From "The DC-10's Defective Doors" prepared by Barbara Himes and Tom L. Beauchamp, and revised by Cathleen Kavepy, John Cuddihy, and Jeff Greene. Taken from *Case Studies in Business, Society, and Ethics,* 4th edition, by Tom Beauchamp. Upper Saddle River, NJ: Prentice Hall. 1998. Copyright © Tom L. Beauchamp. Reproduced by permission of the authors.

CASE STUDY 6

Tenerife Runway Collision[1]

Karl Weick

Introductory Note by W.M.E. and M.M: *The worst disaster in aviation history occurred on March 27, 1977, at the Tenerife airport in the Canary Islands. Five hundred and eighty-three people perished as two jumbo jets collided on the runway. Human factors such as loss of cognitive efficiency due to excessive stress and anxiety and loss of communicative accuracy between pilots and the air traffic controllers were the major causes of the accident.*

The dynamics of cockpit interactions between pilot and copilot, ambiguous radio communication between air traffic controllers, along with the stressful environment inside the cockpit, led to the inability of the pilots of both airplanes to successfully interpret and make sense of the radio communications.

This technological disaster of the Second Industrial Revolution was caused by human factors.

There is a growing appreciation that large-scale disasters such as Bhopal (Shrivastava, 1987) and Three Mile Island (Perrow, 1981) are the result of separate small events that become linked and amplified in ways that are incomprehensible and unpredictable. This scenario of linkage and amplification is especially likely when systems become more tightly coupled and less linear (Perrow, 1984).

What is missing from these analyses, however, is any discussion of the processes by which crises are set in motion. Specifically, we lack an understanding of the ways in which separate small failures become linked. We know that single cause incidents are rare, but we don't know how small events can become chained together so that they result in a disastrous outcome . . .

Description of Tenerife Disaster

On March 27, 1977, KLM flight 4805, a 747 bound from Amsterdam to the Canary Islands, and Pan Am flight 1736, another 747 bound from Los Angeles and New York to the Canary Islands, were both diverted to Los Rodeos airport at Tenerife because the Las Palmas airport, their original destination, was closed because of a bomb explosion. The KLM landed first at 1:38 p.m., followed by Pan Am, which landed at 2:15 p.m. Because Tenerife is not a major airport, its taxi space was limited. This meant that the Pan Am plane had to park behind the KLM flight in such a way that it could not depart until the KLM plane left. When the Las Palmas airport reopened at 2:30 p.m., the Pan Am flight was ready to depart because its passengers had remained on board. KLM's passengers, however, had left the plane so there was a delay while they reboarded and while the plane was refueled to shorten its turnaround time at Las Palmas. KLM began its taxi for takeoff at 4:56 p.m. and was initially directed to proceed down a runway parallel to the takeoff runway (see Figure 8–4).

This directive was amended shortly thereafter and KLM was requested to taxi down the takeoff runway and at the end, to make a 180-degree turn and await further instruction.

Pan Am was requested to follow KLM down the takeoff runway and to leave the takeoff runway at taxiway C3, use the parallel runway for the remainder of the taxi, and then pull in behind the KLM flight. Pan Am's request to hold short of the takeoff runway and stay off it until KLM had departed was denied. After the KLM plane made the 180-degree turn at the end of the takeoff runway, rather than hold as instructed, it started moving and reported, "we are now at takeoff." Neither the air traffic controllers nor the Pan Am crew were certain what this ambiguous phrase meant, but Pan Am restated to controllers that it would report when it was clear of the takeoff runway, a

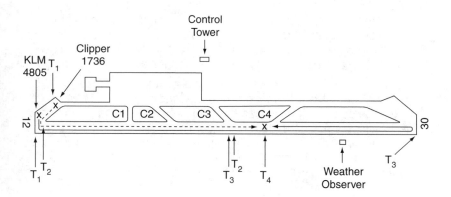

KLM 4805 and PanAm 1736
March 27, 1977
Elevation: 2073 feet
Runway: 3400 × 45 meters

T_1 = 1659:10 (GMT) T_3 = 1705:53 (GMT)
Pan Am on runway   Pan Am passing C3
KLM enters runway   KLM receiving ATC clearance

T_2 = 1702:08 (GMT) T_4 = 1706:49 (GMT)
Pan Am enters runway   Accident between impact point near C4
KLM at C3

Figure 8–4
Tenerife airport diagram.

communiqué heard inside the KLM cockpit. When the pilot of the
KLM flight was asked by the engineer, "is he not clear then, that Pan
Am?" the pilot replied "yes" and there was no further conversation.
The collision occurred 13 seconds later at 5:06 p.m. None of the 234
passengers and 14 crew on the KLM flight survived. Of the 380 passengers and 16 crew on the Pan Am plane, 70 survived, although 9
died later, making a total loss of 583 lives.

A brief excerpt from the Spanish Ministry of Transport and Communication's investigation of the crash describes interactions among the
KLM crewmembers immediately before the crash. These interactions,
reconstructed from the KLM cockpit voice recorder (CVR), are the focus of the remainder of our analysis.

As the time for the takeoff approached, the KLM captain "seemed a
little absent from all that was heard in the cockpit." He inquired several times, and after the copilot confirmed the order to backtrack, he
asked the tower if he should leave the runway by C-1, and subsequently asked his copilot if he should do so by C-4. On arriving at
the end of the runway, and making a 180-degree turn in order to
place himself in takeoff position, he was advised by the copilot that
he should wait because they still did not have an Air Traffic Control
(ATC) clearance. The captain asked him to request it, and he did, but
while the copilot was still repeating the clearance, the captain
opened the throttle and started to take off. Then the copilot, instead
of requesting takeoff clearance or advising that they did not yet have
it, added to his read-back, "We are now at takeoff."

The tower, which was not expecting the aircraft to take off because it
had not been given clearance, interpreted the sentence as, "We are
now at takeoff position." (When the Spanish, American and Dutch investigating teams heard the tower recording together and for the first
time, no one, or hardly anyone, understood that this transmission
meant that they were taking off.) The controller replied: "O.K. . . . stand
by for takeoff . . . I will call you." Nor did the Pan Am, on hearing the
"We are now at takeoff," interpret it as an unequivocal indication of
takeoff. However, in order to make their own position clear, they said,
"We are still taxiing down the runway." This transmission coincided
with the "Stand by for takeoff . . . I will call you," causing a whistling
sound in the tower transmission and making its reception in the KLM
cockpit not as clear as it should have been, even though it did not
thereby become unintelligible.

The communication from the tower to the Pan Am requested the latter to report when it left the runway clear. In the cockpit of the KLM,
nobody at first confirmed receiving these communications until the
Pan Am responded to the tower's request that it should report leaving the runway with an "O.K., we'll report when we're clear." On hearing this, the KLM flight engineer asked, "Is he not clear then?" The
captain did not understand him and he repeated, "Is he not clear that
Pan American?" The captain replied with an emphatic "Yes." Perhaps

influenced by his great prestige, making it difficult to imagine an error of this magnitude on the part of such an expert pilot, both the copilot and flight engineer made no further objections. The impact took place about 13 seconds later (Anonymous, 1978: 115).

Tenerife as a Stressful Environment

Stress is often defined as a relation between the person and the environment, as in Holyroyd and Lazarus's statement that "psychological stress requires a judgment that environmental and/or internal demands tax or exceed the individual's resources for managing them" (1982: 22). Their use of the word *judgment* emphasizes that stress results from an appraisal that imposes meaning on environmental demands. Typically, stress results from the appraisal that something important is at stake and in jeopardy (McGrath, 1976).

There were several events impinging on people at Tenerife that are likely to have taxed their resources and been labeled as threatening. These events, once appraised as threatening, had a cumulative, negative effect on performance . . .

Environmental Demands at Tenerife.   The KLM crew felt growing pressure from at least three sources: Dutch law, difficult maneuvers, and unpredictable weather. Because the accident took place near the end of March, members of the KLM crew were very near the limits of time they were allowed to fly in one month. This was more serious than a mere inconvenience because in 1976 the Dutch enacted a tough law on "Work and Rest Regulations for Flight Crews" (Roitsch, Babcock, & Edmunds, 1979: 14) that put strict limits on flight and duty time. The computation of these limits was complex and could no longer be done by the captain nor did the captain have any discretion to extend duty time. Therefore, the KLM crew faced the possibility of fines, imprisonment, and loss of pilot license if further delays materialized. The crew was informed that if they could leave Las Palmas by 7 p.m. their headquarters thought they could make it back to Amsterdam legally, but headquarters would let them know in Las Palmas.

Further pressure was added because the maneuver of turning a 747 around (backtracking) at the end of a runway is difficult, especially when that runway is narrow. It takes a minimum width of 142 feet to make a 180-degree turn in a 747 (Roitsch et al., 1979: 19) and the Tenerife runway was 150 feet wide. Finally, the weather was unpredictable, and at Tenerife that creates some unique problems. Tenerife is 2073 feet above sea level and the sea coast is just a few miles away. This means that clouds rather than fog float into the airport. When KLM's crew backtracked, they saw a cloud 3000 feet down the runway moving toward them at 12 knots (Roitsch et al., 1979: 12), concealing the Pan Am plane on the other side. Pan Am was taxiing inside this cloud and passed its assigned runway exit because it could not see it. KLM entered that same cloud 1300 feet into its takeoff roll and that is where the collision occurred. The tower did not see the collision or the resulting fire because of the cloud, nor could the firefighters find the fire at first when they were summoned. The density of the cloud is further shown by the fact that when the firefighters started to put out the fire on one plane, the KLM plane, they didn't realize that a second plane was on fire nearby because they couldn't see it (Anonymous, 1978: 117–119).

The KLM crew was not the only group that was under pressure. Las Palmas airport had reopened for traffic at 2:30, barely 15 minutes after Pan Am had landed at Tenerife. Pan Am was ready to leave Tenerife immediately except that they were blocked by KLM 4805 and continued to be blocked for another 2-1/2 hours. Reactions of the Pan Am crew to the lengthening delays undoubtedly were intensified by the fact that they had originally asked to circle over Las Palmas because they had sufficient fuel to do so, a request that was denied by Spanish controllers. The Pan Am crew also saw the weather deteriorating as they waited for KLM to leave. They had been on duty 1 hour, although they were not close to the limits of their duty time.

Controllers at Tenerife were also under pressure because they were shorthanded, they did not often handle 747s, they had no ground radar, the centerline lights on the runway were not operating, they were working in English (a less familiar second language), and their normal routines for routing planes on a takeoff and landing were disrupted because they had planes parked in areas they would normally use to execute these routines . . .

The Breakdown of Coordination Under Stress

The phrase "operator error" is misleading in many ways, but among the most subtle problems is the fact that the term is singular (Hayashi, 1985). An operator error is usually a collective error (Gardenier, 1981), but it is only recently that efforts have been made to understand the ways in which team interaction generates emergent potentialities for and remedies of local failures (Hirokawa, Gouran, & Martz, 1988). The crew in the KLM cockpit provides a unique glimpse of some ways in which crises become mobilized when crew interaction breaks down.

Individualism in the Cockpit.   The setting in the KLM cockpit was unusual, not only because the captain was the head of flight training and a member of the top management team, but also because this captain had given the copilot (first officer) his qualification check in a 747 just two months earlier. This recently certified first officer made only two comments to try to influence the captain's decision during the crucial events at the head of the runway. The Airline Pilots Association (ALPA) report of the crash describes those comments this way:

> The KLM first officer was relatively young and new in his position and appeared to be mainly concerned with completing his

tasks so as not to delay the captain's timing of the takeoff. He only made two comments in order to try to influence the captain's takeoff decision. When the captain first began pushing up the thrust levers, he said, "Wait a minute, we do not have an ATC clearance." The captain, rather than admitting to an oversight, closed the thrust levers and responded by saying, "No, I know that, go ahead ask." The second occurrence was at the end of the ATC clearance read back. The KLM first officer observed that the captain had commenced the takeoff and finished the ATC clearance read back by stating, "We are, uh, taking off" or "We are at takeoff" over the radio. After many hours of replaying the tapes, it is difficult to be sure what statement the first officer made. For this reason, we assume that neither the approach controller nor the Pan Am crew were positive about what was said. The Study Group believes that this ambiguous statement by the first officer was an indication that he was surprised by the KLM captain's actions in commencing the takeoff. We believe the first officer thought something was wrong with the takeoff decision by the captain, and tried to alert everyone on frequency that they were commencing takeoff. The KLM captain did not comment on his first officer's radio transmission but rather became immediately involved in setting takeoff power and tracking the runway centerline (Roitsch et al., 1979: 18).

The first officer is not the only person acting in a manner that is more individual than collective (Wagner & Moch, 1986). The same was true for the engineer.

The flight engineer was the first and current President of the European Flight Engineers Organization. There is an odd statement about him in the ALPA documents. It says that he was not in favor of integrating the functions of the engineering position with those of the pilot crewmembers, such as communication, navigation, and general monitoring of the operation of the flight. "He is said to have felt that flight engineering should consist of specialized emphasis on power-plant and systems analysis and maintenance consideration" (Roitsch et al., 1979: 5).

Recall that the engineer was the last point where this accident could have been prevented when he asked, "Is he not clear then, that Pan Am?" Recordings suggest that he made this statement in a "tentative manner" (Roitsch et al., 1979: 22) just as the plane entered the thick cloud and the pilots had their hands full keeping the plane on the runway . . .

Speech-Exchange Systems as Organizational Building Blocks

. . . The Tenerife disaster was built out of a series of small, semi-standardized, misunderstandings, among which were the following:

1. KLM requested 2 clearances in one transmission (we are now ready for takeoff and are waiting for ATC clearance). Any reply could be seen as a comment on both requests.

2. The controller, in giving a clearance, used the words "after take-off" ("maintain flight level niner zero right after takeoff proceed with heading zero four zero until intercepting the three two five radial from Las Palmas"), which could have been heard by the KLM crew as permission to leave. The ATC Manual (7110.650, October 25, 1984) clearly states, under the heading "Departure Terminology" that controllers should, "Avoid using the term "take-off" except to actually clear an aircraft for takeoff or to cancel a takeoff clearance. Use such terms as 'depart,' 'departure,' or 'fly' in clearances when necessary" (heading 4-20: 4-5). Thus, the Tenerife controller could have said "right turn after departure" or "right turn fly heading zero four."

3. As we have seen, the phrase "We are now taking off" is non-standard and produced confusion as to what it meant.

4. When the controller said to KLM, "Okay . . . stand by for takeoff . . . I will call you," a squeal for the last portion of this message changed the timbre of the controller's voice. This may have led the KLM crew to assume that a different station was transmitting and that the message was not intended for them.

5. The controller did not wait to receive an acknowledgement (e.g., "Roger") from KLM after he had ordered them to "standby for takeoff." Had he done so, he might have discovered a misunderstanding (Hurst, 1982: 176).

6. Shortly before the collision, for the first and only time that day, the controller changed from calling the Pan Am plane "Clipper 1736" to the designation "Pappa Alpha 1736." This could sound like the controller is referring to a different plane. (Roitsch et al., 1979: 22).

Interactive Complexity as Indigenous to Human Systems

. . . Human systems are not necessarily protected from disasters by loose coupling and linear transformation systems, because these qualities can change when people are subjected to stress, ignore data, regress, centralize, and become more self-centered.

Thus it would be a mistake to conclude from Perrow's (1984) work that organizations are either chronically vulnerable to normal accidents or chronically immune from them. Perrow's (1984: 63) structural bias kept him from seeing clearly that, when you take people and their limitations into account, susceptibility to normal accidents can change within a relatively short time. Several events at Tenerife show the system growing tighter and more complex:

1. Controllers develop ad hoc routing of 2 jumbo jets on an active runway because they have no other place to put them (Roitsch et al., 1979: 8).

2. Controllers have to work with more planes than they have before, without the aid of ground radar, without a tower operator, and with no centerline lights to help in guiding planes.

3. Controllers keep instructing pilots to use taxiway "Third Left" to exit the active runway, but this taxiway is impossible for a 747 to negotiate. It requires a 148 degree left turn followed by an immediate 148 degree right turn onto a taxiway that is 74 feet wide (Roitsch et al., 1979: 19).

4. Thus, neither the KLM pilot nor the Pan Am pilot are able to do what the controller tells them to do, so both pilots assume that the controller really means for them to use some other taxiway. Nevertheless, the KLM pilot may have assumed that the Pan Am pilot had exited by taxiway third left (Roitsch et al., 1979: 24).

5. The longer the delay at Tenerife, the higher the probability that all hotel rooms in Las Palmas would be filled, the higher the probability that the air corridor back to Amsterdam would be filled with evening flights, occasioning other air traffic delays, and the greater the chance for backups at Las Palmas itself, all of which increased the chances that duty time would expire while the KLM crew was in transit.

6. Throughout the afternoon there was the continuing possibility that the terrorist activities that had closed Las Palmas could spread to Tenerife. In fact, when the tower personnel heard the KLM explosion, they first thought that fuel tanks next to the tower had been blown up by terrorists (Roitsch et al., 1979: 8). . . .

In conclusion, small details can enlarge and, in the context of other enlargements, create a problem that exceeds the grasp of individuals or groups. Interactive complexity is likely to become more common, not less . . . It is not a fixed commodity, nor is it a peculiar pathology confined to nuclear reactors and chemical plants. It may be the most volatile linkage point between micro and macro processes we are likely to find in the next few years.

References

Anonymous. (1978). "Spaniards analyze Tenerife accident," *Aviation Week and Space Technology*, November 20: 113–121.

Gardenier, J.S. (1981). "Ship navigational failure detection and diagnosis." In J. Rasmussen and W.B. Rouse (Eds.), *Human detection and diagnosis of system failures.* New York: Plenum: 49–74.

Hayashi, K. (1985). "Hazard analysis in chemical complexes in Japan—especially those caused by human errors," *Ergonomics* 28: 835–841.

Hirokawa, R.Y, Gouran, D.S., and Martz, A.E. (1988). Understanding the sources of faulty group decision-making: A lesson from the Challenger disaster. *Small Group Behavior* 19: 411–433.

Holroyd, K.A., and Lazarus, R.S. (1982). "Stress, coping, and somatic adaptation." In L. Goldberger and S. Breznitz (Eds.), *Handbook of stress.* New York: Free Press: 21–35.

Hurst, R. (1982). "Portents and challenges." In R. Hurst and L.R. Hurst (Eds.), *Pilot error.* New York: Jason Aronson: 164–177.

McGrath, J.E. (1976). "Stress and behavior in organizations." In M.D. Dunnette (Ed.), *Handbook in industrial and organizational psychology.* Chicago: Rand-McNally: 1351–1395.

Perrow, C. (1981). "Normal accident at Three Mile Island," *Society* 18 (5): 17–26.

Perrow, C. (1984). *Normal accidents.* New York: Basic Books.

Roitsch, P.A., Babcock, G.L., and Edmunds, W.W. (1979). *Human factors report on the Tenerife accident.* Washington, DC: Airline Pilots Association.

Shrivastava, Paul. (1987). *Bhopal: Anatomy of a crisis.* Cambridge, MA: Ballinger.

Wagner, J.A., and Moch, M.K. (1986). "Individualism and collectivism: Concept and measure," *Group and Organization Studies* 11: 280–304.

Endnote

1. From "The Vulnerable System: An Analysis of the Tenerife Air Disaster" by Karl Weick in *Journal of Management* Vol. 9, #3; by permission of JAI Press. All rights retained by JAI Press.

CASE STUDY 7

Santa Barbara Oil Spill[1]

Harvey Molotch

Introductory Note by W.M.E. and M.M: *Two eruptions, one on January 28, 1969, the other on February 12, 1969, occurred off the coast of Santa Barbara, California, spilling crude oil at the rate of 5,000 gallons per day. The spillage continued for over a month until engineers finally sealed the fissure. Miles of beaches were damaged.*

The two eruptions received national attention as news reports showed hundreds of photographs of blackened beaches and oil-soaked birds. The damage to the "ecological chain" was substantial. Damage to wildlife was considerable; thousands of water fowl and other species of animals were adversely affected. The cause is attributed to the corporate power of oil companies, which subordinated public health to profit maximization.

The Santa Barbara oil spill was one of the first great environmental disasters. The enactment of the National Environmental Policy Act, passed in 1969, made environmental protection a matter of national policy with the establishment of the Environmental Protection Agency in 1970.[2]

This technological disaster of the Second Industrial Revolution was caused by organizational systems factors.

More than oil leaked from Union Oil's Platform A in the Santa Barbara Channel—a bit of truth about power in America spilled out along with it. . . . A few historical details concerning the case under examination are in order. For over fifteen years, Santa Barbara's political leaders had attempted to prevent despoliation of their coastline by oil drilling on adjacent federal waters. Although they were unsuccessful in blocking eventual oil leasing (in February 1968) of federal waters beyond the three-mile limit they were able to establish a sanctuary within state waters (thus foregoing the extraordinary revenues which leases in such areas bring to adjacent localities—e.g., the riches of Long Beach).

It was therefore a great irony that the one city which voluntarily exchanged revenue for a pure environment should find itself faced, on January 29, 1969, with a massive eruption of crude oil—an eruption which was, in the end, to cover the entire city (as well as much of Ventura and Santa County coastline as well) with a thick coat of crude oil. The air was soured for many hundreds of feet inland and the traditional economic base of the region (tourism) was under threat. After ten days of unsuccessful attempts, the runaway well was brought under control, only to be followed by a second eruption on February

12. This fissure was closed on March 3, but was followed by a sustained "seepage" of oil—a leakage that continues, at this writing, to pollute the sea, the air, and the famed local beaches. The oil companies had paid $603 million for their lease rights and neither they nor the federal government bear any significant legal responsibility toward the localities which these lease rights might endanger.

If the big spill had occurred almost anywhere else (e.g., Lima, Ohio; Lompoc, California), it is likely that the current research opportunity would not have developed. But Santa Barbara is different. Of its 70,000 residents, a disproportionate number are upper class and upper middle class. They are persons who, having a wide choice of where in the world they might live, have chosen Santa Barbara for its ideal climate, gentle beauty, and sophisticated "culture." Thus a large number of worldly rich, well-educated persons—individuals with resources, spare time, and contacts with national and international elites—found themselves with a commonly shared disagreeable situation: the pollution of their otherwise near-perfect environment. Santa Barbarans thus possessed none of the "problems" which otherwise are said to inhibit effective community response to external threat: they are not urban villagers (cf. Gans, 1962); they are not internally divided and parochial like the Springdalers (Vidich and Bensman, 1960); nor emaciated with self-doubt and organizational naiveté as is supposed of the ghetto dwellers. With moral indignation and high self-confidence, they set out to right the wrong so obviously done to them.

Their response was immediate. The stodgy *Santa Barbara News-Press* inaugurated a series of editorials, unique in uncompromising stridency. Under the leadership of a former State Senator and a local corporate executive, a community organization was established called "GOO" (Get Oil Out!) which took a militant stand against any and all oil activity in the Channel. In a petition to President Nixon (eventually to gain 110,000 signatures), GOO's position was clearly stated:

> . . . With the seabed filled with fissures in this area, similar disastrous oil operation accidents may be expected. And with one of the largest faults centered in the channel waters, one sizeable earthquake could mean possible disaster for the entire channel area. . . . Therefore, we the undersigned do call upon the state of California and the Federal Government to promote conservation by:
>
> 1. Taking immediate action to have present offshore oil operations cease and desist at once.
> 2. Issuing no further leases in the Santa Barbara Channel.
> 3. Having all oil platforms removed from this area at the earliest possible date.

The same theme emerged in the hundreds of letters published by the *News-Press* in the weeks to follow and in the positions taken by virtually every local civic and government body. Both in terms of its volume (372 letters published in February alone) and the intensity of the revealed opinions, the flow of letters was hailed by the *News-Press* as "unprecedented." Rallies were held at the beach; GOO petitions were circulated at local shopping centers and sent to friends around the country; a fund-raising dramatic spoof of the oil industry was produced at a local high school. Local artists, playwrights, advertising men, retired executives and academic specialists from the local campus of the University of California (UCSB) executed special projects appropriate to their areas of expertise.

A GOO strategy emerged for a two-front attack. Local indignation, producing the petition to the President and thousands of letters to key members of Congress and the executive *would* lead to appropriate legislation. Legal action in the courts against the oil companies and the federal government would have the double effect of recouping some of the financial losses certain to be endured by the local tourist and fishing industries while at the same time serving notice that drilling would be a much less profitable operation than it was supposed to be. Legislation to ban drilling was introduced by Cranston in the U.S. Senate and Teague in the House of Representatives. Joint suits by the city and County of Santa Barbara (later joined by the State) for $1 billion in damages were filed against the oil companies and the federal government.

All of these activities—petitions, rallies, court action and legislative lobbying—were significant for their similarity in revealing faith in "the system." . . . There was a muckraking tone to the Santa Barbara response: oil and the profit-crazy executives of Union Oil were ruining Santa Barbara—but once our national and state leaders became aware of what was going on . . . justice would be done . . .

As subsequent events and inexplicable silence of the democratically-elected representatives began to fall into place as part of a more general problem, American democracy came to be seen as a much more complicated affair than a system in which governmental officials actuate the desires of the "people who elected them" once those desires come to be known. Instead, increasing recognition came to be given to the "all-powerful oil lobby"; to legislators "in the pockets of Oil"; to academicians "bought" by Oil; and to regulatory agencies which lobby for those they are supposed to regulate. In other words, Santa Barbarans became increasingly ideological . . . , and in the words of some observers, increasingly "radical" (Mintz, 1969). Writing from his lodgings in the area's most exclusive hotel (the Santa Barbara Biltmore), an irate citizen penned these words in his published letter to the *News-Press:*

> We the people can protest and protest and it means nothing because the industrial and military juntas are the country. They tell

us, the People, what is good for the oil companies is good for the People. To that I say, Like Hell . . . Profit is their language and the proof of all this is their history (Anonymous [a], 1969: A-6).

As time wore on, the editorials and letters continued in their bitterness.

The Executive Branch and the Regulatory Agencies: Disillusionment

From the start, Secretary [of the Interior] [Walter J.] Hickel's actions were regarded with suspicion. His publicized associations with Alaskan Oil interests did his reputation no good in Santa Barbara. When, after a halt to drilling (for "review" of procedures) immediately after the initial eruption, Hickel one day later ordered a resumption of drilling and production (even as the oil continued to gush into the channel), the government's response was seen as unbelievably consistent with conservationists' worst fears. That he backed down within 48 hours and ordered a halt to drilling and production was taken as a response to the massive nationwide media play then being given to the Santa Barbara plight and to the citizens' mass outcry just then beginning to reach Washington.

. . . Hickel's failure to support any of the legislation introduced to halt drilling was seen as an action favoring Oil. His remarks on the subject, while often expressing sympathy with Santa Barbarans (and for a while placating local sentiment), were revealed as hypocritical in light of the action not taken. Of further note was the constant attempt by the Interior Department to minimize the extent of damage in Santa Barbara or to hint at possible "compromises" which were seen locally as near-total capitulation to the oil companies.

Volume of Oil Spillage.   Many specific examples might be cited. An early (and continuing) issue in the oil spill was the volume of oil spilling into the Channel. The U.S. Geological Survey (administered by Interior), when queried by reporters, broke its silence on the subject with estimates which struck as incredible in Santa Barbara. One of the extraordinary attributes of the Santa Barbara locale is the presence of a technology establishment among the most sophisticated in the country. Several officials of the General Research Corporation (a local R & D firm with experience in marine technology) initiated studies of the oil outflow and announced findings of pollution volume at a "minimum" of ten-fold the Interior estimate. Further, General Research provided (and the *News-Press* published) a detailed account of the methods used in making the estimate (Allan, 1969). Despite repeated challenges from the press, Interior both refused to alter its estimate or to reveal its method for making estimates. Throughout the crisis, the divergence of the estimates remained at about ten fold.

The "seepage" was estimated by the Geological Survey to have been reduced from 1,260 gallons per day to about 630 gallons. General

Research, however, estimated the leakage at the rate of 8,400 gallons per day at the same point in time as Interior's 630 gallon estimate. The lowest estimate of all was provided by an official of the Western Oil and Gas Association, in a letter to the *Wall Street Journal.* His estimate: "Probably less than 100 gallons a day" (Anonymous [b], 1969: A-1).

Damage to Beaches. Still another point of contention was the state of the beaches at varying points in time. The oil companies, through various public relations officials, constantly minimized the actual amount of damage and maximized the effect of Union Oil's cleanup activity. What surprised (and most irritated) the locals was the fact that Interior statements implied the same goal. Thus Hickel referred at a press conference to the "recent" oil spill, providing the impression that the oil spill was over, at a time when freshly erupting oil was continuing to stain local beaches. President Nixon appeared locally to "inspect" the damage to beaches, and Interior arranged for him to land his helicopter on a city beach which had been cleaned thoroughly in the days just before, but spared him a close-up of much of the rest of the County shoreline which continued to be covered with a thick coat of crude oil. (The beach visited by Nixon has been oil stained on many occasions subsequent to the President's departure.) Secret servicemen kept the placards and shouts of several hundred demonstrators safely out of Presidential viewing or hearing distance . . .

Damage to Wildlife. Birds ingest oil on feathers; continuous preening thus leads to death. In what local and national authorities called a hopeless task, two bird-cleaning centers were established to cleanse feathers and otherwise administer to damaged wild-fowl. (Oil money helped to establish and supply these centers.) Both spokesmen from Oil and the federal government then adopted these centers as sources of "data" on the extent of damage to wild-fowl. Thus, the number of dead birds due to pollution was computed on the basis of number of fatalities at the wild-fowl center. This of course is preposterous given the fact that dying birds are provided with very inefficient means of propelling themselves to such designated places. The obviousness of this dramatic understatement of fatalities was never acknowledged by either Oil or Interior—although noted in Santa Barbara . . .

With these details under their belts, Santa Barbarans were in a position to understand the sweeping condemnation of the regulatory system as contained in a *News-Press* front page, banner headlined interview with Rep. Richard D. Ottenger (D-NY) quoted as follows: "And so on down the line. Each agency has a tendency to become the captive of the industry that it is to regulate" (Anonymous [c], 1969: A1) . . .

Science and Technology: Disillusionment

From the start, part of the shock of the oil spill was that such a thing could happen in a country with such sophisticated technology. The much-overworked phrase, "If we can send a man to the moon . . ." was even more overworked in Santa Barbara. When, in years previous, Santa Barbara's elected officials had attempted to halt the original sale of leases, "assurances" were given from Interior that such an "accident" could not occur, given the highly developed state of the art. Not only did it occur, but the original gusher of oil spewed forth completely out of control for ten days and the continuing "seepage" which followed it [remained] . . . seven months later. That the government would embark upon so massive a drilling program with such unsophisticated technologies was striking indeed.

Further, not only were the technologies inadequate and the plans for stopping a leak, should it occur, nonexistent, but the area in which the drilling took place was known to be ultra-hazardous from the outset. That is, drilling was occurring on an ocean bottom known for its extraordinary geological circumstances—porous sands lacking a bedrock "ceiling" capable of containing runaway oil and gas. Thus the continuing leakage through the sands at various points above the oil reservoir is unstoppable, and *could have* been anticipated with the data *known to all* parties involved . . .

This striking contrast between the sophistication of the means used to locate and extract oil compared to the primitiveness of the means to control and clean it up was widely noted in Santa Barbara. It is the result of a system that promotes research and development that leads to strategic profitability rather than to social utility. The common sight of men throwing straw on miles of beaches within sight of complex drilling rigs capable of exploiting resources thousands of feet below the ocean's surface made the point clear.

The futility of the clean-up and control efforts was widely noted in Santa Barbara. Secretary Hickel's announcement that the Interior Department was generating new "tough" regulations to control off-shore drilling was thus met with great skepticism. The Santa Barbara County Board of Supervisors was invited to "review" these new regulations—and refused to do so in the belief that such participation would be used to provide the fraudulent impression of democratic responsiveness—when, in fact, the relevant decisions had been already made. In previous years when they were fighting against the leasing of the Channel, the supervisors had been assured of technological safeguards; now, as the emergency continued, they could witness for themselves the dearth of any means for ending the leakage in the Channel. They had also heard testimony of a high-ranking Interior engineer who, when asked if such safeguards could positively prevent future spills, explained that "no prudent engineer would ever make such a claim" (Anonymous [d], 1969: A-1).

Science was also having its non-neutral consequences on the other battlefront being waged by Santa Barbarans. The chief Deputy Attorney General of California, in his April 7, 1969, speech to the blue-ribbon Channel City Club of Santa Barbara, complained that the oil industry is preventing oil drilling experts from aiding the Attorney General's office in its lawsuits over the Santa Barbara oil spill. (Anonymous [e], 1969)

Complaining that his office has been unable to get assistance from petroleum experts at California universities, the Deputy Attorney General further stated:

> The university experts all seem to be working on grants from the oil industry. There is an atmosphere of fear. The experts are afraid that if they assist us in our case on behalf of the people of California, they will lose their oil industry grants. (Anonymous [e], 1969)

At the Santa Barbara Campus of the University, there is little Oil money in evidence and few, if any, faculty members have entered into proprietary research arrangements with Oil. Petroleum geology and engineering is simply not a local specialty. Yet it is a fact that Oil interests did contact several Santa Barbara faculty members with offers of funds for studies of the ecological effects of the oil spill, with publication rights stipulated by Oil. It is also the case that the Federal Water Pollution Control Administration explicitly requested a UC Santa Barbara botanist to withhold the findings of his study, funded by that Agency, on the ecological consequences of the spill (Anonymous [f], 1969: A-3). Except for the Deputy Attorney General's complaint, none of these revelations received any publicity outside of Santa Barbara. But the Attorney's allegation became something of a statewide issue. A professor at the Berkeley campus, in his attempt to refute the allegation, actually confirmed it. Wilbur H. Somerton, Professor of petroleum engineering, indicated he could not testify against Oil:

> because my work depends on good relations with the petroleum industry. My interest is serving the petroleum industry. I view my obligation to the community as supplying it with well-trained petroleum engineers. We train the industry's engineers and they help us. (Anonymous [g], 1969)

Santa Barbara's leaders were incredulous about the whole affair. The question—one which is more often asked by the downtrodden sectors of the society—was asked: "Whose University is this, anyway?" A local executive and GOO leader asked, "If the truth isn't in the universities, where is it?" A conservative member of the State Legislature, in a move reminiscent of SDS demands, went so far as to ask an end to all faculty "moonlighting" for industry. In Santa Barbara, the only place where all of this publicity was occurring, there was thus an opportunity for insight into the linkages between knowledge, the University, Government and Oil and the resultant non-neutrality of science. The backgrounds of many members of the DuBridge

Panel were linked publicly to the oil industry. In a line of reasoning usually the handiwork of groups like SDS, a *News-Press* letter writer labeled Dr. DuBridge as a servant of Oil interests because, as a past President of Cal Tech, he would have had to defer to Oil in generating the massive funding which that institution requires. In fact, the relationship was quite direct. Not only has Union Oil been a contributor to Cal Tech, but Fred Hartley (Union's President) is a Cal Tech trustee. The impropriety of such a man as DuBridge serving as the key "scientist" in determining the Santa Barbara outcome seemed more and more obvious. . . .

References

Allen, Allan A. (1969, 20 May). *Santa Barbara oil spill.* Statement presented to the U.S. Senate.

Anonymous (a). (1969). *Santa Barbara News Press,* February 26.

Anonymous (b). (1969). *Santa Barbara News Press,* August 5.

Anonymous (c). (1969). *Santa Barbara News Press,* March 1.

Anonymous (d). (1969). *Santa Barbara News Press,* February 19.

Anonymous (e). (1969). *Santa Barbara News Press,* August 8.

Anonymous (f). (1969). *Santa Barbara News Press,* July 29.

Anonymous (g). (1969). *Santa Barbara News Press,* April 12.

Gans, Herbert. (1962). *The Urban Villagers.* New York: The Free Press of Glencoe.

Mintz, Morton. (1969). *The Washington Post,* June 29.

Vidich, Arthur, and Bensman, Joseph. (1958). *Small town in mass society.* Princeton, NJ: Princeton University Press.

Endnotes

1. From "Oil in Santa Barbara and Power in America" by Harvey Molotch (1970) in *Sociological Inquiry* 40: (1): 131–145. Reprinted with permission of the author and the University of Texas Press. All rights retained by the author and the University of Texas Press.

2. All efforts by Santa Barbara community organizations to halt offshore drilling failed because of the superior power of the oil companies. President George Bush, however, came to the rescue by imposing a moratorium on offshore drilling in 1990 through 2002. On January 12, 1998, President Clinton announced a 10-year extension of the moratorium on oil drilling off virtually all U.S. ocean coastlines. [Eds.]

CASE STUDY 8

Love Canal Toxic Waste Contamination[1]

Lewis G. Regenstein

Introductory Note by W.M.E. and M.M: *During an 11 year period, from 1942 to 1953, more than 21,000 tons of toxic waste were buried in a 20- to 25-foot deep cement trough (the abandoned "Love Canal"), which was situated in a neighborhood in Niagara Falls, New York. Epidemiological studies revealed cases of cancer, miscarriages, birth defects, and other illnesses among the local inhabitants. Corporate malfeasance and the inability of community leaders, including the municipal school board, to protect the local citizens were the causes of the toxic contamination.*

Community activists were outraged when a New York State court failed to find Occidental Chemical Corporation (OCC) guilty of criminal charges. In 1994, after 14 years of litigation, OCC agreed to reimburse the state $98 million for its cleanup expenses and assume responsibility for the future maintenance of the site. Love Canal became a synonym for environmental disaster and toxic contamination, leading Congress to pass into law on December 12, 1980, the Comprehensive Environmental Response, Compensation, and Liability Act (CERCLA), the so-called Superfund, a multi-billion dollar program established to clean up abandoned toxic waste dumps. Since the creation of the Superfund, the Environmental Protection Agency has put more than 900 sites on its list of the nation's most dangerous dumps.

A technological disaster of the Second Industrial Revolution, Love Canal was the result of socio-cultural factors.

Love Canal is probably the best known, and most infamous, of the nation's toxic waste sites. The name has come to symbolize chemical contamination of areas and neighborhoods.

Between 1942 and 1953, the Olin Corporation and the Hooker Chemical Corporation buried over twenty thousand tons of deadly chemical wastes, including dioxin, in Love Canal, located near Niagara Falls, New York. Many of the compounds dumped there are known to cause cancer, miscarriages, birth defects, and other illnesses and disorders.

In 1953, Hooker Chemical donated the land to the local board of education for a token payment of one dollar, but did not clearly warn of the dangerous nature of the chemicals buried there, even when a school, homes, and playgrounds began to be built in the area.

In 1976, after years of unusually heavy rains raised the water table and basements became flooded, problems became apparent in the neighborhood. Homes began to reek of chemicals; children and pets came home with chemical burns on their feet and hands, and some pets even died, as did trees, flowers, and vegetables. Soon, people in the neighborhood began to experience an extraordinarily high number of serious and unexplainable illnesses, including higher than normal rates of cancer, miscarriages, and deformities in newborns. In the spring of 1978, alarmed by the situation and frustrated by the lack of action on the part of local, state, and federal authorities, a twenty-seven-year-old housewife named Lois Gibbs took matters into her own hands. She organized her neighbors into the Love Canal Homeowners Association and began a two-and-a-half-year-long fight to have the government relocate them to another area. In the process, she turned Love Canal into a household name across the country and helped focus nationwide attention on the growing problem of toxic waste disposal, which culminated in the passage of the so-called Superfund Law in December of 1980.

As the number of people experiencing health problems in the neighborhood rose, it became increasingly apparent that the community might not be safe for human habitation. Under pressure from neighborhood activists and as a result of extensive coverage in the local and national media, government officials finally began to recognize the need for action. In August 1978, the New York State health commissioner, Dr. Robert P. Whalen, recommended that pregnant women and children under the age of two be evacuated, saying that there was "growing evidence . . . of subacute and chronic health hazards, as well as spontaneous abortions and congenital malformations" at the site. When the state tested the air, water, soil, and homes for toxic chemicals later that month, over eighty different compounds were found, many of which were thought to be capable of causing cancer. Chemical pollution of the air was measured at two-hundred-and-fifty to five thousand times the levels considered safe.

The 1978 study found an unusually high miscarriage rate of 29.4 percent in the neighborhood, with five of the twenty-four children born in the area listed as having birth defects. State health officials estimated that women in the area had a 50 percent higher-than-normal risk of miscarriage. Another report found that in 1979, only two of seventeen pregnant women in Love Canal gave birth to normal children. Four had miscarriages, two had stillbirths, and nine had babies born with defects.

Epidemiological studies of the affected population revealed an alarming pattern of illness among exposed residents. For example, on Ninety-sixth Street, where fifteen homes were located, eight people developed cancer in the twelve-year period between 1968 and 1980; six women had cancerous breasts removed; one man contracted bladder cancer; and another developed throat cancer. In addition, a seven-year-old boy experienced convulsions and died of kidney failure, and a pet dog had to be destroyed after developing cancerous tumors.

Some of the most alarming health data were gathered by Dr. Beverly Paigen of the Roswell Cancer Institute, who found a much higher incidence of illness among people who lived in houses that were located above moist ground or wet areas—those most prone to contamination by rising ground water. On March 21, 1979, testifying before the House Subcommittee on Oversight and Investigations, she described the tragic history of several families who lived in one such house located directly over liquid wastes that were seeping out of the ground. Among the four families, there existed three cases of nervous breakdown, three hysterectomies due to uterine bleeding, cancer, or both, and numerous cases of epilepsy, asthma, and bronchitis.

Dr. Paigen also found a significant excess of childhood disorders among youngsters born to residents of wet areas, including nine instances of birth defects among the sixteen children born in such areas between 1974 and 1978. She determined that the overall incidence of birth defects was 20 percent; the miscarriage rate was estimated at 25 percent, compared with just 8.5 percent for women moving into the area. She also reported that eleven out of thirteen hyperactive children lived in wet areas, and that 380 percent more asthma occurred there than in the "dry" areas of Love Canal. The incidence of urinary disease and convulsive disorders was almost triple that of dry areas, and the rates of suicides and nervous breakdowns almost quadruple. Significantly, Dr. Paigen found that Love Canal residents suffering from illnesses ranging from migraines to pneumonia to severe depression reported marked health improvements when they moved out of the area and away from the contamination. Eventually, New York State authorities termed the area "a grave and imminent peril" to the health of those living nearby. Several hundred families were moved out of the neighborhood, and the others were advised to leave; the school was closed and a barbed wire fence placed around it.

In May 1980, further testing revealed high levels of genetic damage among neighborhood residents, resulting in an additional seven hundred and ten families being evacuated at a cost estimated to run between $3 million and $60 million. In October 1980, President Jimmy Carter declared the neighborhood a disaster area. In the end, some sixty families decided to remain in their homes and reject the government's offer to buy them out. The total cost for the cleanup has been estimated at $250 million. Twelve years after the neighborhood was abandoned, the state of New York approved plans to allow families to move back into some parts of the area, and homes were permitted to be sold.

The major impact of the massive publicity generated by Love Canal was to draw national attention to the dangers of toxic chemicals and hazardous waste, and to pressure Congress and the White House to pass laws to address the problems caused by such pollution. Public outrage was particularly fueled by revelations that Hooker had engaged in a cover-up of the dangers of the situation at Love Canal.

When Hooker deeded the land to the Niagara Falls Board of Education for one dollar, the company did not warn the board about the lethal nature of the chemicals it had buried there. But Hooker did take pains to protect itself: the legal document transferring the property disclaims liability for any deaths or injuries that might occur on the land and specifies that the school board assume responsibility for any claims that result from exposure to the buried chemicals. In its letter agreeing to donate the land to the new school, Hooker's executive vice president adopted the tone of the good corporate citizen, stressing the company's desire to serve the needs of its community.

Even when a neighborhood began to be built in the area, Hooker failed to issue warnings about the dangers there. On June 18, 1958, a company memo observed that "the entire area is being used as a playground," and that "3 or 4 children had been burned by material at the old Love Canal property." But concerned about legal repercussions, Hooker remained silent. Ten years later, road workers building a highway near Love Canal uncovered leaking drums full of toxic chemicals that Hooker analyzed and discussed in a March 21, 1968, internal company memo. Again, no action was taken to alert the public about the presence of these dangerous chemicals.

Indeed, the chemical industry continued to downplay the threat to public health documented at Love Canal long after evidence was uncovered. For example, on October 14, 1979, the late Armand Hammer, then chairman of Hooker's parent company, Occidental Petroleum, said on NBC's "Meet the Press" that the Love Canal problem had "been blown up out of context." And when Hammer chaired Occidental's annual stockholder's meeting in May 1980, the company rejected a resolution calling on the firm to adopt policies designed to prevent future Love Canals. Even the vice president of the Chemical Manufacturers Association, Geraldine Cox, suggested during a July 2, 1980, television interview on the "MacNeil/Lehrer News Hour" that Love Canal residents were comparable to hypochondriacs with imaginary or exaggerated illnesses.

By 1980, shocking information was being released concerning thousands of other toxic waste dumps scattered across the country that were potential Love Canals. The U.S. Environmental Protection Agency (EPA) estimated that there existed thirty-two thousand to fifty thousand hazardous waste disposal sites in the United States, of which twelve hundred to two thousand might pose "significant risks to human health or the environment." And new chemical waste sites were being discovered at a rate of two hundred a month. At the time, the EPA estimated that only about 10 percent of the one hundred and fifty million tons of hazardous wastes being generated each year were disposed of in a safe and legal manner; 90 percent were being dumped illegally or disposed of in a way that posed a potential threat to humans or the environment. The agency called the situation "the most serious environmental problem in the U.S. today." (More recent estimates put the amount of hazardous waste being produced each year at 300 million tons—roughly a ton for every man, woman, and child in the country.) On May 16, 1979, Assistant Attorney General James Moorman testified before the House Subcommittee on Oversight and Investigations that toxic waste dumping was the "first or second most serious environmental problem in the country." He pointed to a lack of effective environmental regulations and a lack of enforcement of existing anti-dumping laws.

Since Love Canal, and partly in response to the uproar it caused, laws designed to protect the public from toxic chemicals have been passed or strengthened. In November 1980, a provision of the Resource Conservation and Recovery Act (RCRA) went into effect, theoretically requiring that toxic wastes be tracked "from cradle to grave." And on December 11, 1980, President Carter signed into law a diluted version of the Environmental Emergency Response Act—the so-called Superfund Law—that creates funds to pay for the cleanup of hazardous waste sites and makes owners and operators of waste disposal sites, as well as producers and transporters of hazardous materials, liable for cleanup costs. But while these laws have greatly reduced the improper disposal of dangerous chemicals, weak government enforcement has significantly weakened their effectiveness. For example, as of the beginning of 1993, only 149 of 1,256 priority Superfund sites had been cleaned up. . . .

It may well prove impossible to completely clean up the nation's dump sites at any price, but the anticipated costs for steps that must be taken in the next few years are staggering. One EPA study estimated that it would cost over $44 billion just to clean up the most dangerous sites, with the public having to pick up half the tab. Other projections put the figure for addressing Superfund waste sites at over a trillion dollars—which would require half a century to accomplish.

Endnote

1. From "Love Canal" by Lewis G. Regenstein. Taken from *When Technology Fails*. Edited by Neil Schlager. Copyright © 1994 Gale Research. Reprinted by permission of The Gale Group.

CASE STUDY 9

Apollo I Fire[1]

Leonard Bruno

Introductory Note by W.M.E. and M.M: *On January 27, 1967, three astronauts strapped into the command module died while they were rehearsing a countdown during the United States' preparations to fly to the moon. The flames explosively engulfed the pure oxygen atmosphere of the interior of the capsule. It was seven hours before the capsule cooled enough to remove the bodies. The cause of the failure was an in-capsule fire due to faulty wiring design and installation. Apollo's board of inquiry, a panel of 21 experts charged with investigating specific systems, subsystems, and events that could have a potential relationship to the fire, issued a report disapproving of engineering design and testing, both in the National Aeronautics and Space Administration (NASA) and North American Aviation, Inc., the major contractor for the Apollo I spacecraft. Among other things, the board found poor workmanship by North American, inferior quality control by NASA of the work done, and lack of adequate escape mechanisms for the astronauts in case of in-capsule failure.*

This disaster led to numerous design changes, manufacturing requirements, and safety standards on the Apollo spacecraft, especially in regard to wiring, use of combustible materials in the capsule, and redesign of the escape hatch.

This technological disaster of the Third Industrial Revolution was caused by technical design factors.

On January 27, 1967, at the NASA (National Aeronautics and Space Administration) base in Cape Canaveral, three astronauts were killed during a routine ground test of the Apollo command module. Although the exact cause of the fire remained undetermined, it probably began with an electrical arc caused by poor wiring design and installation. Once some combustible materials ignited, the fire was fed by the pure oxygen under pressure in the module. The astronauts had no equipment to suppress the fire, and were unable to open the six-bolt escape hatch in the short time that elapsed before the fire blazed.

With the successful completion of NASA's two-man Gemini program, AS-204 was to be the first manned Apollo mission. The United States was in its race with the Soviet Union to land men on the Moon, and it was with the three-man Apollo capsules that the Americans planned to accomplish this feat.

In August 1966, the AS-204 crew was named. It consisted of two space veterans, Virgil "Gus" Grissom and Edward H. White, and one new astronaut Roger B. Chaffee who had not yet flown in space. NASA planned for the crew to make a December 1966 "shakedown" flight of up to two weeks in Earth's orbit.

This flight would test the Apollo's new command and service modules (CSM) built by North American Aviation, which had not been deployed as a manned mission. Following a series of mechanical problems, however, the flight date was postponed until February 1967.

In preparation for this flight, the astronauts were to complete a four-phase ground test of the CSM's systems. One phase of this process required that the astronauts enter the command module, which was perched on a Saturn 1B launch vehicle and emptied of fuel; seal and pressurize the module; and then begin a "plugs out" test to see if the spacecraft could run on its internal power system.

The command module was designed to function in space, and therefore to be subject to greater internal than external pressure. To simulate this condition as closely as possible, NASA's engineers planned to pressurize the cabin with pure oxygen at approximately 16 pounds per square inch, rather than the normal 5 pounds per square inch of air. Although pure oxygen itself will not ignite, it can, when pressurized, rapidly feed an existing fire. No one, however, considered a cabin fire likely. There was no fire extinguisher in the capsule, and, since the launch vehicle was not fueled, the fire crews were on standby rather than maximum alert. Furthermore, no one considered excessive the 90 seconds it took to open the new six-bolt escape hatch; most experts were convinced that the alternate quick-release hatches were accident-prone and dangerous.

NASA's apparent complacency about safety precautions stemmed from the fact that, in six years of manned space flights, no astronaut had died in the course of a mission. The "plugs-out" test, conducted with an unfueled launch vehicle, seemed both routine and low-risk.

On January 27, 1967, "Gus" Grissom, Edward White, and Roger Chaffee undertook the "plugs-out" test on the new Apollo command and service modules. By 1:00 p.m. on that Friday afternoon, the three-man crew had crawled through the open hatch and assumed their flight positions. After two hours of tests conducted with the hatch open, the capsule was sealed and the cabin pressurized to 16.2 pounds per square inch of pure oxygen. The crew then went through a practice countdown and ran simulation tests for more than three hours, regularly interrupted by minor problems.

Shortly after 6:00 p.m., fifteen minutes before lift-off was to be simulated, the spacecraft switched to internal power. NASA engineers called a hold to check on some problem, and the astronauts waited yet again. The astronauts had been in their couches some five and one-half hours; it had been a long day. Suddenly, at 6:31 p.m., telemetry from the spacecraft indicated that a major short had occurred somewhere in the nearly twelve miles of electrical wiring packed into the command module. Less than ten seconds later, Roger Chaffee made an almost casual report: "Fire, I smell fire."

The spacecraft was not equipped with internal cameras, but a camera was focused on its porthole. At the first report of fire, all the camera operator could see was a sudden bright glow. The operator then saw flames flickering across the porthole and Edward White's hands reaching above his head to get at the bolts securing the hatch. He saw a lot of movement, and then another pair of arms struggling with the hatch. Soon, dark smoke completely obscured the scene. The last sound from the astronauts was a now frantic cry from Roger Chaffee: "We've got a bad fire—let's get out . . . we're burning up!" Seconds after this last transmission, the tremendous pressure inside the cabin split the capsule open and a blaze of flame gushed out. Only 18 seconds had passed from the first call of fire to this explosion.

Help was close at hand, but the control personnel in the White Room were momentarily held back by the explosion. Due to the thick smoke, it then took five men working in shifts five and a half minutes to remove three separate hatches: the boost protective cover that shielded the command module during launch, the ablative hatch, and the inner hatch. Inside the capsule, once the smoke thinned, they found the three dead men. Chaffee was still strapped onto his couch, and the bodies of White and Grissom were lying close together below the hatch. White's handprint was outlined in ash on the hatch.

Official autopsies would later identify the cause of the deaths as asphyxiation, observing that although each astronaut had suffered serious burns, these were not fatal. However, the heat during those few terrible seconds had been so intense—the holes burned in aluminum tubing indicate temperatures of at least 760 degrees Celsius—that the astronaut's suits had melted and fused with the molten nylon and Velcro inside the capsule, forming a synthetic liquid that solidified as it cooled. Doctors arrived fourteen minutes after the first alarm of fire, but it took them seven hours to remove the bodies; those of White and Grissom had been welded to the capsule floor.

NASA's entire multi-billion dollar effort to put a man on the Moon virtually came to a standstill as a special Board of Inquiry sought explanations for the accident. On February 22, 1967, almost a month after the disaster, a seven-man review board issued an interim report stating that although no definite cause of the fire could be established, the most likely origin was an electrical malfunction. A fourteen-volume report of some three thousand pages came out in early April; this report specified that the fire was probably caused by an electrical arc that occurred in the vicinity of the environmental control equipment under Grissom's couch.

The full report was highly critical of the conditions that enabled the fatal accident. It indicated that some of the wiring unaffected by the fire revealed "numerous examples of poor installation, design, and workmanship." Moreover, the capsule was loaded with highly combustible materials such as Velcro and the nylon netting used to prevent loose objects from floating around in the zero gravity of space. In its conclusion, the report implied that the disaster may have been inevitable given the substandard manufacturing procedures and the lack of safety measures.

The critique of NASA at least implicit in this report was not lost on Congress, and a House space subcommittee opened hearings on April 7, 1967, to assess the space program and the Apollo accident. Most observers decided that NASA had been pushing the Apollo program too hard and too fast; there simply had not been enough time to thoroughly test all its systems. The Apollo spacecraft was still an unproven and evolving craft, having undergone 623 changes between August 1966, when it was delivered by North American Aviation to NASA, and the fatal accident in January 1967. By the end of the Congressional hearings, the prevailing view was that making limited adjustments to the Apollo capsule and allowing the space program to continue on its timetable would be ill-advised. An exhaustive review of the entire spacecraft was necessary, however long it took.

NASA was therefore obligated to step back and reassess the full scope of its systems and procedures, particularly in terms of safety standards. NASA took good advantage of this enforced break from its rivalry with the Soviet space program, and the outcome of a year and a half's worth of reevaluation was a completely redesigned Apollo command module. Some fifteen hundred modifications were made to the command module, resulting in a considerably more secure and fireproof vehicle.

First, NASA installed high-quality wiring. Flameproof coatings were applied over all wire connections, plastic switches were replaced by metal ones, and soldering became more meticulous. Almost all flammable materials inside the module were removed. A new, fire-resistant material known as Beta cloth was developed for spacesuits. Instead of igniting at 500 degrees Celsius, as did the old suits made of Nomex, the Beta cloth suits could withstand temperatures of more than 800 degrees Celsius.

A considerable debate over the use of pure oxygen resulted in a compromise in the favor of safety. When they were in space, the crew would breathe pure oxygen at five pounds per square-inch. For ground testing and launch however, the cabin would be filled with a mixture of oxygen and nitrogen at sea-level pressure. To prevent this nitrogen from causing "bends" (nitrogen narcosis or poisoning), the astronauts would breathe only through their spacesuits, which contained pure oxygen, until the cabin had been purged of nitrogen.

One irony of the Apollo accident was that Grissom, White, and Chaffee's workday was scheduled to end with a test of the new six-bolt escape hatch. This system had replaced the quick-release, explosively charged hatch that had been used on both Mercury and Gemini spacecraft, but which in 1961 had been blamed for prematurely blowing and almost sinking Grissom and his *Mercury Liberty Bell 7* when he landed in the Atlantic. The new Apollo design returned to a quick-escape system; this improved hatch took only twelve seconds to release and opened outward so that internal pressure would not affect its functioning. Among the revelations of the full report on the Apollo accident was that even if the astronauts had managed to undo the six-bolt hatch, their efforts would have been futile. When the internal pressure exceeded the external pressure by more than 0.25 pounds per square inch, the hatch, which swung inward, became impossible to open.

If good can be said to come of such a disaster, it was that the American government and industry coalition became conscious of the need to raise design, workmanship, and safety standards. The fire both enabled and forced NASA to step back from its politically-controlled timetables and methodically build the high-performance spacecraft that ultimately proved capable of putting a man on the Moon.

Endnote

1. From "Apollo I" by Leonard Bruno. Taken from *When Technology Fails*. Edited by Neil Schlager. Copyright © 1994 Gale Research. Reprinted by permission of The Gale Group.

CASE STUDY 10

Three Mile Island[1]
Charles Perrow

Introductory Note by W.M.E. and M.M.: *The worst disaster in the history of American commercial nuclear power occurred on March 28, 1979, in Middletown, Pennsylvania. The disaster caused a partial meltdown of the reactor core. Experts debate the actual number of curies of radiation released into the environment, but the 2,200 lawsuits filed by area residents attest to the concern about the actual human exposure to harmful ionizing radiation. The financial toll was enormous. The cleanup alone cost approximately $975 million. The total lifetime cost will be close to $2 billion without ever producing any significant amount of electricity. Causes of the malfunction are many, but numerous operator mistakes contributed significantly to the failure.*

The disaster left an indelible mark on public consciousness and on the finances of the nuclear industry as well. It left the industry's public image in tatters and prompted a number of costly new regulations that further hindered the growth of the nuclear power industry. In fact, no new nuclear power plants have been ordered constructed since 1978. The U.S. Department of Energy has recently conceded that the accident at Three Mile Island was far more serious than the government or the utilities managing the plant originally reported. In the years following the partial meltdown, more than 2,000 local residents filed personal injury lawsuits against General Public Utilities, owner of the Three Mile Island plant during the events, for radiation-related injuries. The company has subsequently settled a number of claims.

This technological disaster of the Third Industrial Revolution is primarily attributed to human factors.

. . . The accident started in the cooling system. There are two cooling systems. The primary cooling system contains water under high pressure and at high temperature that circulates through the core where the nuclear reaction is taking place. This water goes into a steam generator, where it bathes small tubes circulating water in a quite separate system, the secondary cooling system, and heats this water in the secondary system. This transfer of heat from the primary to the secondary system keeps the core from overheating, and uses the heat to make steam. Water in the secondary system is also under high pressure until it is called upon to turn into steam, which drives the turbines that generate the electric power. The accident started in the secondary cooling system.

The water in the secondary system is not radioactive (as is the water in the primary system), but it must be very pure because its steam drives the finely precisioned turbine blades. Resins get into the water and have to be removed by the condensate polisher system, which removes particles that are precipitated out.

The polisher is a balky system, and it had failed three times in the few months the new unit had been in operation. After about eleven hours of work on the system, at 4:00 a.m. on March 28, 1979, the turbine tripped (stopped). Though the operators did not know why at the time, it is believed that some water leaked out of the polisher system—perhaps a cupful—through a leaky seal.

Seals are always in danger of leaking, but normally it is not a problem. In this case, however, the moisture got into the instrument air system of the plant. This is a pneumatic system that drives some of the instruments. The moisture interrupted the air pressure applied to two valves on two feedwater pumps. This interruption "told" the pumps that something was amiss (though it wasn't) and that they should stop. They did. Without the pumps, the cold water was no longer flowing into the steam generator, where the heat of the primary system could be transferred to the cool water in the secondary system. When this flow is interrupted, the turbine shuts down, automatically, by an automatic safety device, or ASD.

But stopping the turbine is not enough to render the plant safe. Somehow, the heat in the core, which makes the primary cooling system water so hot, has to be removed. If you take a whistling teakettle off the stove and plug its opening, the heat in the metal and water will continue to produce steam, and if it cannot get out, it may explode. Therefore, the emergency feedwater pumps came on (they are at H in Figure 8–5; the regular feedwater pumps which stopped are above them in the figure).

They are designed to pull water from an emergency storage tank and run it through the secondary cooling system, compensating for the water in that system that will boil off now that it is not circulating. (It is like pouring cold water over your plugged tea kettle.) However, these two pipes were unfortunately blocked; a valve in each pipe had been accidentally left in a closed position after maintenance two days before. The pumps came on and the operator verified that they did, but he did not know that they were pumping water into a closed pipe.

The President's Commission on the Accident at Three Mile Island (the Kemeny Commission) spent a lot of time trying to find out just who was responsible for leaving the valves closed, but they were unsuccessful. Three operators testified that it was a mystery to them how the valve had gotten closed, because they distinctly remembered opening them after the testing. You probably have had the same problem with closing the freezer door or locking the front door; you are sure you did, because you have done it many times.

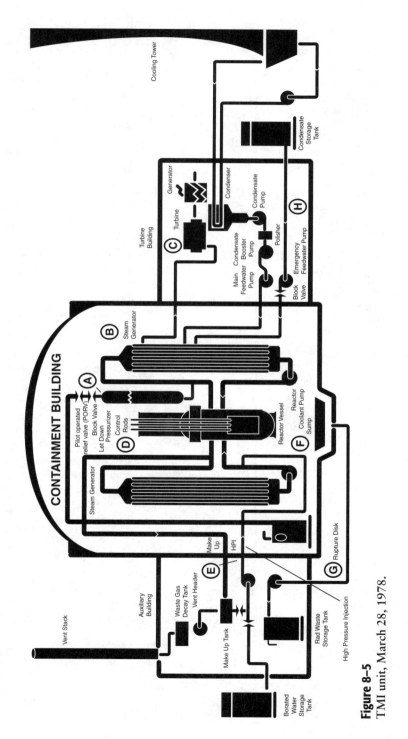

Figure 8-5
TMI unit, March 28, 1978.

Failure #1	{ Clogged condensate polisher line Moisture in instrument air line False signal to turbine	ASD	Reactor coolant pumps come on Primary coolant pressure down, temperature up Steam voids in coolant pipes and core, restricting flow forced by coolant pumps, creating uneven pressures in system
ASD*	Turbine stops		
ASD	Feedwater pumps stop		
ASD	Emergency feedwater pumps start		
Failure #2	Flow blocked; valves closed instead of open	ASD	High Pressure Injection (HPI) starts, to reduce temperature
	No heat removal from primary coolant		Pressurizer fills with coolant as it seeks outlet through PORV
	Rise in core temperature and pressure	"Operator error"	Operators reduce HPI to save pressurizer, per procedures
ASD	Reactor scrams		Temperature and pressure in core continue to rise because of lack of heat removal, decay heat generation, steam voids, hydrogen generation from the zirconium-water reaction, and uncovering of core
	Reactor continues to heat, "decay heat"		
	Pressure and temperature rise		
ASD	Pilot Operated Relief Valve (PORV) opens		
ASD	PORV told to close		
Failure #3	PORV sticks open		Reactor coolant pumps cavitate and must be shut off, further restricting circulation
Failure #4	PORV position indicator signifies it has shut		

*ASD (automatic safety device)
Source: Kemeny, John, et al. *Report of the President's Commission on the Accident at Three Mile Island*. Washington D.C.: Government Printing Office, 1979.

Figure 8–5 (continued)

Operators testified at the Commission's hearings that with hundreds of valves being opened or closed in a nuclear plant, it is not unusual to find some in the wrong position—even when locks are put on them and a "lock sheet" is maintained so the operators can make an entry every time a special valve is opened or closed. . . . The Three Mile Island (TMI) operators finally had to concede reluctantly that large valves do not close by themselves, so someone must have goofed.

There were two indicators on TMI's gigantic control panel that showed that the valves were closed instead of open. One was obscured by a repair tag hanging on the switch above it. But at this point the operators were unaware of any problem with emergency feedwater and had no occasion to make sure those valves, which are always open except during tests, were indeed open. Eight minutes later, when they were baffled by the performance of the plant, they discovered it. By then much of the initial damage had been done. Apparently our knowledge of these plants is quite incomplete, for while some experts thought the closed valves constituted an important operator error, other experts held that it did not make much difference whether the valves were closed or not, since the supply of emergency feedwater was limited and worse problems were happening anyway.

With no circulation of coolant in the secondary system, a number of complications were bound to occur. The steam generator boiled dry. Since no heat was being removed from the core, the reactor "scrammed." In a scram the graphite control rods, 80 percent silver, drop into the core and absorb the neutrons, stopping the chain reaction. (In the first experiments with chain reactions, the procedure was the same—"drop the rods and scram;" thus the graphic term scram for stopping the chain reaction.) But that isn't enough.

The decaying radioactive materials still produce some heat, enough to generate electricity for 18,000 homes. The "decay heat" in this 40-foot-high stainless steel vessel, taller than a three-story building, builds up enormous temperature and pressure. Normally there are thousands of gallons of water in the primary and secondary cooling systems to draw off the intense heat of the reactor core. In a few days this cooling system should cool down the core. But the cooling system was not working.

There are, of course, ASDs to handle the problem. The first ASD is the pilot-operated relief valve (PORV), which will relieve the pressure in the core by channeling the water from the core through a big vessel called a pressurizer, and out the top of it into a drain pipe (called the "hot leg"), and down into a sump. It is radioactive water and is very hot, so the valve is a nuisance. Also, it should only be open long enough to relieve the pressure; if too much water comes through it, the pressure will drop so much that the water can flash into steam, creating bubbles of steam, called steam voids, in the core and the primary cooling pipes. These bubbles will restrict the flow of coolant, and allow certain spots to get much hotter than others, in particular, spots by the uranium rods, allowing them to start fissioning again. The PORV is also known by its Dresser Industries' trade name of "electromatic relief valve" . . . It is expected to fail once in every fifty usages, but on the other hand, it is seldom needed.

The President's Commission turned up at least eleven instances of it failing in other nuclear plants (to the surprise of the Nuclear Regulatory Commission and the builder of the reactor, Babcock and Wilcox, who only knew of four) and there had been two earlier failures in the short life of TMI-Unit 2. Unfortunately, it just so happened that this time, with the block valves closed and one indicator hidden, and with the condensate pumps out of order, the PORV failed to reseat, or close, after the core had relieved itself sufficiently of pressure.

This meant that the reactor core, where the heat was building up because the coolant was not moving, had a sizeable hole in it—the stuck open relief valve. The coolant in the core, the primary coolant system, was under high pressure, and was ejecting out through the stuck valve into a long curved pipe, the "hot leg," which went down to a drain tank. Thirty-two thousand gallons, one third of the capacity of the core, would eventually stream out. This was no small pipe break someplace as the operators originally thought; the thing was simply uncorked, relieving itself when it shouldn't.

Since there had been problems with this relief valve before (and it is a difficult engineering job to make a highly reliable valve under the conditions in which it must operate), an indicator had recently been added to the valve to warn operators if it did not reseat. The watchword is "safety" in nuclear plants. But, since nothing is perfect, it just so happened that this time the indicator itself failed, probably because of a faulty solenoid, a kind of electromagnetic toggle switch. Actually, it wasn't much of an indicator, and the utility and supplier would have been better off to have had none at all. Safety systems, such as warning lights, are necessary, but they have the potential for deception. If there had been no light assuring them the valve had closed, the operators would have taken other steps to check the status of the valve, as operators did in a similar accident at another plant a year and a half before. But if you can't believe the lights on your control panel, an army of operators would be necessary to check every part of the system that might be relevant. And one of the lessons of complex systems and TMI is that any part of the system might be interacting with other parts in unanticipated ways. The indicator sent a signal to the control board that the valve had received the impulse to shut down. (It was not an indication that the valve had actually shut down; that would be much harder to provide.) So the operators noted that all was fine with the PORV, and waited for reactor pressure to rise again, since it had dropped quickly when the valve opened for a second. The cork stayed off the vessel for two hours and twenty minutes before a new shift supervisor, taking a fresh look at the problems, discovered it.

We are now, incredibly enough, only thirteen seconds into the "transient," as engineers call it. (It is not a perversely optimistic term meaning something quite temporary or transient, but rather it means a rapid change in some parameter, in this case, temperature.) In these few seconds there was a false signal causing the condensate pumps to fail, two valves for emergency cooling out of position and the indicator obscured, a PORV that failed to reseat, and a failed indicator of its position. *The operators could have been aware of none of these.*

Moreover, while all these parts are highly interdependent, so that one affects the other, they are not in direct operational sequence. Direct operational sequence is a sequence of stages as in a production line, or an engineered safety sequence. The operator knows that a block in the condensate line will cause the condensate pump to trip, which will stop water from going to the steam generator and then going to the turbine as steam to drive it, so the turbine will shut down because it will have no source of power to turn it. This is quite comprehensible. But connected to this sequence, although not a part of its production role, is another system, the primary cooling system, which regulates the amount of water in the core. The water level in the core was judged to have fallen, which it had, because of the drop in the pressure and temperature in the primary cooling system. But for the operators there was no obvious connection between this drop and a turbine "trip" (shutdown). Unknown to them, there was an intimate connection because of the interactive complexity of the system. The connection is through the PORV, but that also has no production sequence or safety sequence connection to the trip of the turbines, or to the failure of the condensate polisher system, even had the operators been able to ascertain that this was the cause of the turbine trip. The PORV is expected to operate on the basis of core pressure, regardless of the functioning of the turbine, the secondary cooling system (feedwater to the steam generators and turbine), or the emergency core cooling pumps.

Even if there is a part of the system that is in direct operational sequence, an information failure in any part of that sequence can render the connection opaque, if not invisible. For example, the PORV is connected in a direct sequence to a drain pipe, then to a drain tank, and when that overflows, to a sump. A couple of readings of excessive radioactive water will appear along the way. But for the operators, this was water from an "unknown origin," since they were assured, by the signal light, that the PORV was closed. Since they assumed a pipe break somewhere and since the piping system in the plant is so complex that a member of the Presidential Commission had to use a magnifying glass to try to follow it on the drawings, there was reason to believe that the water could have come from any number of places. Indeed, later in the accident they found that radioactive water was not traveling to the tank they intended, but because of complex flow and pressure interactions, was going to a different, wrong tank, which also overflowed, this time in the auxiliary building.

Here we have the essence of the normal accident: the interaction of multiple failures that are not in a direct operational sequence. You could underline this definition, but there is one other ingredient we have not explored in detail—incomprehensibility . . . most normal accidents have a significant degree of incomprehensibility . . .

The Kemeny Commission thought the operators should have known, and berated them in its report—they were "oblivious" to the danger; two readings "should have clearly alerted" them to the LOCA (Loss of Coolant Accident); "the major cause of the accident was due to inappropriate actions by those who were operating the plant," they said in their final report (Kemeny et al., 1979: 2). Babcock and Wilcox agreed; this was the sole cause of the accident, they argued in a press conference. The British Secretary of State for Energy was less diplomatic—the accident was caused by "stupid errors," he said.

Actually, there were three readings that should have indicated a LOCA to the operator, and it is a lesson in the fate of warnings to examine them. First, we should note that a LOCA is the most feared of the probable accidents in a plant, for it means the core can melt, and in what are called worst-case analyses could cause a steam explosion and rupture the vessel, spewing radioactivity. Even without a steam explosion, the extreme heat of open fissioning could breach containment. LOCA will occur when the water level drops below the level of the fuel rods and they overheat. But there is no direct measure of water level in the core in the Babcock and Wilcox reactors. One could be put on, said a Babcock and Wilcox official during a press conference, but it would be hard to provide and would create other complications. One hesitates to penetrate the core more than needed, and it would be hard to measure surging water under high pressure, about to flash into steam. So, let's examine the indirect measures.

One device measured drain tank pressures. But, it is not considered a particularly vital indicator by the designers, and is located on the backside of a 7-foot high control panel, near the bottom. Not suspecting they were in a LOCA, no one bothered to examine it (though the record is vague on this question). Another indicator showed the temperature of the drain tank. With hundreds of gallons of hot coolant spewing out and going to the drain tank, that temperature reading should be way up. It was indeed up. But they had been having trouble with a leaky PORV for some weeks, meaning that there was always some coolant going through it, so it was usual for it to be higher than normal. It did shoot up at one point, they noted, but that was shortly after the PORV opened, and when it didn't come down fast that was comprehensible, because the pipe heats up and stays hot. "That hot?" a commissioner interrogating an operator asked, in effect. The operator replied, in effect, "Yes; if it were a LOCA I would expect it to be much higher." It was not the LOCA they were trained for on the simulators that are used for training sessions, since it had some coolant coming in through an emergency system, and some coming in through High-Pressure Injection (HPI), which was only throttled back, not stopped. Their training never imagined a multiple accident with a stuck PORV and blocked valves. Well, what about the drop in pressure in the core itself? Surely this would indicate that the coolant was getting out somehow. But the operators discounted that indicator as erroneous or simply mysterious because it contradicted the one next to it, the pressurizer indicator, which was rising. A supervisor testified:

> I think we knew we were experiencing something different, but I think each time we made a decision it was based on something we knew about. For instance: pressure was low, but they had opened the feed valves quickly in the steam generator, and they thought that might have been "stuck." There was logic at that time for most of the actions, even though today you can look back and say, well, that wasn't the cause of that, or, that shouldn't have been that long. (President's Commission, 1979, 57)

... Besides, about this time—just four or five minutes into the accident—another more pressing problem arose. The reactor coolant pumps that had turned on started thumping and shaking. They could be heard and felt from far away in the control room. Would they withstand the violence they were exposed to? Or should they be shut off? A hasty conference was called, and they were shut off. (It could have been, perhaps should have been, a sign that there were further dangers ahead, since they were "cavitating"—not getting enough emergency coolant going through them to function properly.)

In the control room there were three audible alarms sounding, and many of the 1,600 annunciator lights (little rectangles of plastic with some code numbers and letters on them) were on or blinking. The operators did not turn off the main audible alarm because it would cancel some of the annunciator lights. The computer was beginning to run far behind schedule; in fact it took some hours before its message that something might be wrong with the PORV finally got its chance to be printed. Radiation alarms were coming on. The control room was filling with experts; later in the day there were about forty people there. The phones were ringing constantly, demanding information the operators did not have.

Two hours and twenty minutes after the start of the accident, a new shift came on. The record is unclear, but either the new shift supervisor decided to check the PORV, or an expert talking with a supervisor over the telephone questioned its status, and the operators discovered the stuck valve, and closed a block valve to shut off the flow to the PORV. The operator testified at the Kemeny Commission hearings that it was more of an act of desperation to shut the block valve than an act of understanding. After all, he said, you do not casually block off a safety system. It was fortunate that it occurred when it did; incredible damage had been done, with substantial parts of the core melting, but had it remained open for another thirty minutes or so, and HPI remained throttled back, there would probably have been a complete meltdown, with the fissioning material threatening to breach containment. But the accident was far from over. New dangers appeared every few hours. Thirty-three hours into the accident another unexpected and mysterious interaction occurred. Confusion still reigned when the first sign of the famous hydrogen bubble appeared; the bubble threatened the integrity of the plant for the next few days. Again we have a lesson in the meaning of warnings, and in the difficulty that even experts have in understanding such a complex human-made system as a nuclear plant. Here is the background: The fuel rods—36,816 of them—contain enriched uranium in little pills, all stacked within a thin liner, like the cigarette paper around tobacco, only about 12 feet long. Water circulates through the stacks of rods and cools them so they won't burn too fast. When they get too hot, though, the liner, or "cladding," can react with the water in a zirconium-water reaction. This consumes oxygen, thus freeing hydrogen, making hydrogen bubbles, which then can make pockets of hydrogen gas if there is room for them, and a dandy explosion if there is also a bit of oxygen and a spark . . .

Of such complexities is the normal accident made. For all but one operator, presumably, and for all the experts, the pressure spike and the hydrogen bubble were incomprehensible. To understand the accident, they would have had to know that the core was seriously uncovered, and that a zirconium-water reaction was likely (a possibility disputed by an expert), and would have had to recall that the PORV had been open, allowing the hydrogen to get out of the core vessel into the building that contained it. These are not expected sequences in a production or safety system; they are multiple failures that interacted in an incomprehensible manner for all but at least one person, who, incredibly enough, wasn't talking, or didn't examine the implications of his hunch. A warning such as the spike is only effective if it fits into our mental model of what is going on . . .

References

Kemeny, John, et al. (1979). *The need for change: The legacy of TMI. Report of the President's Commission on the Accident at Three Mile Island.* Washington, DC: Government Printing Office.
President's Commission on Three Mile Island. (1979). *Hearings,* May 30: 30, 57.

Endnote

1. From "Normal Accident at Three-Mile Island," Taken from Chapter 1 of *Normal Accidents: Living With High-Risk Technologies* by Charles Perrow. Copyright © 1984 Charles Perrow. Reprinted with permission of author.

CASE STUDY 11

Challenger Disaster[1]

Russell Boisjoly, Ellen Curtis, and Eugene Mellican

Introductory Note by W.M.E. and M.M: *The Challenger space shuttle 51-L exploded on January 28, 1986, 73 seconds after lift-off, killing its crew of seven, including the first potential civilian in space. In addition to the tragic loss of life, tens of millions of taxpayer dollars went up in smoke along with the shuttle and rocket. A presidential commission headed by former Secretary of State William Rogers investigated the disaster and found numerous failures at NASA, including faulty risk assessment and flawed group decision making.*

During the investigation it was discovered that expert engineers knew of design and performance flaws in the "O-rings," the technical components that were the immediate cause of the explosion. Data from previous flights identified the failure of the O-rings to seal when at colder than normal temperatures. The night before the fateful launch, Morton Thiokol engineers tried to convince Morton Thiokol management of the dangers of launching the space shuttle in below freezing temperatures The failure of engineers to convince top management of the dangers and the managers'ultimate rejection of the engineers' concerns also exhibited organizational miscommunication and flawed decision making.

This technological disaster of the Third Industrial Revolution was the result of organizational systems factors.

On January 28, 1986, the space shuttle Challenger exploded 73 seconds into its flight killing the seven astronauts aboard. As the nation mourned the tragic loss of the crewmembers, the Rogers Commission was formed to investigate the causes of the disaster. The Commission concluded that the explosion occurred due to seal failure in one of the solid rocket booster joints. Testimony given by Roger Boisjoly, Senior Scientist and acknowledged rocket seal expert, indicated that top management at NASA and Morton Thiokol had been aware of problems with the O-ring seals, but agreed to launch against the recommendation of Boisjoly and other engineers. Boisjoly had alerted management to problems with the O-rings as early as January, 1985, yet several shuttle launches prior to the Challenger had been approved without correcting the hazards. This suggests that the management practice of NASA and Morton Thiokol had created an environment which altered the framework for decision-making, leading to a breakdown in communication between technical experts and their supervisors, and top level management, and to the acceptance of risks that both organizations had historically viewed as unaccept-able. With human lives and national interest at stake, serious ethical concerns are embedded in this dramatic change in management practice . . .

By reason of the courageous activities and testimony of individuals like Roger Boisjoly, the Challenger disaster provides a fascinating il-lustration of the dynamic tension between organizational and indi-vidual responsibility . . .

Preview for Disaster

On January 24, 1985, Roger Boisjoly, Senior Scientist at Morton Thiokol, watched the launch of Flight 51-C of the space shuttle pro-gram. He was at Cape Canaveral to inspect the solid rocket boosters from Flight 51-C following their recovery in the Atlantic Ocean and to conduct a training session at Kennedy Space Center (KSC) on the proper methods of inspecting the booster joints. While watching the launch, he noted that the temperature that day was much cooler than recorded at other launches, but was still much warmer than the 18 degree temperature encountered three days earlier when he arrived in Orlando. The unseasonably cold weather of the past several days had produced the worst citrus crop failures in Florida history.

When he inspected the solid rocket boosters several days later, Boisjoly discovered evidence that the primary O-ring seals on two field joints had been compromised by hot combustion gases (i.e., hot gas blow-by had occurred) which had also eroded part of the primary O-ring. This was the first time that a primary seal on a field joint had been penetrated. When he discovered the large amount of blackened grease between the primary and secondary seals, his concern height-ened. The blackened grease was discovered over 80 degree and 110 degree arcs, respectively, on two of the seals, with the larger arc indi-cating greater hot gas blow-by. Post-flight calculations indicated that the ambient temperature of the field joints at launch time was 53 degrees. This evidence, coupled with his recollection of the low temperature the day of the launch and the citrus crop damage caused by the cold spell, led to his conclusion that the severe hot gas blow-by may have been caused by, and related to, low temperature. After reporting these find-ings to his superiors, Boisjoly presented them to engineers and man-agement at NASA's Marshall Space Flight Center (MSFC). As a result of his presentation at MSFC, Roger Boisjoly was asked to participate in the Flight Readiness Review (FRR) on February 12, 1985, for Flight 51-E which was scheduled for launch in April, 1985. This FRR represents the first association of low temperature with blow-by on a field joint, a con-dition that was considered an "acceptable risk" by Larry Mulloy, NASA's Manager for the Booster Project, and other NASA officials.

Roger Boisjoly had twenty-five years of experience as an engineer in the aerospace industry. Among his many notable assignments were the performance of stress and deflection analysis on the flight control equipment of the Advanced Minuteman Missile at Aeronautics, and serving as a lead engineer on the lunar module of Apollo at Hamilton Standard. He moved to Utah in 1980 to take a position in the Applied Mechanics Department as a Staff Engineer at the Wasatch Division of Morton Thiokol. He was considered the leading expert in the United States on O-rings and rocket joint seals and received applause for his work on the joint seal problems from Joe C. Kilminster, Vice President of Space Booster Programs, Morton Thiokol (Kilminster, July, 1985). His commitment to the company and the community was further demonstrated by his service as Mayor of Willard, Utah, from 1982 to 1983.

The rough questioning he received at the February 12th FRR convinced Boisjoly of the need for further evidence linking low temperature and hot gas blow-by. He worked closely with Arnie Thompson, Supervisor of Rocket Motor Cases, who conducted subscale laboratory tests in March, 1985, to further test the effects of temperature on O-ring resiliency. The bench tests that were performed provided powerful evidence to support Boisjoly's and Thompson's theory—low temperatures greatly and adversely affected the ability of O-rings to create a seal on solid rocket booster joints. If the temperature was too low (and they did not know what the threshold temperature would be), it was possible that neither the primary or secondary O-rings would seal!

One month later the post-flight inspection of Flight 51-B revealed that the primary seal of a booster nozzle joint did not make contact during its two minute flight. If this damage had occurred in a field joint, the secondary O-ring may have failed to seal, causing the loss of the flight. As a result Boisjoly and his colleagues became increasingly concerned about shuttle safety. This evidence from the inspection of Flight 51-B was presented at the FRR for Flight 51-F on July 1, 1985; the key engineers and managers at NASA and Morton Thiokol were now aware of the critical O-ring problems and the influence of low temperature on the performance of the joint seals.

During July 1985, Boisjoly and his associates voiced their desire to devote more effort and resources to solving the problems of O-ring erosion. In his activity reports dated July 22 and 29, 1985, Boisjoly expressed considerable frustration with the lack of progress in this area, despite the fact that a Seal Erosion Task Force had been informally appointed on July 19th. Finally, Boisjoly wrote the following memo, labeled "Company Private," to R. K. (Bob) Lund, Vice President of Engineering for Morton Thiokol, to express the extreme urgency of his concerns. Here are some excerpts from that memo:

> This letter is written to insure that management is fully aware of the seriousness of the current O-ring erosion problem . . . The mistakenly accepted position on the joint problem was to fly without fear of failure . . . is now drastically changed as a

result of the *SRM 16A* nozzle joint erosion that eroded a secondary O-ring with the primary O-ring never sealing. If the same scenario should occur in a field joint (and it could), then it is a jump ball as to the success or failure of the joint . . . The result would be a catastrophe of the highest order—loss of human life . . .

It is my honest and real fear that if we do not take immediate action to dedicate a team to solve the problem, with the field joint having the number one priority, then we stand in jeopardy of losing a flight along with all the launch pad facilities. (Boisjoly, 1985a)

On August 20, 1985, R. K. Lund formally announced the formation of the Seal Erosion Task Team. The team consisted of only five full-time engineers from the 2500 employed by Morton Thiokol on the Space Shuttle Program. The events of the next five months would demonstrate that management had not provided the resources necessary to carry out the enormous task of solving the seal erosion problem.

On October 3, 1985, the Seal Erosion Task Force met with Joe Kilminster to discuss the problems they were having in gaining organizational support necessary to solve the O-ring problems. Boisjoly later stated that Kilminster summarized the meeting as a "good bullshit session." Once again frustrated by bureaucratic inertia, Boisjoly wrote in his activity report dated October 4th:

. . . NASA is sending an engineering representative to stay with us starting Oct. 14th. We feel that this is a direct result of their feeling that we (MTI) are not responding quickly enough to the seal problem . . . upper management apparently feels that the SRM program is ours for sure and the customer be damned. (Boisjoly, 1985b)

Boisjoly was not alone in his expression of frustration. Bob Ebeling, Department Manager, Solid Rocket Motor Igniter and Final Assembly, and a member of the Seal Erosion Task Force, wrote in a memo to Allan McDonald, Manager of the Solid Rocket Motor Project, "HELP! The seal task force is constantly being delayed by every possible means . . . We wish we could get action by verbal request, but such is not the case. This is a red flag" (McConnell, 1987).

At the Society of Automotive Engineers (SAE) conference on October 7, 1985, Boisjoly presented a six-page overview of the joints and the seal configuration to approximately 130 technical experts in hope of soliciting suggestions for remedying the O-ring problems. Although MSFC had requested the presentation, NASA gave strict instructions not to express the critical urgency of fixing the joints, but merely to ask for suggestions for improvement. Although no help was forthcoming, the conference was a milestone in that it was the first time that NASA allowed information on the O-ring difficulties to be expressed in a public forum. That NASA also recognized that the O-ring problems were not receiving appropriate attention and manpower considerations from Morton Thiokol management is further evidenced by Boisjoly's October 24 log entry, ". . . Jerry Peoples (NASA) has informed his people that our group needs more authority and people to do the job. Jim Smith (NASA) will corner Al McDonald today to attempt to implement this direction."

The October 30 launch of Flight 61-A of the Challenger provided the most convincing, and yet to some the most contestable, evidence to date that low temperature was directly related to hot gas blow-by. The left booster experienced hot gas blow-by in the center and aft field joints without any seal erosion. The ambient temperature of the field joints was estimated to be 75 degrees at launch time based on post-flight calculations. Inspection of the booster joints revealed that the blow-by was less severe than that found on Flight 51-C because the seal grease was a grayish black color, rather than the jet black hue of Flight 51-C. The evidence was now consistent with the bench tests for joint resiliency conducted in March. That is, at 75 degrees the O-ring lost contact with its sealing surface for 2.4 seconds, whereas at 50 degrees the O-ring lost contact for 10 minutes. The actual flight data revealed greater hot gas blow-by for the O-rings on Flight 51-C which had an ambient temperature of 53 degrees than for Flight 61-A which had an ambient temperature of 75 degrees. Those who rejected this line of reasoning concluded that temperature must be irrelevant since hot gas blow-by had occurred even at room temperature (75 degrees). This difference in interpretation would receive further attention on January 27, 1986.

During the next two and one-half months, little progress was made in obtaining a solution to the O-ring problems. Roger Boisjoly made the following entry into his log on January 13, 1986, "O-ring resiliency tests charts . . . requested on September 24, 1985 are now scheduled for January 15, 1986."

The Day before the Disaster

At 10 a.m. on January 27, 1986, Arnie Thompson received a phone call from Boyd Brinton, Thiokol's Manager of Project Engineering at MSFC, relaying the concerns of NASA's Larry Wear, also at MSFC, about the 18 degree temperature forecast for the launch of Flight 51-L, the Challenger, scheduled for the next day. This phone call precipitated a series of meetings within Morton Thiokol, at the Marshall Space Flight Center; and at the Kennedy Space Center that culminated in a three-way teleconference involving three teams of engineers and managers, that began at 8:15 p.m. E.S.T.

Joe Kilminster, Vice President, Space Booster Programs, of Morton Thiokol began the telecom by turning the presentation of the engineering charts over to Roger Boisjoly and Arnie Thompson. They presented thirteen charts which resulted in a recommendation against the launch of the Challenger. Boisjoly demonstrated their concerns with the performance of the O-rings in the field joints during the initial phases of Challenger's flight with charts showing the effects of primary O-ring erosion, and with temperature, on the ability to maintain a reliable secondary seal. The tremendous pressure and release of power from the rocket boosters create rotation in the joint such that the metal moves away from the O-rings so that they cannot maintain contact with the metal surfaces. If, at the same time, erosion occurs in the primary O-ring for any reason, then there is a reduced probability of maintaining a secondary seal. It is highly probable that as the ambient temperature drops, the primary O-ring will not seal, that there will be hot gas blow-by and erosion of the primary O-ring, and that a catastrophe will occur when the secondary O-ring fails to seal.

Bob Lund presented the final chart that included the Morton Thiokol recommendation that the ambient temperature including wind must be such that the seal temperature would be greater than 53 degrees to proceed with the launch. Since the overnight low was predicted to be 18 degrees, Bob Lund recommended against launch on January 28, 1986, or until the seal temperature exceeded 53 degrees.

NASA's Larry Mulloy bypassed Bob Lund and directly asked Joe Kilminster for his reaction. Kilminster stated that he supported the position of his engineers and he would not recommend launch below 53 degrees.

George Hardy, Deputy Director of Science and Engineering at MSFC, said he was "appalled at that recommendation," according to Allan McDonald's testimony before the Rogers Commission. Nevertheless, Hardy would not recommend to launch if the contractor was against it. After Hardy's reaction, Stanley Reinartz, Manager of Shuttle Project Office at MSFC, objected by pointing out that the solid rocket motors were qualified to operate between 40 and 90 degrees Fahrenheit.

Larry Mulloy, citing the data from Fight 61-A which indicated to him that temperature was not a factor, strenuously objected to Morton Thiokol's recommendation. He suggested that Thiokol was attempting to establish new Launch Commit Criteria at 53 degrees and that they couldn't do that the night before a launch. In exasperation Mulloy asked, "My God, Thiokol, when do you want me to launch? Next April?" (McConnell, 1987). Although other NASA officials also objected to the association of temperature with O-ring erosion and hot gas blow-by, Roger Boisjoly was able to hold his ground and demonstrate with the use of his charts and pictures that there was indeed a relationship: The lower the temperature the higher the probability of erosion and blow-by and the greater the likelihood of an accident. Finally, Joe Kilminster asked for a five-minute caucus off-air.

According to Boisjoly's testimony before the Rogers Commission, Jerry Mason, Senior Vice President of Wasatch Operations, began the caucus by saying that "a management decision was necessary." Sensing that an attempt would be made to overturn the no-launch decision, Boisjoly and Thompson attempted to re-review the material previously presented to NASA for the executives in the room. Thompson took a pad of paper and tried to sketch out the problem with the joint, while Boisjoly laid out the photos of the compromised joints from Flights 51-C and 61-A. When they became convinced that no one was listening, they ceased their efforts. As Boisjoly would later testify, "There was not one positive pro-launch statement ever made by anybody" (*Report of the Presidential Commission,* 1986, IV, p. 792, hereafter abbreviated as R.C.).

According to Boisjoly, after he and Thompson made their last attempts to stop the launch, Jerry Mason asked rhetorically, "Am I the only one who wants to fly?" Mason turned to Bob Lund and asked him to "take off his engineering hat and put on his management hat." The four managers held a brief discussion and voted unanimously to recommend Challenger's launch . . .

Aside from the four senior Morton Thiokol executives present at the teleconference, all others were excluded from the final decision. The process represented a radical shift from previous NASA policy. Until that moment, the burden of proof had always been on the engineers to prove beyond a doubt that it was safe to launch. NASA, with their objections to the original Thiokol recommendation against the launch, and Mason, with his request for a "management decision," shifted the burden of proof in the opposite direction. Morton Thiokol was expected to prove that launching Challenger would not be safe (R.C., IV, p. 793).

The change in the direction so deeply upset Boisjoly that he returned to his office and made the following journal entry:

> I sincerely hope this launch does not result in a catastrophe. I personally do not agree with some of the statements made in Joe Kilminster's written summary stating that SRM-25 is okay to fly (Boisjoly, 1987).

The Disaster and its Aftermath

On January 28, 1986, a reluctant Roger Boisjoly watched the launch of the Challenger. As the vehicle cleared the tower, Bob Ebeling whispered, "we've just dodged a bullet." (The engineers who opposed the launch assumed that O-ring failure would result in an explosion almost immediately after engine ignition.) To continue in Boisjoly's words, "At approximately T + 60 seconds Bob told me he had just completed a prayer of thanks to the Lord for a successful launch. Just thirteen seconds later we both saw the horror of the destruction as the vehicle exploded" (Boisjoly, 1987).

Challenger Disaster, January 28, 1986. Copyright Corbus. Used by permission.

Morton Thiokol formed a failure investigation team on January 31, 1986, to study the Challenger explosion. Roger Boisjoly and Arnie Thompson were part of the team that was sent to MSFC in Huntsville, Alabama. Boisjoly's first inkling of a division between himself and management came on February 13 when he was informed at the last minute that he was to testify before the Rogers Commission the next day. He had very little time to prepare his testimony. Five days later, two Commission members held a closed session with Kilminster, Boisjoly, and Thompson. During the interview, Boisjoly gave his memos and activity reports to the Commissioners. After that meeting, Kilminster chastised Thompson and Boisjoly for correcting his interpretation of the technical data. Their response was that they would continue to correct his version if it was technically incorrect.

Boisjoly's February 25th testimony before the Commission, rebutting the general manager's statement that the initial decision against the launch was not unanimous, drove a wedge further between him and Morton Thiokol management. Boisjoly was flown to MSFC before he could hear the NASA testimony about the pre-flight telecom. The next day, he was removed from the failure investigation team and returned to Utah.

Beginning in April, Boisjoly began to believe that for the previous month he had been used solely for public relations purposes. Although given the job of Seal Coordinator for the redesign effort, he was isolated from NASA and the seal redesign effort. His design information had been changed without his knowledge and presented without his feedback. On May 1, 1986, in a briefing preceding closed sessions before the Rogers Commission, Ed Garrison, President of Aerospace Operations for Morton Thiokol, chastised Boisjoly for "airing the company's dirty laundry" with the memos he had given the Commission. The next day, Boisjoly testified about the change in his job assignment. Commissioner Rogers criticized Thiokol management,

> . . . if it appears that you're punishing the two people or at least two of the people who are right about the decision and objected to the launch which ultimately resulted in criticism of Thiokol and then they're demoted or feel that they are being retaliated against, that is a very serious matter. It would seem to me, just speaking for myself, they should be promoted, not demoted or pushed aside. (R.C., V, p. 1586)

Boisjoly now sensed a major rift developing within the corporation. Some co-workers perceived that his testimony was damaging the company image. In an effort to clear the air, he and McDonald requested a private meeting with the company's three top executives, which was held on May 16, 1986. According to Boisjoly, management was unreceptive throughout the meeting. The CEO told McDonald and Boisjoly that the company "was doing just fine until Al and I testified about our job reassignments" (Boisjoly, 1987). McDonald and Boisjoly were nominally restored to their former assignments, but Boisjoly's position became untenable as time passed. On July 21, 1986, Roger Boisjoly requested an extended sick leave from Morton Thiokol.

Ethical Analysis

It is clear from this case study that Roger Boisjoly's experiences before and after the Challenger disaster raise numerous ethical questions that are integral to any explanation of the disaster and applicable to other management situations, especially those involving highly complex technologies. The difficulties and uncertainties involved in the management of these technologies exacerbate the kind of bureaucratic syndromes that generate ethical conflicts in the first place. In fact, Boisjoly's experiences could well serve as a paradigmatic case study for such ethical problems, ranging from accountability to corporate loyalty and whistleblowing. Underlying all these issues, however, is the problematic relationship between individual and organizational responsibility. Boisjoly's experiences graphically portray the tensions inherent in this relationship in a manner that discloses its importance in the causal sequence leading to the Challenger disaster . . . By focusing on the problematic relationship between individual and organizational responsibility, this analysis reveals that the organizational structure governing the space shuttle program became the locus of responsibility in such a way that not only did it undermine the responsibilities of individual decision makers within the process, but it also became a means of avoiding real, effective responsibility throughout the entire management system . . .

Although fragmentary and tentative in its formulation, this set of considerations points toward the conclusion that however complex and sophisticated an organization may be, and no matter how large and remote the institutional network needed to manage it may be, an active and creative tension of responsibility must be maintained at every level of the operation. Given the size and complexity of such endeavors, the only way to ensure that tension of attentive and effective responsibility is to give the primacy of responsibility to that ultimate principal of all moral conduct: the human individual . . .

References

Boisjoly, Roger M. (1985a). Applied mechanics memorandum to Robert K. Lund. Vice President, Engineering, Wasatch Division Morton Thiokol, Inc., July 31.

Boisjoly, Roger M. (1985b). Activity Report, SRB Seal Erosion Task Team Status, October 4, 26: 50.

Boisjoly, Roger. (1987). "Ethical dimensions: Morton Thiokol and the shuttle disaster." Speech given at Massachusetts Institute of Technology, January 7.

Kilminster, J.C. (1985). Memorandum (E000-FY86-003) to Robert Lund, Vice-President, Engineering, Wasatch Division, Morton Thiokol, Inc., July 5.

McConnell, Malcolm. (1987). *Challenger, a major malfunction: A true story of politics, greed, and the wrong stuff.* Garden City, NY: Doublesday and Company.

Report of the Presidential Commission on the Space Shuttle Challenger Accident. (1986). (Rogers Commission). Washington, DC: U.S. Government Printing Office.

Endnote

1. From "Roger Boisjoly and the *Challenger* Disaster: The Ethical Dimensions," by Russell Boisjoly, Ellen Curtis, and Eugene Mellican in *Journal of Business Ethics* 8: 217–230, 1989. © Kluwer Academic Publishers. Printed in the Netherlands. Reproduced by permission of Kluwer Academic Publishers.

CASE 12

Bhopal Poison Gas Release[1]

B. Bowonder, L. Kasperson, and R. Kasperson

Introductory Note by W.M.E. and M.M.: *On December 3, 1984, a poisonous cloud of methyl isocyanate, a chemical compound used to make pesticides, escaped and passed over the town of Bhopal, India, eventually causing the deaths of 14,000 people. In addition, over 30,000 permanent injuries, including blindness, 20,000 temporary injuries, and 150,000 minor injuries were reported. Lax governmental controls, incompetent operator training and inadequate social infrastructures were causes of the disaster. In addition, the depreciated value of life in Third World countries resulted in negligence in plant design, including safety standards.*

The Bhopal poison gas release is the worst industrial disaster in history. The case points up the dangers of technology transfer and the often-irresponsible actions of multinational corporations operating in Third World countries. For example, after legal battles raged for years, Union Carbide settled out of court with the Indian Government for a fraction of the compensation demanded in the lawsuits filed by the victims of the disaster. Ultimately, the victims received little compensation, leading to a 1999 lawsuit filed against Union Carbide and its CEO Warren Anderson. The lawsuit charges the company and CEO with violations of international law and the fundamental human rights of the victims and survivors of the disaster. In addition to the legal morass, long-term health effects are still being investigated such as respiratory illnesses resulting from exposure to the poisonous gas.

This technological disaster of the Third Industrial Revolution was the result of socio-cultural factors.

Runaway chemical reactions are rare events, particularly in this heyday of the redundant and "defense-in-depth" safety design for complex, high-risk technologies. Yet during the chill of night between December 2 and 3, 1984, a statistically improbable worst-case scenario moved from the computer simulations of the risk assessors and played itself out on the unsuspecting citizens of Bhopal, India. A parade of failures—in design, in maintenance, in operation, in emergency response, and in management—conspired with a southerly wind and a temperature inversion to push a lethal cloud of methyl isocyanate (MIC) out to kill and injure thousands of people, animals, and plants in the area. By sunrise, the unprecedented horror had catapulted Bhopal to the head of history's roll of industrial disasters (see Table 8–1).

Table 8-1
Major industrial disasters in the 20th century.

Year	Accident	Site	Number of Fatalities
1921	Explosion in chemical plant	Oppau, Germany	561
1942	Coal-dust explosion	Honkeiko, Colliery, China	1,572
1947	Fertilizer ship explosion	Texas City, USA	562
1956	Dynamite truck explosion	Cali, Colombia	1,100
1974	Explosion in chemical plant	Flixborough, UK	28[a]
1975	Mine explosion	Chasnala, India	431
1976	Chemical leak	Seveso, Italy	0(?)[b]
1979	Biological/chemical warfare plant accident	Novosibirsk, USSR	300
1984	Natural gas explosion	Mexico City	452+[c]
1984	Poison gas leak	Bhopal, India	14,000[d]

[a] 3,000 evacuated
[b] 700 evacuated, hundreds of animals killed, 200 cases of skin disease
[c] 4,258 injured, 31,000 evacuated
[d] 100,000 evacuated, 50,000 severely impaired
Sources: Patrick Lagadec, *Major Technological Risk: An Assessment of Industrial Disasters* (New York: Pergamon Press, 1982); Saliesh Kottary, "Whose Life Is It Anyway?" *The Illustrated Weekly of India* (December 30, 1984–January 5, 1985), 8; "Union Carbide Halts Production of Pesticide Gas," *Financial Times* (London) December 5, 1984, 1.

. . . Union Carbide Corporation, the parent company involved in the disaster at Bhopal, has more than twenty years experience in the safe manufacture, use, transportation, and storage of MIC (to say nothing of a host of other hazardous products). With a cadre of scientists and technicians and an institutional structure for environmental protection, India is better equipped than other developing countries to manage hazardous technologies. Given this framework, other industries in other places are more likely candidates for catastrophic disasters. Thus it is essential to understand how and why this particular surprise occurred at Bhopal if we are to ward off future similar tragedies.

Union Carbide was scarcely an unwelcome intruder in Bhopal. The Indian government promoted the siting of industries in less developed states such as Madhya Pradesh where Bhopal is located. Eager to attract major industries, Madhya Pradesh leaders offered incentives to companies that would bring jobs and indigenous manufacturing to its unindustrialized cities; Union Carbide, for example, built on government land for an annual rent of less than $40 an acre. A plant that would manufacture the carbaryl pesticides to fuel India's ongoing Green Revolution was particularly welcome as another step toward self-sufficient food production. Hence, the 1970 decision of Union Carbide of India Limited (UCIL) to manufacture the pesticide Sevin in an advanced facility in central India was met with great fanfare.

Sevin, manufactured from MIC, had received the endorsement of the Indian Council of Agricultural Research. Use of the pesticide decreases insect damage of cotton, lentils, and other vegetables by as much as 50 percent. Even in the wake of the accident, few serious observers have suggested that India do away with Sevin and other carbaryl pesticides, which, ironically, are substitutes for "more dangerous" DDT and organophosphates. Given the high toxicity (Kimmerle & Iben, 1964) of MIC, however, it is clear that the chemical requires, at all stages, special handling commensurate to the risk.

It is easy to contend that high-risk facilities have no place in densely populated urban areas. Yet such a facility is apt to attract squatter settlements to its gates, whether it be a liquefied-natural gas facility in Mexico City, a petrochemical complex in Cubatao, Brazil, or a pesticides factory in Bhopal. The showpiece UCIL factory and other industries that set up shop in Bhopal surely contributed to the staggering rise in population—from 350,000 in 1969, to 700,000 in 1981, to over 800,000 in 1984. As a Union Carbide official recently put it:

> In India, land is scarce and the population often gravitates towards areas that contain manufacturing facilities. That's how so many people came to be living near the fences surrounding our property. (Browning, 1984: 3)

It is also, of course, how risks come to fall so disproportionately on the poor (Susman, O'Keefe, & Wisner, 1983; Wijkman & Timberlake, 1985) . . .

The Bhopal operation, never very profitable, broke even in 1981 but thereafter began to lose money. By 1984 the plant produced less than 1,000 of a projected 5,000 tons and lost close to $4 million. UCIL, contemplating selling the operation, began to issue incentives for early retirement and cut back on its workforce. Many of the skilled workers left for securer pastures. Things were not going well.

Early Warnings

Whether cost-cutting measures and the departure of skilled personnel caused lapses in safety is difficult to ascertain. Nevertheless, the Bhopal plant experienced six accidents—at least three of which involved the release of MIC or phosgene, another poisonous gas, between 1981 and 1984. These accidents scarcely presaged the catastrophic release, but taken together they surely could have pointed to safety problems at the plant. Indeed, a phosgene leak that killed one worker on December 26, 1981, generated an official inquiry, but the findings (filed three years later) gathered dust in the Madhya Pradesh labor department until after the Bhopal accident, when two officials lost their jobs for having failed to act upon the report's safety recommendations (Anonymous [a], 1985).

Meanwhile, a local journalist warned that the plant's proximity to Bhopal's most densely populated areas was inviting disaster. In 1982 Rajkumar Keswani took on UCIL in a series of articles in the Hindi press. "Sage, please save this city," "Bhopal on the mouth of a volcano," and "If you don't understand, you will be wiped out," the headlines warned (Keswani, 17 September 1982; Keswani, 1 October 1982; Keswani, 8 October 1982). On June 16, 1984, he tried again, this time with what he calls "an exhaustive report on the Union Carbide threat." "The alarm fell on deaf ears," he wrote one week after the Bhopal accident (Keswani, 1984).

A 1982 safety audit by an inspection team from the parent company cited a number of safety problems, including the danger posed by a manual control on the MIC feed tank, the unreliability of certain gauges and valves, and insufficient training of operators (Kail, Poulson, & Tyson, 1982). UCIL claims to have corrected the deficiencies, but auditors have never confirmed the corrections.

Just before the accident in Bhopal, Union Carbide Corporation's safety and health survey of its MIC unit II Plant in Institute, West Virginia, cited 34 less serious and 2 major concerns, the first of which was the "potential for runaway reaction in unit storage tanks due to a combination of contamination possibilities and reduced surveillance during block operation" (Union Carbide, 1984). Why the parent company, which owns 50.9 percent of the Bhopal plant, failed to share with its subsidiary its two major concerns (the second was the serious potential for overexposure to chloroform) is unclear. Some Union Carbide officials contend that the different cooling systems— brine at Institute and freon at Bhopal—made the *hazard* communication unnecessary, but this is difficult to square with the recommendation:

> The fact that past incidences of water contamination may be warnings, rather than examples of successfully dealing with problems, should be emphasized to all operating personnel. (Union Carbide, 1984, 462)

Equally puzzling is the parent company's earlier overriding of an alleged UCIL protest against the installation of such large, 15,000 gallon, storage tanks at Bhopal (Brushnan & Subramaniam, 1988).

In any event, MIC sat in storage at the Bhopal plant for at least three months prior to the accident. Such storage invites disaster, for the tiniest ingress of water, caustic soda, or even MIC itself is sufficient to set in motion an exothermic (heat-producing) chemical reaction (Worthy, 1985). One Indian scientist has even hypothesized that the re-action at Bhopal began slowly and imperceptibly at least two weeks before the fateful night in which it reached violent runaway propor-tions (Munilidbaran, 1985).

Accident Analysis

Some time shortly before the 10:45 p.m. shift change at the Bhopal plant on December 2, 1984, water and/or another contaminant en-tered MIC storage tank 610, thereby triggering a violent chemical re-action and a dramatic rise in temperature and pressure. It is not known whether the incoming control room operator was aware that the 10:20 p.m. tank pressure read 2 psi (pounds per square inch), but the 11:00 p.m. reading of 10 psi does not seem to have struck anyone as unusual. Nor should it have, since normal operations ran at pressures between 2 and 25 psi.

By the time the operator did take notice of the rising pressure—from 10 psi at 11:00 p.m. to 30 psi at 12:15 a.m.—the reading was racing to the top of the scale (55 psi). Escaping MIC vapor ruptured a safety disc and popped the safety valve. On the heels of this initial release came a series of compromises and failures of virtually all the safety systems designed to prevent release. The deadly gas spewed out over the slums of Bhopal . . .

Consequences of the Accident

More than six months after the accident, even the early (acute) con-sequences of the accident are not well defined and the longer-term (chronic) effects are very uncertain. Estimates of human fatalities range from 1,400 to 14,000. The Indian government's official count, as of June 1985, of 1,762—based on death certificates—undoubtedly underestimates the toll because many people fled from the city and died in outlying regions, the deaths occurred disproportionately among people living in the nearby squatter settlements about whom little information exists, and the upcoming elections provided incen-tive for official minimization of the number of fatalities.

More recently, Asoke K. Sen, the Indian law minister, put the fatali-ties at more than 2,000 (Diamond, 1985), closer to the more widely accepted estimate of 2,500. This figure, based on body counts conducted by members of the press, is subject to errors in observation and tabulation and also misses the deaths that oc-curred outside the city. Because the exodus from the city was ex-tensive within 24 hours of the release, a significant number of fatalities may have escaped count.

The Medico Friends Circle estimated at least 4,000 deaths, based on the sale of death shrouds in the week following the accident. Other estimates have reached 14,000. . . . Estimation is con-founded further because over 80 percent of the deaths occurred

Bhopal Poison Gas Release, December 3, 1984. Copyright Corbus. Used by permission.

outside the hospitals. Only 438 deaths are recorded in the various hospitals on December 3 and 4 (Anonymous [b], 1985). Although a precise breakdown of age among the fatalities is unavailable, the deaths were disproportionately concentrated among children and especially infants.

Information on the number of people exposed to MIC as well as on the long-term health effects from exposure is even less available. (Vegetation analyses are under way and may improve understanding of MIC's effects.) The most widely accepted estimates indicate that 200,000 persons were exposed, and that 50,000 to 60,000 received substantial exposure (Waldhoz, 1985). Law Minister Sen recently indicated that doctors have treated 200,000 persons for exposure and 17,000 persons in Bhopal have been permanently disabled, largely from lung ailments (Diamond, 1985). Evidence of continuing physiological effects—including abnormally high blood levels of carboxyhemoglobin and methemoglobin, low vital lung capacity, neurological abnormalities, widespread gastritis, and vomiting—is accumulating. At the time of the accident, the German toxicologist Max Daunderer warned about the stages of MIC's effects—irritated eyes, skin, and lungs during the first four to seven days, serious central nervous system effects developing after three to four weeks, and then delayed central nervous system disorders, including paralysis (Rout, 1985). The degree of reversibility or irreversibility of these effects, however, is not known.

The Indian Council of Medical Research is coordinating a massive data-gathering effort on long-term morbidity conducted by the All-India Institute of Medical Science, the VP Patel Chest Institute, the Industrial Toxicology Research Centre, and the KEM Hospital in Bombay. Meanwhile, widespread distribution (often without medical supervision) of antibiotics and corticosteroids, as well as widespread malnutrition and chronic diseases among many victims, have complicated the assessment of MIC's effects on morbidity. Studies aimed at defining potential genetic and carcinogenic effects are only beginning and results will require another three to five years. Although MIC passed the Ames test—a short-term test for mutagenicity—the U.S. National Toxicology Program is planning an ambitious series of animal studies to elucidate long-term effects on respiratory, reproductive, and immune systems (Dagani, 1985).

Special concern exists over possible damage to women's health. Amidst accounts of abnormally high levels of uncommon vaginal discharge, excessive menstrual bleeding, retroverted position of the uterus with severe restricted mobility, and other disorders, the press has alleged that government and medical teams are avoiding women's health complaints. Junior gynecologists and midwives in hospital maternity wards have reported unusually high numbers of premature or underweight babies and physical deformities among the 20 to 30 infants born daily in Bhopal (Anonymous [b], 1985). Despite the sketchy evidence, some local doctors had advised a number of the 1,000 women who were pregnant at the time of the accident to undergo abortions (Anonymous [b], 1985), advice which a government services department has contested. The Indian Council of Medical Research recently reported that a study of the effects of MIC exposure on fetal growth in some 500 babies does not indicate fetal damage.

Effects of the accident on mental health are acknowledged but are the least studied. Reports of mental trauma and other psychiatric effects persist, yet there is no systematic program for monitoring or treating mental health problems. A distinguished group of psychiatrists who visited Bhopal several weeks after the accident acknowledged that people were indeed suffering from anxiety and depression, but it was difficult to attribute these symptoms to the accident (Anonymous [a], 1985). More recently the King George Medical College has found widespread mental disorders among the Bhopal population.

Other damages and burdens add to the human health problems. Some 70,000 to 100,000 persons left the city for distances up to 50 kilometers, with resulting disruption and economic loss. An estimated

1,600 animals died on the first and second days after the accident, posing a serious disposal problem—eventually solved by digging a 1-acre burial pit 5 kilometers from the city. Ecological effects—among the least understood of the accident's long-term effects—included apparent damage to certain vegetation, animal, and fish species but not to others are under study through the Indian Council of Agricultural Research.

What of the relief efforts to avert future health and ecological effects? Despite the remarkable emergency performance of the Indian medical system many victims of Bhopal are suffering further harm. As often happens in disasters, the non-affected residents of Bhopal show a general lack of interest for the victims, many of whom are poor immigrants, not from Bhopal.

Numerous private relief organizations appeared quickly on the scene, administered their aid, and departed. The Indian and Madhya Pradesh governments preside over an uneven relief program, consisting of small doles of money, recently scheduled to be revised to $180 per affected family (from $833 for each death and $12–$240 for each person suffering injury or illness, depending upon the severity), and a free ration package of 12 kilograms of oil, and one-half kilogram of sugar per adult each month plus 200 milliliters of milk daily per child. But of the 18,000 families expected to receive compensation, only half have been identified by June 1985 and of these only 4,000 had actually received payment (Khandekar, 1985).

The Indian government announced in June 1985 that 1,500 housing units and a 100-bed hospital would be built to accommodate the most seriously ill Bhopal survivors (Diamond, 1985). But despite the substantial efforts by the Indian government and a reported $27 million expenditure by the Madhya Pradesh state government, the relief effort falls short of what is needed . . .

References

Anonymous (a). (1985). "Probe report gathered dust for 3 years," *Times of India,* January 2.

Anonymous (b). (1985). "The crime continues." *Sunday Calcutta,* April 7: 25.

Browning, J.B. (1984). Director of health safety and environmental affairs, Union Carbide Corporation, in "News release," P-0082-84, December 6: 3.

Brushnan, B., and Subramaniam, A. (1988). "Bhopal: What really happened?" *Business India* 7 (182): 109.

Dagani, Ron. (1985). "Data on MIC's toxicity are scant, leave much to be learned," *Chemical and Engineering News* 40, February.

Diamond, Stuart. (1985). "India, Carbide trade charges," *The New York Times,* June 20: D5.

Kail, L.K., Poulson, J.M., and Tyson, C.S. (1982). *Operational safety survey.* Danbury, CT: Union Carbide Corporation.

Keswani, Rajkumar. (1982). "Sage, please save this city," *Hindi Weekly, Saptahik Report,* September 17.

Keswani, Rajkumar. (1982). "Bhopal on the mouth of a volcano," *Hindi Weekly, Saptahik Report,* October 1.

Keswani, Rajkumar. (1982). "If you don't understand, you will be wiped out," *Hindi Weekly, Saptahik Report,* October 8.

Keswani, Rajkumar. (1984). "Bhopal's killer plant," *Indian Express* (Delhi), December 9.

Khandekar, Sreekant. (1985). "An area of darkness," *India Today* 134, June 30.

Kimmerle, G., and Iben, A. (1964). Toxicitat von Methylisocyanat und dessen quantitativer Bestimmung in der Luft, *Arkive für Toxicologie* 20, 235–241.

Munilidbaran, Sukumar. (1985). "The Bhopal tragedy: Carbide's counter-offensive, special report," *The Herald Review* (Bangalore) 3, April 7.

Rout, M.K. (1985). *The Bhopal tragedy: Analysis and related issues.* Bhubaneswar: Orrisa State Prevention and Control of Pollution Board.

Susman, Paul, O'Keefe, Phil, and Wisner, Ben. (1983). "Global disasters: A radical interpretation." In K. Hewitt (Ed.), *Interpretations of calamity.* Boston: Allen and Unwin: 263–293.

Union Carbide Corporation Engineering and Technology Services, Central Engineering Department, Safety/Health Group. (1984). *Operational safety/health survey.* Charleston, WV: Union Carbide Corporation. Reprinted in U.S. Congress, House Committee on Energy and Commerce, Subcommittee on Health and the Environmental, Hazardous Air Pollutants: Hearing. December 14, 1984, 8th Congress, 2nd sess., 1984, Serial 192, 458–476.

Waldhoz, Michael. (1985). "Bhopal death toll, survivor problems still debated," *Wall Street Journal,* March 21: 22.

Wijkman, A., and Timberlake, L. (1985). *Natural disasters: Acts of God or acts of man?* London: Earthscan.

Worthy, Ward. (1985). "Methyl isocyanate: The chemistry of a hazard," *Chemical and Engineering News* 11 (28): February 11.

Endnote

1. *Environment,* Vol. 27, #7, 1985. "Avoiding Future Bhopals" by B. Bowonder, J. Kasperson, and R. Kasperson. Reprinted with permission of the Helen Dwight Reid Educational Foundation. Published by Heldref Publications, 1319 Eighteenth St., NW, Washington, DC 20036-1802. Copyright © 1985.

CHAPTER *9*

Lessons Learned from the Case Studies of Technological Disasters

"It is important to acknowledge mistakes and make sure you draw some lessons from them."

—Bill Gates

A startling fact emerges from the 12 exemplary case studies of technological disasters presented in Chapter 8: Many of the cases of technological disaster considered in that chapter, as well as those discussed in Chapter 5, could actually have been prevented! Indeed, an argument can be made that many of the technological disasters could have been anticipated on the basis of what experts and executives in the various industries involved already knew. According to the civil engineer Henry Petroski, "the greatest tragedy underlying design errors and the resultant failures is that many of them do indeed seem avoidable, yet one of the potentially most effective means of improving reliability in engineering appears to be most neglected" (Petroski, 1994: ix)—namely, the critical importance of studying disasters.

SPECIFIC LESSONS LEARNED

An analysis of the various case studies of technological disaster presented in Chapters 5 and 8 yields sets of specific lessons to be learned (see Figures 9–1 and 9–2). By identifying and classifying

Case Studies	Lessons Learned
1. USS *Princeton* Explosion	1. Marked deviation from design standards for naval guns led to weaknesses in the barrel, resulting in an explosion during firing.
2. *Titanic* Sinking	2. Overconfidence in technology creates a sense of omnipotence and lack of vigilance.
3. Aisgill Train Wreck	3. Human resource management failure and rigid management scheduling stemmed from preoccupation with maximizing profits.
4. Johnstown Flood	4. Shoddy construction standards reflected callous attitudes of owners of fishing club toward subordinate social classes.
5. DC-10 Crash	5. FAA's regulatory failures to control design problems of aircraft manufacturers and reluctance to issue airworthiness directives grounding defective aircraft until repairs were made.
6. Tenerife Runway Collision	6. Miscommunication among pilots and air flight controllers and inability to control stress and anxiety paved the way for the collison.
7. Santa Barbara Oil Spill	7. Corporate power of oil companies subordinated public considerations of health and safety to its own calculus of profits.
8. Love Canal Toxic Waste Contamination	8. School board and community organizations failed to appreciate health hazards of toxic chemicals, and the polluting corporation covered up its irresponsible behavior to limit liability and malfeasance.
9. Apollo I Fire	9. Faulty wiring by contractors combined with failure of NASA personnel to use proper methods of cabin pressurization.
10. Three Mile Island	10. Complexity of technology increased the fallibility of human judgment.
11. Challenger Disaster	11. Conflicting agendas of NASA and Morton Thiokol management and engineers led to faulty risk assessment and decision making.
12. Bhopal Poison Gas Release	12. Depreciation of the value of life in a Third World country justified cutting corners in the design of the plant, including laxness of safety standards.

Figure 9–1

Specific lessons learned from the case studies of technological disasters presented in chapter 8.

Case Studies	Lessons Learned
1. Dee Bridge Collapse	1. Extension of design beyond known limits and past successes in design breed failure.
2. Hyatt Regency Walkway Collapse	2. Engineers fail to conform to acceptable practice in communicating "design intent" to contractors; imprudent deviation from original design of connection structures linking the two walkways.
3. Therac-25 Computer Malfunction	3. Design defects in the computer system; insufficient testing failed to identify avoidable bugs in the computer program.
4. SS *Mendi* Collision	4. Reckless behavior of the captain on the high seas caused two ships to collide; noncompliance with regulations to come to the aid of distressed ships at sea caused the death of 800 military personnel.
5. USS *Vincennes* Failure	5. Human-computer interaction failure; operational difficulties in the interpretation of the hyper-complex computer system AEGIS, under stress of battle, led to operator anxiety, task fixation, and unconscious distortion of facts.
6. Chernobyl Nuclear Catastrophe	6. A poorly designed graphite-moderated nuclear reactor, accompanied by numerous violations of operating procedures, careless disablement of safety systems, and reckless operator judgments led to the disaster.
7. *Sultana* Sinking	7. Laxity in the enforcement of safety standards provided a climate of carelessness, negligence, and bad judgments: 1,800 men are loaded onto a steamboat with the capacity of 376. Excessive weight put stress on the boilers, which exploded, sinking the boat. Ignorance of boiler safety standards was also responsible for the failure.
8. B.F. Goodrich Brake Scandal	8. Elite engineers were involved in a conspiracy to falsify data and develop a scheme of deception to protect firm's contract with the U.S. Air Force.

Figure 9–2
Specific lessons learned from the case studies of technological disasters presented in chapter 5.

Case Studies	Lessons Learned
9. BART System Malfunctions	9. Automated control system was found to have major design flaws; testing and operator training was inadequate; major software bugs plagued the computerized transit system.
10. Monongah Mine Explosion	10. Ignorance of safety issues and lack of enforceable safety regulations were primarily responsible for the mine disaster.
11. Triangle Shirtwaist Factory Fire	11. Gross violations of building safety codes and callous disregard for immigrant workers were significant causes of the fire.
12. Dow Corning Breast Implants	12. Distorted cultural values concerning the female body image, inadequate regulatory control over medical devices, and the company's withholding of data of known dangers of breast implants all contributed to the implant failure.

Figure 9–2 (*continued*)

the causes and lessons of technological disasters, we create a fund of ideas for better technology assessment and technology management.

If an analysis of the case studies of technological disasters helps identify known hazards associated with various technologies, the obvious questions arise: Why were the known risks not adequately taken into account? Why were steps not taken to prevent them? These case studies reveal that time and again, the same or similar mistakes have led to technological disasters. This insight has led us to develop our four root causes of technological disaster. It is our thesis that the analysis of case studies is the key to understanding the nature of technological disasters. Studies of failures provide opportunities for design engineers and organizational analysts concerned with technology management to heed the lessons learned and hence to avoid making similar errors in the future. If design engineers, organizational planners, and policymakers—those who decide how organizational structures, machines, and other technological systems are designed and built—would study technological disasters with the same systematic approach given to the studies of technological successes, they could greatly improve the reliability of the technological systems they design, implement, and manage.

GENERAL LESSONS LEARNED

One obvious lesson is that we *can* learn from technological disasters, provided that design engineers and technology management specialists share important information about past failures. Unfortunately, this seemingly obvious point is often neglected. Also neglected are human factors, such as those identified by Glegg, a specialist in engineering design:

> . . . we may forget that a 'designer' or 'engineer' is first and foremost a human being. A human being who designs or engineers has, in additon, a wide range of complicated and interacting emotions. It is easy to become emotionally involved in a new invention and so resent criticism of it even when that criticism, if heeded, would turn a failure into a success. (Glegg, 1971: 92–93)

Organizations that manage high technology all too often focus on narrow and disparate goals, overlooking the perspectives of the full cast of stakeholders or constituencies involved. Commercial goals of economy and efficiency often trump values of safety and the public good. As Roger McCarthy, principal engineer for Failure Analysis Associates—one of the premier engineering firms that investigates technological failures—points out, "The systems that involve huge energy reservoirs and therefore the highest potential for catastrophic accidents are also the ones where huge amounts of capital are tied up" (Roush, 1993: 52). In other words, since billions of dollars are tied up in the industrial production of oil, chemicals, transportation, electricity, nuclear power, etc., it is often difficult—sometimes practically impossible—to hold these industries fully accountable for their neglect of safety because of the power and influence they wield in our industrialized civilization. This fact leads to "organizational inertia" concerning the proper management of technology and the "institutional neglect of low-probability, high consequence events" (Roush, 1993: 54).

A few examples will advance the argument that lack of shared knowledge concerning failures impedes the ability of stakeholders to learn from past mistakes and, as a result, leaves them unprepared to fully understand how and why technological disasters happen. The reason why so many people perished in the *Titanic* tragedy was the inadequate number of lifeboats. In fact, the *Titanic* lacked a sufficient number of lifeboats decades *after* most of the passengers and crew of the steamship *Arctic* perished because of the same problem (Martin and Schinzinger, 1996: 80). On May 15, 1980, a ship rammed the Tampa Bay Skyline Bridge. The bridge subsequently collapsed, eventually killing 13 people. This disaster, "was the largest and most tragic of a growing number of incidents of ships colliding with bridges over navigable waterways" (Martin and Schinzinger, 1996: 83). Other notorious ship-bridge collisions are the Maracaibo Bridge (Venezuela, 1964) and the Tasman Bridge (Australia, 1975). The Tampa Bay Skyline Bridge collapsed when hit by a ship because the bridge was not designed with horizontal impact forces in mind—the same engineering cause at the center of all of the bridge failures mentioned (Martin and Schinzinger, 1996: 82). According to Martin and Schinzinger, horizontal impacts were not considered simply because the relevant codes did not require it.

It is well known, of course, that valves are notoriously unreliable components of hydraulic systems. Yet, a pressure relief valve, and "lack of definitive information regarding its open or shut state," contributed to the nuclear reactor accident at Three Mile Island on March 28, 1979 (Martin and Schinzinger, 1996: 83). Similar malfunctions had occurred with identical valves on another nuclear reactor just six months earlier at the Toledo, Ohio, nuclear power plant. The required reports of malfunction were filed with Babcock and Wilcox, the manufacturer of both the Toledo reactor and the reactor at Three Mile Island. However, the manufacturers failed to share the information with the managers at Three Mile Island (Martin and Schinzinger, 1996: 84).

Problems of communication and/or mismanagement need to be overcome and replaced with open discussion and debate about design and other engineering failures in order to prevent repeating similar mistakes in the future. The few examples just discussed are hardly an exhaustive account of past disasters that tragically did not prompt technologists, managers, and policy makers to heed important lessons. They are illustrative of a telling point that "ignorance is the father of disaster, a father whose progeny multiply hideously when powerful, complex technologies are involved" (Mark, 1987: 49).

In fact, John Kemeny, Chairman of the Presidential Commission on TMI, points to individual and organizational failures to learn from past disasters as crucial factors that led to the TMI accident.

> Of the three 'people problems' [discussed earlier], I saved the Nuclear Regulatory Commission for last. I have to report to you that the agency . . . was a total disaster. It was clearly not part of the solution but a serious part of the problem . . . they had no systematic way—I mean that absolutely literally and I am repeating sworn testimony by senior NRC officials—*they had no systematic way of learning from experience.* It was an agency convinced that the equipment was so foolproof that nothing bad could possibly happen . . . (Kemeny, 1980: 69)

Our analyses of case studies in Part IV, as well as the discussions of numerous case studies throughout the book, provide an opportunity for improving the safety of technological innovations. This is achieved by alerting design engineers and organizational and policy planners to common pitfalls in the design, management, and diffusion of technological systems—something that must begin with the study of failures. According to Petroski, "the concept of failure is central to the design process, and it is by way of thinking in terms of obviating failure that successful designs are achieved" (Petroski, 1985:1). Petroski's point is that an engineering or technological disaster often can serve as a powerful stimulus for the development and creation of new engineering knowledge.

The examination of a failed technology frequently brings to the fore previously unknown, underestimated, or neglected variables concerning the materials and components. These generally remain concealed or even suppressed until a system failure forces a reexamination of existing scientific and technical knowledge. This can induce a reevaluation of the technology, followed by adjustments and corrections of errors. While technological failures and disasters have a variety of negative consequences, in some instances the lessons they generate, if heeded, may improve the standards and safety of technological systems.

The *Titanic* sinking, for example, led to unprecedented international regulations and the development of international organizations concerned with safety on the high seas. Starting in 1913, two years after the *Titanic*'s sinking, a series of annual conferences were convened on the Safety of Life at Sea (SOLAS). Also in 1913, the International Ice Patrol (IPP) was formed. The IPP is still in operation, now using sophisticated equipment such as aerial surveillance, satellite images, and radio-equipped oceanographic drifter buoys to detect icebergs (Tenner, 1997: 330).

Consider another example: After the Apollo I fire, NASA made hundreds of modifications, including the installation of higher quality wiring and flameproof coatings over all wire connections, as well as the removal of virtually all flammable materials from inside the module. A new fire-resistant material known as Beta cloth was developed for astronauts' spacesuits. Finally, adjustments were made that required the cabin to be filled with a combination of nitrogen and oxygen, instead of pure oxygen, as was the case with the Apollo I (Van Duyne, 1994).

Such case studies are rife with lessons to be learned from technological disasters. A new technology, once developed and used, is first evaluated for success. The accompanying perceptions are that a technology, once developed and accepted as safe, becomes a standard and permanent addition to our technological civilization. In fact, this was the perception of NASA engineers and scientists when they evaluated the success of the unmanned Gemini flights and decided to send manned missions into space with the introduction of the Apollo program. When a technological innovation unexpectedly proves to be defective, it shatters our convictions about the dependability of our handiwork. It forces us to reevaluate the components of the technological achievement. In the course of reevaluating the technological components, scientists and engineers are compelled to redevelop or reengineer the innovation. This suggests a cycle, as shown in Figure 9–3.

The cycle of technological development—success→failure→ reevaluation→reengineering—is illustrated in many of our cases involving technical design defects. A successful engineering development is often generalized and becomes a standard for engineering practice—until an unexpected failure occurs. Engineers and applied scientists need to become sensitive to such cycles because, to paraphrase the philosopher George Santayana: those who ignore the lessons of technological disasters are bound to repeat them.

It is crucial that such learning from technological disasters becomes routine in industry operations and engineering school education. As Roush points out:

> A society facing a technological disaster is presented with a
> choice: whether to repair the technology in question and get on

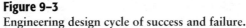

Figure 9–3
Engineering design cycle of success and failure.

with life as quickly as possible, or whether to use the facts brought to light to map out the ways in which the society depends on that technology, the extent to which these needs are legitimate, and how they might be met more safely or fairly. (1993: 50)

Hence, it behooves technologists and policy makers to take heed of the engineering design cycle presented in Figure 9–3. Only if we can identify patterns that lead to technological disaster can we learn to prevent them. With the aid of such knowledge we can come to a better collective understanding about the social role of technology and its positive, as well as negative impacts. Along with Roush (1993), we would like to point out that we have at least two choices. We can submit the lessons learned from technological disasters to scientific scrutiny, cataloguing and classifying, developing theories as to their origins, causes, and mitigation. We could also participate in widespread public debate over their significance, involving legislators, corporate executives, public interest groups, and concerned citizens. However, we can also simply close our eyes to the consequences and lessons of technological disasters. We can decide to write them off as unavoidable "externalities" associated with technological "progress." We can ignore the lessons learned, or we can realize how crucial such lessons can be for the proper management of technology.

Now, as we argued previously, the functioning of our sociotechnical systems involves a complex web of factors—technical, human, organizational, and socio-cultural—each involving perhaps hundreds of separate items and all subject to failure. In order to identify the relevant actors involved, failure analysis must extend beyond mere technical causes. Taking full advantage of the lessons learned from technological failures and disasters would require extending the analysis to the human factors, organizational factors, and socio-cultural factors.

As technology becomes more complex and its scope becomes wider, human factors become increasingly more crucial in terms of a technology's success or failure. Large-scale technologies are so complex that simple inattention, human error, poor training, or miscommunication may become causes of potentially destructive failure. This was evident in the case of Three Mile Island, where the complexity of technology increased greatly the fallibility of human judgment, especially when coupled with human inattentiveness, inadequate training, confusion, and miscommunication. The Presidential Commission acknowledged that human causes of the Three Mile Island accident were evident at all levels: operational, managerial, and regulatory. As the report states:

> We are convinced that if the only problems were equipment problems, this Presidential Commission would never have been created . . . but wherever we looked, we found problems with the human beings who operate the plants, with the management that runs the key operation, and with the agency that is charged with assuring the safety of nuclear power plants. . . . To prevent nuclear accidents as serious as Three Mile Island, fundamental changes will be necessary in the organization, procedures, and practices, and above all, in the attitudes of the Nuclear Regulatory Commission; and to the extent that the institutions we investigated are typical, of the nuclear industry. (Peterson, 1982: 36)

Additional case studies provide ample lessons to be learned about the role of human factors in generating technological disasters. The Tenerife runway collision illustrates the disastrous consequences of individuals' inabilities to handle the stress and anxiety that often accompany the operation of complex technologies and teach us about the vulnerabilities of human attempts to manage such technologies. The same lessons can be learned from the shooting down of the Iranian airliner by the USS *Vincennes,* as well as the negligent operations and management of the Chernobyl nuclear power plant.

Technological disasters also teach us that we must be alert to the unanticipated side effects of technological innovations (Tenner, 1997). Some of the unintended consequences of technology have the potential for mass destruction of natural habitats and other types of environmental and ecological destruction. Chernobyl, Exxon *Valdez,* Bhopal, Love Canal, and the Santa Barbara

Oil Spill are all cases in point. Less dramatic, but certainly more potentially destructive, are the phenomena of global warming and the depletion of the ozone layer (Anonymous, 1998; Kerkin, 2000; Houlder, 2000).

Such unintended consequences often point to lessons that can be learned if the organizations involved in the development of technology, along with the influence of the profit motive on their operations, are made transparent (Tenner, 1997: 310). In the Love Canal and Bhopal cases, for example, we discover that instead of being designed and operated to be safety-promoting, the systems in question end up being error-inducing. Tenner concludes that:

> The real question is not whether new disasters will occur. Of course they will. It is whether we gain or lose ground as a result. It is whether our apparent success is part of a long-term and irreversible improvement of the human condition or a deceptive respite in a grim and open-ended Malthusian pressure of human numbers and demands against natural limits. (Tenner, 1997: 348)

As we have pointed out in Chapter 4, human beings do indeed have the capacities to successfully manage technology in a variety of ways—as in the case of High Reliability Organizations. One way is to focus on what lessons can be learned from the mismanagement of technological systems. An adequate understanding of how the lessons of failed technologies can lead to more reliable design and management strategies is possible only if the analysis is extended, as Jasanoff points out, "to the organizational, social, political and moral dimensions" of technological disasters (1991: 2). According to Jasanoff:

> There is a new emphasis on the organizational context in which the failed technology was embedded and, with this, a recognition that corrective policies have to address not only the design of artifacts but also (indeed perhaps even more so) the human practices and presuppositions that determine their management and use. Seen from this perspective, a serious technological mishap ceases to be merely accidental, for it opens windows onto previously unsuspected weaknesses in the social matrix surrounding the technology. (1991: 2)

In addition to the potential for organizational learning, technological disasters also teach us about the functioning of technology in society, the support systems surrounding it and the socio-cultural matrix of values that sustain it. The destruction following the methyl isocyanate leak at Bhopal provided important lessons about the strengths and weaknesses of communication systems, social welfare systems, and emergency preparedness operations, including search and rescue facilities (Wexler, 1990; Sethi, 1992). The potentially avoidable tragedy of Bhopal has also forced us to consider global questions of corporate responsibility, relationships between governments and corporations, and the complex socio-cultural factors involved in the transfer of sophisticated technologies from industrialized countries to Third-World countries.

Multinational corporations have an important and constructive role to play in promoting economic growth and environmental protection in the Third World. This proves to be difficult in practice, however, given the various and often inconsistent cultural value systems and legal structures that exist between different nations. Nevertheless, one lesson from the Bhopal disaster is clear:

> If multinational corporations want access to markets and materials in developing countries, and if these countries want the benefit of direct foreign investment on a sustainable basis, then the public and private sectors will have to set new precedents in cooperating to achieve environmental protection. Unless this happens, further tragedies like Bhopal and other smaller and less visible incidents . . . will eat away the fabric of confidence on which progress depends. (Speth, 1988: 15)

One of the lessons to be learned from the Bhopal case is that the cause of technological disasters may involve factors that transcend organizational boundaries, such as social, political, and cultural variables. Kapitza (1993) identifies related kinds of socio-cultural causes involved in the failure at Chernobyl such as negligent regulation by the Soviet government of its commercial nuclear power industry. Valerii Legasov, chief deputy director of the Kurchatov nuclear energy institute, criticized Chernobyl management for not being more concerned with problems of safety in the overall operations of the plant (Rich, 1988: 285).

Neither the executive branch of the federal government nor the legislative, administrative, or judicial branches are adequately equipped with the knowledge and resources required to handle complex questions about the development and deployment of large-scale sociotechnical systems. This is one significant lesson

John Kemeny, Chairman of the Kemeny Commission on Three Mile Island, draws from the accident at Three Mile Island:

> The message I want to give you, after a hard and long reflection, is that I'm very much afraid it is no longer possible to muddle through. The issues we deal with do not lend themselves to that kind of treatment alone. Therefore, I conclude that our democracy must grow up. (Kemeny, 1980: 74)

Kemeny has two suggestions for democracy's maturity in an age of technology. First is "the existence of respected, nonpartisan, interdisciplinary teams" of specialists involved at all levels of public policy concerning decision making about technology (Kemeny, 1980: 74). Kemeny's second suggestion emphasizes the proper training and education of scientists and technologists, an education that goes beyond the simple dissemination of technical facts and mathematical theories. There is a need to "educate the next generation of leaders so that they can directly understand and come to grips with the monumental issues of our time" (Kemeny, 1980: 74). Kemeny's main lessons are reserved for technologists and politicians. These two groups, along with corporations, are primarily responsible for the development and deployment of potentially hazardous technologies.

Who is responsible for safeguarding the public's interests when dangerous technologies are developed and implemented? Included among relevant watchdog organizations are government agencies and public interest groups, along with various professional societies responsible for setting safety standards. There are indeed a variety of government agencies with a mandate to protect the public against harms produced by different industries. The same can be said for professional and educational organizations. However, the diffusion of responsibility among the multiple organizations all too often results in the actual neglect of the social and economic harms caused by failures of technology.

As we will see in Part V, where we take up these policy issues in more detail, the present state of our educational, professional, and political institutions reveals that serious hurdles must be overcome in order to protect ourselves from technological hazards.

In sum, failures of technology teach us about the limitations of human knowledge, the tragedy of hubris, the complacency with which we treat the powers of technology, and the failure of our institutions to control and manage individual and societal risks. "We cannot allow our fascination with the power of what we can do to blind us to what we cannot. It is no longer a matter of humility. It is a matter of survival" (Dumas, 1999: 328).

Understanding gained from a close analysis of technological disasters, suggests the identification of a set of general lessons—such as those listed in Figure 9–4—which, if heeded, could lead to the design, development, and management of safer and more reliable technology.

1. Despite expert engineering design and the application of "state-of-the-art knowledge," there is no total immunity from risk. A frequent disjuncture between science and technology increases the probability of failure.

2. No matter how sophisticated the hardware, the human element is present in setting requirements, setting designs to meet them, in production, testing, installation, maintenance, repair, and operation. Moreover, human beings are vulnerable to ignorance, error, and greed.

3. Malfunctions due to human error, especially in decision making, are more likely when psychological factors such as stress, anxiety, or "groupthink" are operative.

4. Organizational complexity places special demands on communication linkages that are often inadequate, fragmented, and nonintegrated.

5. The routine management of technology is too often focused on narrow organizational objectives that seek to maintain the status quo, meet deadlines and quotas, and reduce costs. Such objectives tend to compromise the safety of technological systems.

6. In almost all of the cases discussed, failure to properly assess the potential negative impact of technology consequently led to the resulting tragic effects.

7. Technology acts as an organizing force that requires massive natural, human, and financial resources, all of which have the tendency to concentrate both economic and political power in the hands of the few people who control it, often at the expense of vulnerable groups in society.

8. Complex technologies are supposedly monitored and controlled by public policy implemented at the governmental level. These policies are intended to safeguard both people and property. However, both administrative agencies, and citizen "watchdog" groups are often bureaucratized and/or underfunded, rendering them unable to achieve their objectives.

9. It is crucial that communication between the private and public sectors be made visible and publicly accountable. This would lead to more democratic governance over decisions regarding technological development.

10. Every failure of technology raises ethical issues because society is technology's ultimate patron and is thus affected by all of technology's impacts and consequences.

Figure 9–4
General lessons learned from the analysis of case studies of technological disasters.

CONCLUSION

The case studies presented throughout this book have drawn attention to the various engineering, economic, organizational, political, and cultural constraints that impede effective learning from the lessons provided by technological disasters. Due to such constraints, learning is all too often incremental and unsystematic rather than comprehensive and systematic. Would that Petroski's observations were true that "the more case histories a designer is familiar with or the more general the lessons he or she can draw from the cases, the more likely the patterns of erroneous thinking can be recognized and generalizations reached about what to avoid" (Petroski, 1994: 6). Current learning from technological disasters can be best characterized as "muddling through" (Lindblom, 1957). However, muddling through is inadequate in that it fails to lead to a "paradigm shift" in organizational, institutional, and social learning essential for the kind of collective wisdom needed for the design, development, and diffusion of safer, more reliable, and more humane technologies. The set of general lessons enumerated in Figure 9–4 attempt to provide the first steps toward such a systematic paradigm shift.

References

Anonymous. (1998). "Ukraine tallies sharp rise in illnesses near Chernobyl," *The New York Times,* April 23: A5.

Dumas, Lloyd. (1999). *Lethal arrogance: Human fallibility and dangerous technologies.* New York: St. Martin's Press.

Glegg, Gordon L. (1971). *The design of design.* Cambridge, England: Cambridge University Press.

Houlder, V. (2000). "Hole in ozone layer could be closed within 50 years," *Financial Times,* December 4: 16.

Jasanoff, Sheila. (1991). "Introduction: Learning from disaster." In *Learning from disaster: Risk management after Bhopal.* Philadelphia: University of Pennsylvania Press: 1–21.

Kapitza, Sergei. (1993). "Lessons of Chernobyl: The cultural causes of the meltdown," *Foreign Affairs* 72 (3): 7–12.

Kemeny, John G. (1980). "Saving American democracy: The lessons of Three Mile Island," *Technology Review* 83 (7): 65–75.

Kerkin, A. (2000). "Treaty talks fail to find consensus in global warming," *The New York Times,* November 26: 1.

Lindblom, Charles. (1957). "The science of muddling through," *Public Administration Review* 19: 79–88.

Mark, Hans. (1987). "The Challenger and Chernobyl: Lessons and reflections." In Hans Mark, Tom L. Beauchamp, Jesse Luton, Martin Marty, and Andrew Cecil (Eds.), *Traditional moral values in the age of technology.* Dallas: University of Texas Press: 31–57.

Martin, M., and Schinzinger, R. (1996). *Ethics in engineering.* New York: McGraw-Hill.

Peterson, Russell W. (1982). "Three Mile Island: Lessons learned for America." In Christoph Hohenemser and Jeanne X. Kasperson (Eds.), *Risk in the technological society.* Boulder, CO: Westview Press: 35–45.

Petroski, Henry. (1985). *To engineer is human: The role of failure in successful design.* New York: St. Martin's Press.

Petroski, Henry. (1994). *Design paradigms: Case histories in error and judgment in engineering.* Cambridge, England: Cambridge University Press.

Rich, Vera. (1988). "Legasov's indictment of Chernobyl managment," *Nature,* 333 (26): 285.

Roush, Wade. (1993). "Learning from technological disasters," *Technology Review,* August/September: 50–58.

Sethi, Praskash S. (1992). "The inhuman error: Lessons from Bhopal," *New Management* 8: 40–46.

Speth, James. (1988). "What we can learn from Bhopal," *Environment* 27 (1): 15.

Tenner, Edward. (1997). *Why things bite back: Technology and the revenge of unintended consequences.* New York: Vintage Books.

Van Duyne, S. (1994). "Apollo I capsule fire." In N. Schlager (Ed.), *When technology fails: Significant technological disasters, accidents, and failures of the twentieth century.* Detroit, MI: Gale Research: 580–586.

Wexler, Mark. (1990). "Learning from Bhopal," *The Midwest Quarterly* (31): 106–129.

PART V

Strategic Responses to Technological Disasters

10

The Responsibilities of Engineers and Scientists

". . . One characteristic of a professional is the ability and willingness to stay alert while others doze. Engineering responsibility should not require the stimulation that comes in the wake of catastrophe."

—Samuel C. Florman

With the advent of modern industrial society, it has become increasingly evident that engineers and scientists, through their work, largely define the course of our technological civilization that is radically altering our world. Yet, while colleges and universities strive to provide their students with the technical and scientific skills necessary to master complex technologies, they rarely prepare them for the potential role they play as agents of social transformation. Most engineers, according to Nandagopal, leave college uninformed and unconcerned about the societal issues relevant to the practice of engineering (Nandagopal, 1990). This predicament arises because of the narrow scope of engineering curricula: social and ethical implications of engineering are, for the most part, ignored. Since the students' education is predominantly technical, students often come to believe that engineering is divorced from human values and the larger societal impacts of technology (Durbin, 1989). Engineering education fails to help engineers become self-conscious and responsible social actors because both educators and practicing engineers often view engineering education and practice in restricted terms (Wacker, 1990).

THE ROLE OF ENGINEERING SCHOOLS

There is a growing need to inculcate, through engineering education, an awareness of the complex set of interrelated issues of engineering design, public safety, and ethics—in particular the potential dysfunctional effects of engineering and technology on society (Glagola, Kam, Loui, and Whitbeck, 1997). For many engineering professors, "engineering is essentially the application of technical expertise to technical problems, and education is primarily the process by which this expertise is reproduced for new generations of engineers" (Glagola *et al.*, 1997). Contrary to this conventional wisdom, engineering is a social as well as a technical practice, and engineering education should promote an understanding of *both* dimensions. Since technology creates potential risks as well as great benefits, engineers, as experts in technology, should assume a more positive stance toward their social responsibilities. They should "not adopt the stereotypical hyper-rational, calculating character of the technocrat, but should apply humanistic values along with technical insights" (Wacker, 1990).

Most engineering educators would agree that two important steps need to be taken: (1) a reevaluation of the role of design and safety in engineering curricula, and (2) the introduction of a comprehensive engineering ethics program in engineering curricula. These two steps are closely related; engineering design courses, as opposed to theory courses, can more readily incorporate ethical and social factors—the social aspects of design—into their curricula (Unger, 1982; Carpenter, 1984; Vanderburg, 1995).

Responsibility for Safety in Engineering Design

The first step engineering schools can take to mitigate the effects of technological disasters is to successfully integrate ethical and social issues in courses dealing with engineering design. According to a study by Lichter (1989), engineers in their educational experiences, in professional societies, and in actual employment are inclined to consider safety only as a necessary constraint on efficient engineering decisions. Such a constraint tends to be viewed *reactively* rather than *proactively*. In his investigations, Lichter found that concern for engineering safety is "tacked on" to a primary concern for "technical" design; safety-related activities are viewed as essentially "non-productive" (Lichter, 1989). As Lichter points out, "this is best reflected in the budgeting of engineering projects where safety is seen as an added cost which, in the absence of a public mandate or evidence of profitability, would normally be avoided in order to maintain a competitive position in the marketplace" (p. 217). Because of such economic constraints, the design engineer might be tempted to use an unconventional technology in order to comply with financial constraints. It has finally come to the attention of engineering educators that unwarranted pressures are often placed on the young engineer in industry. Now, more than ever, educators should stress safety as a central element in all engineering designs (Moonashingha, 1996). Designs must be infused with safety and ethical values, not just with financial values.

Albert Flores has documented the general failure of engineering education to expose future engineers to any serious treatment of safety-related topics (Flores, 1982). Approximately 250 engineers responded to a survey on a wide range of safety-related issues. Three-quarters of those engineers reported never having had a course on safety, and, for those who had been exposed to some treatment of safety, most claimed that the treatment was superficial and rarely the primary focus of the course. Similar results were obtained from a survey of 450 engineers at Otis Elevator, Boeing, and NASA (Flores, 1982).

The fact that Flores's survey was conducted in 1982 ought not lead readers to suppose that the current status of design safety in engineering curricula has changed significantly. While researchers recognize that design safety is becoming more and more important to engineering education, they also point out that engineering schools still have a long way to go to increase the salience placed on design safety (Bryan, 1999; Lau, 1998; Slaymaker and Bates, 1993; Vesilind, 1992).

Rapid developments in engineering innovation are constantly being introduced into engineering materials, structures, and systems. Because of these rapid advances, which often stretch the limits of engineering knowledge and lead into relatively unknown and untried areas of research and development, Bryan argues that "all aspects of the design stage should be examined closely for possible safety and health hazards" (Bryan, 1999: 34). Prominent examples are nuclear power, telecommunications, computer networking, refineries and industrial chemicals, and bioengineering. One field of engineering, however, has begun to take the lead in focusing on the central role of design safety in classes and laboratories. This field is chemical engineering (Akgerman, Anthony, and Darby, 1999; King, 1998; Willey and Piece, 1998). The relatively new field of bioengineering would be the next likely area to start integrating ethical and social values into engineering safety design courses (Grundfest and Scott, 1998).

Stephen Unger was one of the first engineer-ethicists to stress the need to promote design safety in the engineering curriculum (Unger, 1982). In his 1982 publication, Unger outlined the nature of the responsibility that engineers bear for producing safe designs and safe products, both consumer and industrial. He suggests that engineering schools introduce ideas about safety and responsibility through courses on engineering ethics and technology-society studies.

In recent years, numerous engineers have faced ethical dilemmas in which engineering judgment ultimately led to technological disaster. What is more, these cases have received considerable attention and publicity (Pletta and Hon, 1987). Examples include the problem of safely positioning the gas tank in the Ford Pinto case, the accidents of the Bay Area Rapid Transit system in San Francisco, the cargo door latch on the DC-10 aircraft, the falsification of test data in the B.F. Goodrich aircraft brake case, technical miscalculation in the Hyatt Regency walkway collapse, and faulty risk assessment and flawed decision making in the Challenger disaster. All of these cases involved engineers and/or engineering decisions and have become the subject of analyses by engineers, educators, philosophers, and sociologists.

In the DC-10 case, design flaws in the rear cargo door were well known to top engineers and management, as well as to the Federal Aviation Administration. Neglect of safety and design issues in engineering practice, management decisions, and governmental regulation led to a few serious incidents, but the Paris crash was the most serious. All 346 passengers on board lost their lives due to the defectively designed door. In the B.F. Goodrich case, engineers were involved in a conspiracy to commit fraud in a cover-up that included the falsification of technical data to make an unsafe braking system appear safe. In the BART case, engineers were fired for speaking out against what, in their professional judgments, were serious design and safety defects in the newly installed and computerized Bay Area Rapid Transit System. In the Hyatt Regency case, engineers were found guilty by investigators for "failing to conform to accepted custom and practice of engineers for proper communication of the engineer's design intent."

In the Chernobyl case, engineers and other technical personnel, presumed to be competent, violated numerous operational safety rules, causing the worst nuclear power plant disaster to date. Moreover, the odd design of the Soviet reactor was also implicated as a secondary cause of the catastrophe. Long-term neglect of safety requirements at the Bhopal plant, as well as broken-down safety equipment, were partly responsible for the release of tons of deadly methyl isocyanate into the air. The Board of Inquiry in the Apollo I case found that there were "numerous examples of poor installation, design, and workmanship." After the fire, hundreds of modifications were made to improve the safety of the module to make it more fireproof.

Technological disasters sometimes occur when the engineer's judgment is overruled—as in the Challenger and Ford Pinto cases. Sometimes, however, crises arise when engineers fail to report hazardous practices—as in the DC-10 and B.F. Goodrich cases. Finally, some failures result from the direct actions of engineers—as in the Chernobyl and Hyatt Regency disasters. At a minimum, the engineer's obligation with respect to guarding against technological hazards is the conscious effort to produce effective and safe technology in their designs. Engineers are responsible for such technological hazards in many ways, and hence, they have individual and professional duties to design products and technologies that do not harm the public (Schinzinger, 1986; Unger, 1982).

How is it possible to restructure the activity of engineers so that design safety and human welfare, and not only efficient production, are the principal concerns of engineering practice? This would entail, according to Lichter, the modification of the "culture of engineering" as it relates to engineering pedagogy and numerous professional engineering organizations. To accomplish this difficult task, however, engineering educators would have to "institutionalize the notion that the practice of a merely technical activity cannot be divorced from a theory of the normative dimension of engineering activity that helps define the morally proper goals, obligations, and responsibilities of engineering" (Lichter, 1989: 217).

The American engineering profession took positive steps toward these ideals with the establishment of the Accreditation Board of Engineering Education and Technology's Engineering Criteria (ABET, 1998). To become accredited by ABET, engineering schools are required to have at least 24 credits of design content in the four-year undergraduate curriculum. In addition, more and more educators are realizing the importance of introducing ethical and social issues into engineering design courses. However, as Vesilind points out:

> Engineering ethics is often neglected in engineering design because most faculty do not see the value of ethics in engineering. On the contrary, there are many instances in engineering practice when values will be involved and engineering students and professionals will be faced with ethical problems. (1992: 215)

Clearly, what is needed are courses that incorporate and integrate problems of safety, design, and ethics (Manion and Kam, 2000; Rabins, 1998; Unger, 1995; Vanderburg, 1997). Such courses would draw upon actual experiences of practicing engineers to illustrate the social factors of design—aesthetic, economic, political, ethical, environmental, and usage—along with technical factors (Vanderburg, 1995; Whitbeck, 1995). Some engineering schools have recently experimented with such courses. Lau (1998) describes a course that is multidisciplinary and "attempts to increase engineering students' awareness of the relationships between technology and society, and how they can design technology that improves people's lives." McLean (1993) develops a rationale for an ethics of engineering course that "tries to incorporate design, systems management, ethical theory, and insights from sociology and psychology that are completely integrative and interdisciplinary" (p. 20).

Engineering as a Social Practice

Engineering should be understood as a social "practice" for a number of reasons (Wilson, 1990: 117). First, it is through engineering that new technologies are embodied in products and systems developed in order to satisfy the needs and problems facing society. Technology and industry, one might say, fuel the "wealth of nations." Engineering is practiced in corporate organizations that are embedded in larger social structures of economic and power relations. Corporate organizations are critical to the implementation, operation, and maintenance of technological systems. As Wilson points out, "social and economic relations are found at all levels of engineering practice and technology development, hence making engineering a genuine social practice" (1990: 117).

Second, the products of engineering are used in diverse social contexts. To be effective, these products must be designed with their social context in mind. Hence, responsibility for safety and avoidance of public harm are two social goals that make engineering central to almost all social contexts. In fact, the moral imperative to "do no harm" is embedded in many codes of engineering ethics. Identifying this ethical mandate in the very conception of engineering underscores the importance of understanding engineering as a social practice.

Third, engineering has an impact on society in both intended and unintended ways. "New technologies often alter social relations, power relations, working and living conditions, and cultural values" (Wilson, 1990: 118).

Finally, engineering is a social practice because it is often used to tackle and ameliorate social conditions such as urban slums, world hunger, illiteracy, disease, and pollution (Wilson, 1990: 117).

The roots of an engineer's duties and responsibilities to society can be said to be derived from a "social contract" that holds, at least implicitly, between the engineering profession and society. Harris, Pritchard, and Rabins (1995) argue that this social contract between the engineering profession and society at large contains at least four provisions. On the one hand, engineering professionals agree (1) to devote themselves to rendering service to society or to advance the public good and (2) to regulate themselves in the provision of these services. On the other hand, society agrees (3) to give professionals a place of honor and above-average livelihood and (4) to allow them an unusual degree of autonomy in the performance of their professional duties (Harris, Pritchard, and Rabins, 1995). In fact, professional codes have an important role to play in establishing the social contract between the engineering profession and society at large. The code of ethics stands as a "promissory note," made by the engineering profession, to uphold its part of the contract.

Strong commitments to the social contract can be found in almost all contemporary engineering codes of ethics, such as those of the National Society of Professional Engineers (NSPE), the Institute of Electrical and Electronic Engineers (IEEE), the American Society of Mechanical Engineers (ASME), the American Society of Civil Engineers (ASCE), and the American Institute of Chemical Engineers (AIChE), as well as other professional societies. In fact, the emphasis on the obligations to the public—the hallmark of the social contract professional societies model—seems to be accorded an increasingly prominent role in engineering codes of ethics. In 1974, the code of the Engineering Council of Professional Development was revised to state that "engineers shall hold paramount the safety, health and welfare of the public in the performance of their professional duties." Within the last 20 years, this "paramountcy" provision has been adopted by all of the major engineering societies.

THE ROLE OF ENGINEERING SOCIETIES

Engineering societies have an important role to play in managing and controlling hazardous technologies (Unger, 1987). These professional societies can perform at least three major functions in facilitating the management of technological risk and disaster: (1) the advancement of engineering knowledge; (2) the protection of ethical engineers; and (3) the promotion and enforcement of codes of ethics.

The Advancement of Knowledge

As with engineering schools, one important way professional engineering societies can help guard against the hazards of technology is by advancing safety in design. This is accomplished by facilitating the development and dissemination of engineering knowledge, thereby bolstering the engineer's capacity for building safer devices, structures, and systems. In the past, engineering societies have, for the most part, almost exclusively focused their attention on the advancement of technical knowledge. Undoubtedly, the major function remains to promote the discovery of technical knowledge and its dissemination. To this end, engineering societies publish scholarly journals and organize meetings at which technical papers are presented and discussed. The sharing of important knowledge is made possible through the actions of professional engineering societies, such as the Institute of Electrical and Electronics Engineers (IEEE), the American Society of Mechanical Engineers (ASME), the American Society of Civil Engineers (ASCE), the American Institute of Chemical Engineering (AIChE), and the National Society of Professional Engineers (NSPE). For example, relatively recent advances in electronics resulted in improved instrumentation for airlines, thus contributing to safer air travel. A better understanding of materials science has given us more rugged automobile tires, thereby eliminating a major cause of automobile accidents nationwide (Unger, 1987).

An important function of engineering societies that promote the dissemination of engineering knowledge about safety is the development of industry standards. The establishment of the boiler safety codes in the 19th Century set the stage for the role engineering societies have come to play in the development of safety standards. The ASME, for example, administers the National Elevator Code. All in all, more than 200 organizations—professional societies, trade associations, and safety organizations—develop and promulgate voluntary engineering standards in the United States (Sherr, 1982).

Moreover, these standards, voluntarily formulated by the professional organizations, become the de facto standards in industry, in virtually all governmental regulatory agencies, even in legal disputes over technological issues. Professional engineers are responsible for almost all of the technical standards that undergird our entire sociotechnical infrastructure—from telecommunications to computerized networking to the entire built urban environment. This places a huge responsibility for public safety and welfare on the shoulders of professional engineering societies.

The fact that engineering standards are produced voluntarily—that is, they are developed and promulgated by professional organizations themselves—is an example of a self-regulated professional practice. This creates the necessity for institutional trust between engineers, the citizenry, and the government. Hence, this is also part of the social contract existing between engineers and society. Society places high trust in the engineering profession and permits it to develop and abide by its own technical standards, which, as we have noted, become the de facto norms for industry, government, and society at large. The citizenry receives the technological systems it so eagerly adopts, with the tacit promise that the technology will be safe and reliable. In return, engineers are accorded the principle of professional autonomy—at least as far as technical standards are concerned. Thus, a "social compact" is created between the profession of engineering and the citizenry.

Engineering societies coordinate their activities through a loosely organized entity known as the American National Standards Institute (ANSI). For the most part, the ANSI standards are well defined and categorized by technological areas. These areas are as follows:

(1) American Society of Mechanical Engineers (ASME)—Mechanical, Pressure Vessels, Boilers
(2) Institute of Electronic and Electrical Engineers (IEEE)—Electrotechnology, Communications, Power Apparatus and Generation, Nuclear Instrumentation and Control
(3) Society for Automotive Engineering (SAE)—Automotive and Aeronautics
(4) National Fire Prevention Association (NFPA)—Fire and Safety
(5) The American Society for Testing and Materials (ASTM)—Materials and Testing (Sherr, 1982).

Some professional societies have produced literally hundreds of standards. For example, the IEEE has more than 400 standards in print. The American Society for Testing and Materials (ASTM) has published more than 48 volumes of technical standards (Sherr, 1982).

What has been consistently missing from the traditional function of reporting and advancing engineering knowledge, however, is *the reporting and advancing of engineering knowledge about failures*. Lack of established channels of communication, misplaced pride in not asking for information, embarrassment at failure, or fear of litigation often impede the flow of such important information and hence lead to many repetitions of past mistakes.

In a series of influential books, Henry Petroski (1985, 1994) argues that knowledge about failures is essential for good, sound engineering design practices. Failing to present information about failures is a critical deficiency, both for engineering schools and engineering societies. More attention needs to be paid to research and knowledge of failures and disasters in engineering societies' professional conferences and professional trade journals.

Protection of Ethical Engineers

Engineering societies seek to provide support for engineers who come into conflict with corporate management over issues involving the development and management of hazardous technologies. Engineering societies could assist ethical professionals whose conduct has led to retaliation by employers. They could establish funds to pay legal expenses of engineers who are contesting discharges or other types of lawsuits. These funds would not only assist engineers in times of financial and emotional stress but would also signal the support of their professional colleagues. Professional societies could even assist engineers, who have been unfairly dismissed from their jobs, to find new employment. They could support engineers who take strong positions on behalf of worker and public safety, thus leading to greater workplace safety and, in turn, to a safer society. Hence, professional support for socially responsible conduct, the "right" to be an ethical engineer, could play a crucial role in mitigating the effects of technological risk and harm (Unger, 1994). After considering the most likely objections to professional societies supporting individuals and speaking out in the public interest, Unger concludes that:

> Providing support for the ethical practice of engineering is a very worthwhile and appropriate endeavor for engineering societies . . . It is up to engineers, standing together in their professional societies, to see to it that they are no longer subjected to agonizing choices between sacrificing either conscience or career. (1987: 20–21)

The exemplary actions of the Institute of Electric and Electronic Engineers (IEEE) on behalf of the three engineers dismissed at the Bay Area Rapid Transit system (BART) during the 1970s serves as a good example of how engineering societies can come to the aid of ethical engineers.

Two engineers working on the BART project, systems analyst Holger Hjortsvang and programming analyst Max Blankenzee, became concerned with what they perceived to be major design defects in the Automated Train Control (ATC) system, deficiencies they thought would almost certainly compromise public safety. In addition, a third engineer, Robert Bruder, an electrical engineer monitoring various phases of the construction, also became concerned about the unprofessional manner in which the installation and testing of control and communications equipment was being conducted (Unger, 1994: 22). They brought their concerns to management, but management was not responsive to their apprehensions and professionally based judgments concerning safety.

In November 1971, after many attempts to communicate their concerns to their superiors, the three engineers decided they must express their concerns in order to protect the public interest. They contacted Daniel Helix, a sympathetic member of the BART Board of Directors. Unbeknownst to the three engineers, however, Helix released their complaints to the press. When BART management identified all three dissenting engineers as the source of the complaint, they were given the option of resigning or being fired. All three engineers refused to resign, so they were all fired. Some time after their dismissal the three engineers brought a lawsuit against BART, charging breach of contract and violation of their constitutional rights. An extensive report of the BART case was published in the *newsletter* of the IEEE Committee on Social Implications of Technology in September 1973. This report, and many others, including findings of an independent contractor, confirmed the charges made by the three engineers. Extensive redesigns of the ATC and higher safety standards of construction and maintenance were instituted after the BART investigation (Unger, 1994).

Following the publication of the BART report, the case was discussed extensively within the IEEE. In March 1974, a resolution was passed to set up procedures to support engineers who, acting in compliance with ethical principles articulated in the society's code of ethics, may find themselves in professional jeopardy. This resolution eventually led IEEE to file an *amicus curiae* brief on behalf of the three engineers in their civil suit against BART. The brief deals not only with the facts of the case, "but with the broad ethical principles involved; it urged the court to rule that an engineer's employment contract includes an implicit provision that he or she will protect public safety and that discharging an engineer for adherence to this provision constituted a breach of contract by the employer" (Unger, 1994: 26). In other words, IEEE implored the court to acknowledge that engineers have a professional right to implement the provisions of their code of ethics.

One unfortunate consequence of the BART case was that the chance to establish a legal precedent that professionals have a right to implement provisions of their code of ethics was thwarted since the three engineers, because of financial hardships, eventually agreed to an out-of-court settlement with BART management.

Promotion and Enforcement of Codes of Ethics

One of the principal functions of engineering societies has been the promulgation and enforcement of codes of ethics. Professional codes present an image of what professionals understand as their ethical and social obligations. Hence, in analyzing such codes, we can acquire an understanding of how the profession perceives its own purpose as regards its social responsibilities.

The increasing promotion of codes of ethics in professional engineering societies and their growing role in lending support for the right to be a responsible engineer—as is evident in the BART case—demonstrates that engineering professionals *do* indeed have reciprocal responsibilities and rights as regards the functioning of engineering in society. Recognizing the importance of such reciprocal responsibilities and duties to serve the public in their professional practice, physicians long ago developed the Hippocratic Oath. It is noteworthy that not all medical schools require their graduates to take the Hippocratic Oath. According to a study by Orr and Pang (1993), in 1928 only 26 percent of medical schools administered some form of the oath. However, in 1993, the percentage had markedly risen to 98 percent of all medical schools administering some form of the oath.

Requiring that all engineers, upon graduating from engineering schools, take the equivalent of the medical profession's Hippocratic Oath would perhaps provide the clearest statement possible of an engineer's self-proclaimed social responsibilities. Such an "Engineer's Hippocratic Oath" has, in fact, been articulated. The Oath reads as follows:

> I solemnly pledge myself to consecrate my life to the service of humanity. I will give to my teachers the respect and gratitude which is their due; I will be loyal to the profession of engineering

and just and generous to its member; I will lead my life and practice my profession in uprightness and honor; whatever project I shall undertake, it shall be for the good of mankind to the utmost of my power; I will keep far aloof from wrong, from corruption, and from tempting others to vicious practice; I will exercise my profession solely for the benefit of humanity and perform no act for a criminal purpose, even if solicited, far less suggest it; I will speak out against evil and unjust practice wheresoever I encounter it; I will not permit considerations of religion, nationality, race, party politics, or social standing to intervene between my duty and my work; even under threat, I will not use my professional knowledge contrary to the laws of humanity; I will endeavor to avoid waste and the consumption of nonrenewable resources. I make these promises solemnly, freely, and upon my honor. (Susskind, 1973: 118)

In addition to such oaths, almost all engineering societies have formulated a code of ethics for their members. For example, the first canon of the NSPE Code of Ethics states: "The engineer shall hold *paramount* the health, safety, and welfare of the public in the performance of his professional duties" (National Society of Professional Engineers, 2000). This requires members to adopt this concern not as something over and above the profession's requirements and not as remedial, but as integral to the requirements of the profession *per se*—integral to what the profession is all about. As we have seen, this provision is found in almost all engineering codes. Embedded in such prescriptions is the assumption that the goal of engineering is to uphold the values of public health, safety, and welfare above *all* other values. On this view, values such as safety, health, and welfare are overriding values. They "trump" other values such as market competitiveness, profit, efficiency, cost, and benefit-burden trade-offs.

How should engineers interpret the paramountcy clause? If we choose to interpret it as establishing the "do no harm" principle, this is consistent with present engineering practice. Hence, one responsibility engineers owe to society is to design and build safe products, machines, and systems. As Flores (1980) points out: "It is clear that the products, bridges, dams, computer systems, etc. that engineers design should be safe enough to avoid unnecessary harm to life and limb, and perhaps to liberty and property" (p. 212). Moreover, developments in tort law providing for strict product liability have, in many ways, helped to establish this principle in a more formal way (Tribe, 1971a, 1971b). In sum, engineers must abide by the do no harm principle. This entails a duty to always make products and systems that are safe and reliable.

However, no engineering project is totally free of risk, and hence most engineering projects can be interpreted as "experiments" carried out on human subjects. Martin and Schinzinger (1996) argue for such a view. They characterize engineering as social experimentation and, on the analogy with medical experimentation, they argue that the doctrine of informed consent must be recognized and honored between the experimenters (engineers and their corporations) and their subjects (the general public). On this model, an engineering project is seen as an "experiment"—that is, a potentially risky undertaking—on a societal scale (Long, 1983).

Inasmuch as people will be affected by the experiment, their well-being should be considered in the decision-making process of engineering projects. From this it follows that the moral relationship existing between engineers and the public should be grounded along the lines of an ethic of informed consent. Knowledge about technology must be disclosed and discussed. Not just technical knowledge *per se,* but knowledge of the risks and the potential harms, as well as of the benefits, need to be debated in a public forum.

What are we to make of engineers who believe they function as merely applied physical scientists operating in an environment where decisions are made objectively and values do not enter the decision-making process? The hazards associated with the experimental nature of engineering projects—as opposed to the controlled and often less complex nature of scientific investigations, for example—present challenges distinct to engineers, creating for them certain ethical obligations and social responsibilities unique to their profession.

For one thing, safety and risk are crucial variables for engineers, variables from which the purely theoretical scientist is generally immune. To render scientific judgments is, for all intents and purposes, to make decisions based on sound, law-like principles. Making engineering judgments, as opposed to scientific judgments, is to make judgments about safety and risk, namely, seeing the prospective engineering design project in terms of trade-offs between costs and benefits (Lowrance, 1976).

Hence, engineering design judgments cannot always be grounded in sound scientific theory. Engineering methodology consists not of scientific principles alone but of scientific principles along with a constellation of nonscientific "heuristics." Engineers often do not have all of the scientific facts on hand when problems are to be solved and cannot always perform full-scale controlled tests. They extrapolate from proven principles using engineering heuristics and engineering judgment. Sometimes, unfortunately, the extrapolations and judgments fail.

Comparing engineering with science helps us identify some of the social responsibilities of engineers and their profession. What of the people who have a stronger claim to objectivity, disinterestedness, and value neutrality, namely scientists who have been involved in engineering, applied science, and research and development?

THE ROLE OF SCIENCE AND SCIENTISTS

Unlike engineers, who see their role as technological innovators, scientists, concerned with contributing to a body of scientific knowledge, are inclined to develop an "ivory tower" conception of their role. Pursuing knowledge for its own sake is an intrinsic value for scientists. Up until World War II, scientists tended to view their work as value-free and morally neutral, and hence they were generally oblivious to the social consequences of their discoveries. The only norms ideally guiding their research are those identified by Merton, a world-renowned sociologist:

1. Universalism: empirically verified knowledge transcends national boundaries.
2. Organized skepticism: a scientist is obligated to subject his or her research results as well as those of other scientists to the most rigorous scrutiny.
3. Communality: a scientist is obligated to share his or her research results with others since scientific progress requires the free flow of ideas.
4. Disinterestedness: a scientist pursues science for "its own sake." Knowledge and discovery should be pursued for their own sake and not to enhance a scientist's prestige or authority (Merton, 1973: 267–278).

While these ideal norms describe the role of academic scientists engaged in basic research, they clearly do not describe the role of applied scientists employed in government or industrial laboratories operating with explicit missions. The latter are concerned with *instrumental,* as opposed to *intrinsic,* values of knowledge; they are also governed by norms pertaining to proprietary and classified information.

Developments in 20th Century science, however, have tended to blur the distinction between pure and applied research. Examples from atomic and nuclear physics will bear out this point. In the first decade of the 20th Century, Niels Bohr, the renowned Danish physicist, was pursuing atomic research when he formulated his "planetary" model of the atom. In the late 1930s, Hahn and Strassmann conducted experiments on neutron-irradiated uranium, which yielded the surprising result that the uranium nucleus had split into two smaller nuclei (Beckman et al., 1989: 20–26). These basic research findings eventually paved the way for an applied research project to develop an atomic bomb, known under the code name of the "Manhattan Project."

Knowledge and Power

Francis Bacon, the apostle of modern science, anticipated the relationship between basic and applied research when he formulated his well-known aphorism: "Knowledge is power."

Unlike some other values human beings create that are perishable, scientific knowledge tends to endure and to accumulate, enriching the lives of succeeding generations. This is indeed the premise upon which universities were first founded in the Middle Ages and the reason they have flourished for centuries, whereas many other types of organizations have been ephemeral.

Valuable as scientific knowledge may be, is it powerful? In other words, does knowledge guide the fateful decisions taken at national and international levels that affect the lives of millions of human beings? Does knowledge contribute to the wise use of power? In uttering his aphorism, was Bacon describing reality at the dawn of the 17th Century, or was he projecting a vision of a utopia in which knowledge *would* be power, as he subsequently did in his *New Atlantis?* These questions point to the uneasy and perplexing relationship between knowledge and power. On the one hand, knowledge may be entirely divorced from power; it may be pursued and developed without any concern for its impact on the exercise of power. On the other hand, knowledge may be cultivated for the express purpose of undergirding the exercise of power.

When knowledge is applied—that is, converted to power by public policy decision makers—it is never a direct or an automatic process: intervening between knowledge and power are an implicit or an explicit affirmation of values and a commitment to specific interests. Moreover, the values prompting the application of scientific knowledge may be benevolent as well as malevolent, altruistic as well as selfish, humane as well as inhumane. Since the social roles of knowledge producers—scientists—and the public policy decision makers are distinct, there may or may not be consensus among the people performing these roles regarding the values and interests underlying the use of knowledge (Evan, 1981: 11–13).

Examples of the Relationship between Knowledge and Power

The Manhattan Project provides ample data on the complexities of the relationship between knowledge and power. Leo Szilard, a Hungarian physicist, upon learning of the Hahn-Strassman experiments in 1938, quickly inferred that bombarding a uranium nucleus would trigger a chain reaction releasing an immense amount of atomic energy. He further inferred that Nazi Germany would exploit this knowledge by developing an atomic bomb in order to win the forthcoming world war. He thus decided to travel to the United States to confer with his fellow Hungarian physicists of great renown, Eugene Wigner and Edward Teller. Convinced of the political urgency of their knowledge, they met with Albert Einstein and drafted a letter, which Einstein signed, and which they then forwarded, through an intermediary, to President Franklin Roosevelt in 1939. After several years of a feasibility study initiated by President Roosevelt, the Manhattan Project was established in 1942. This is indeed a historic example of a successful linkage between knowledge and power.

Two unsuccessful examples of the relationship between knowledge and power will now be considered. By the time the first nuclear weapon was successfully detonated on July 16, 1945, at Alamogordo, New Mexico, Nazi Germany was already defeated, thus eliminating the primary motivation of many scientists who were recruited to the Manhattan Project. Some of these scientists now had moral misgivings about using the bomb to force the Japanese to surrender. James Franck, a refugee German Nobel Laureate in chemistry, chaired a committee of concerned scientists, including Leo Szilard, which urged that the United States conduct a demonstration test of an atomic bomb on an unoccupied Japanese island. In addition, the scientists advocated putting such weapons under international control to prevent their use and to forestall the emergence of a nuclear arms race between the United States and Soviet Union. The committee submitted its proposal to President Harry Truman, who rejected it.

Niels Bohr also made an effort to link knowledge with power. Convinced that nuclear weapons would challenge the traditional concepts of national sovereignty and concerned about the possible outbreak of a nuclear weapons arms race, he urged Winston Churchill and Franklin Roosevelt in 1944 to inform Stalin of the Manhattan Project and to negotiate a mutual security pact. "The only security from the bomb would be political: negotiation toward an open world, which would increase security by decreasing national sovereignty and damping out violence that attended it. The consequence of refusing to negotiate would be a temporary monopoly followed by an arms race" (Rhodes, 1988: 783). Bohr's scenario of the postwar world, ushered in by the invention of the atomic bomb, was, however, rejected by Winston Churchill and Franklin Roosevelt as "dangerously naïve" (Rhodes, 1988: 783).

These failures on the part of the Manhattan Project scientists to translate knowledge into power prompted them to establish the Federation of Atomic Scientists, whose mission is to end the worldwide arms race, to achieve complete nuclear disarmament, and to avoid the use of nuclear weapons. In cooperation with the members of the Federation of Atomic Scientists, the atomic scientists of the Metallurgical Laboratories of the University of Chicago established in December 1945 *The Bulletin of the Atomic Scientists,* dedicated to addressing the problems of the nuclear age.

Given the foregoing cases of the relationship between knowledge and power, is Francis Bacon's proposition valid? Does knowledge lead to power? As stated, it is generally *not* true. Knowledge *does* indeed lead to power provided three essential resources are available: human, organizational, and financial. The Manhattan Project is an exemplary manifestation of the validity of our revised proposition about knowledge and power.

In the 1930s, nuclear physics attracted the most gifted scientists in the world such as Bethe, Fermi, Feynman, Oppenheimer, Rabi, Rotblat, Teller, and Wigner (Zuckerman, 1977). The scientists recruited to the Manhattan Project by Oppenheimer were truly outstanding, many of whom became Nobel Laureates or were their intellectual peers (Bernstein, 2000).

The organizational structure of the Manhattan Project was also functionally well designed, with Oppenheimer, a nuclear theoretical physicist, appointed as the technical director and General Groves appointed as the administrative director with liaison functions with the War Department.

As for the third essential resource—financial—for a successful relationship between knowledge and power, Roosevelt, in the interest of safeguarding the security of the Manhattan Project, bypassed Congress in his decision to allocate $2 billion—probably one of the largest budget allocations for a single federal project in the first half of the 20th Century.

Professionalization and Internationalization of Science

In the course of developing bodies of knowledge, scientists tend to form professional societies or organizations as mechanisms for disseminating their theories and research findings. These societies give rise to formal and informal networks of communication and a "scientific community" based on a set of shared values and norms.

In the early stage of development of scientific fields, national societies tend to emerge. As sciences grow and develop in different countries, the tendency is for international scientific societies to arise, encouraging the further growth of informal and formal communication networks. These networks stimulate professional collaboration across national boundaries and accelerate the growth of scientific knowledge. The more developed a scientific field, the greater the probability that the values and norms are internalized by its members and the higher the frequency of international collaboration (Evan, 1975: 386–387).

As early as 1847, the American Association for the Advancement of Science (AAAS) was founded for the purpose of furthering the work of scientists, facilitating cooperation among them, improving the effectiveness of science in the promotion of human welfare, and increasing public understanding of the role of science in human progress. By the late 20th Century, AAAS represented all the major fields of science; its annual conferences continue to coordinate the activities of approximately 300 scientific societies. It is also noteworthy that in 1976 the AAAS established the Science and Human Rights Program, the objectives of which include protecting the human rights of scientists and fostering greater understanding of and support for human rights among scientists throughout the world (Rogers, 1981).

With the growth of the scientific community and the multiplication of specialties in the 19th Century, international scientific societies have proliferated. One of the oldest international scientific associations, the International Meteorological Committee, was founded in 1872. Numerous other international associations were subsequently founded, so that by the end of World War I there was a perceived need for a new association to coordinate the multitude of scientific societies, thus giving rise in 1919 to an umbrella organization called the International Council of Scientific Unions. After World War II, the United Nations Educational, Scientific, and Cultural Organization (UNESCO) was instrumental in coordinating the activities of all the social sciences under another umbrella organization known as the International Social Science Council.

Notwithstanding the proliferation of professional societies at national and international levels, there has been a dearth of organizational developments focused on the social responsibility of scientists, namely their ethical concerns about the social impact of technological innovations. In addition to the pioneering work of the Federation of American Scientists mentioned previously, several other organizations deserve consideration.

A kindred national organization in the United States is the Union of Concerned Scientists (UCS). The purpose of this organization of approximately 50,000 concerned scientists is to combine rigorous scientific analysis with citizen advocacy for "a cleaner, healthier environment and a safer world" (Union of Concerned Scientists: www.ucsusa.org). Founded in 1969 by faculty members of the Massachusetts Institute of Technology, its mission is to redirect scientific research to environmental and social problems. Its members collaborate on technical studies on renewable energy options, the impact of global warming, risks of genetically engineered crops and related problems. Results of technical studies are shared with policy makers, news media, and the public.

In 1955, at the height of the Cold War, Bertrand Russell and Albert Einstein issued a humanist manifesto for peace addressed to the participants of the first Pugwash conference:

> We are speaking . . . not as members of this or that nation, continent, or creed, but as human beings, members of the species Man, whose continued existence is in doubt. . . .

> We have to learn to think in a new way. We have to learn to ask ourselves, not what steps can be taken to give military victory to whatever group we prefer, for there no longer are such steps; the question we have to ask ourselves is: what steps can be taken to prevent a military contest of which the issue must be disastrous to all parties?

> There lies before us, if we choose, continual progress in happiness, knowledge, and wisdom. Shall we, instead, choose death, because we cannot forget our quarrels? We appeal, as human beings, to human beings: Remember your humanity, and forget the rest. If you can do so, the way lies open to a new Paradise; if you cannot, there lies before you the risk of universal death. (Pugwash: 1978: 10–12)

Signed by a number of internationally-distinguished scientists, the manifesto urged scientists, regardless of political persuasion, to assemble for the purpose of discussing the threat to civilization posed by thermonuclear weapons. The first meeting was held in 1957 in the Village of Pugwash, Nova Scotia—hence the name "Pugwash Conference." Meeting as private individuals rather than as government representatives, Pugwash participants explored alternative strategies to arms control and other global problems. The results of the Pugwash Conferences are communicated as policy recommendations to national governments around the world.

> It has . . . been reliably stated that . . . (Pugwash Conferences) made a very significant contribution to decisions by govern-

ments on such a major issue as the Partial Test-Ban Treaty. They have also contributed new ideas towards the control and limitation of armaments such as 'nuclear-free zones' and the 'black-box' methods of monitoring underground test-explosions. They may have been useful in preparing the ground for the recent 'SALT' negotiations. (Pentz and Slovo, 1981: 177)

In 1979, the first Student Pugwash Conference was convened at the University of California, San Diego. This offshoot of the Pugwash Conference seeks to foster dialogue among student participants about major global problems (Leifer, 1979: 3–11).

Professional scientific societies, at national as well as international levels, suffer from several fundamental weaknesses. First is a lack of adequate financial resources to launch scientific studies and disseminate their findings to policy makers and the public. Second is a lack of funds to protect the human rights of scientists who choose to act as whistleblowers in government agencies and private corporations. Andrei Sakharov, the father of the Soviet hydrogen bomb, who, after its development, then foreswore participation in any subsequent military research, became a celebrated dissident and an advocate of human rights. As a result of his courageous challenge to the Soviet totalitarian system, he was harassed by the KGB and eventually banished from Moscow to Gorky. During the years of his privation, national and international scientific organizations were unable to prevail upon the Soviet authorities to release him from internal exile.

Yet another deficiency of professional scientific societies is the lack of funds to establish research institutes to study global problems and provide employment for scientists who choose not to work on military and national security projects.

Finally, professional societies, especially at the international level, lack a code of ethics affirming the principle of "do no harm" embodied in the Hippocratic Oath. In a recent lecture entitled "Science and Humanity at the Turn of the Millennium," Sir Joseph Rotblat, a physicist and Nobel Laureate in Peace, addressed this problem:

> The time has . . . come for some kind of Hippocratic Oath to be formulated and adopted by scientists. A solemn oath, or pledge, taken when receiving a degree in science, would, at the least, have an important symbolic value, but might also generate awareness and stimulate thinking on the wider issues among

young scientists . . ." The text of the pledge adopted by the US Student Pugwash Group seems to me highly suitable. The Pledge reads:

> "*I promise to work for a better world, where science and technology are used in socially responsible ways. I will not use my education for any purpose intended to harm human beings or the environment. Throughout my career, I will consider the ethical implications of my work before I take action. While the demands placed upon me may be great, I sign this declaration because I recognize that individual responsibility is the first step on the path to peace.*" (Rotblat, 1999: 7)

Apart from the individual expression by scientists of their social responsibility, there is a need for organizations of scientists to articulate such a commitment collectively (Lakoff, 1979).

If these four fundamental problems of international scientific societies are ever to be solved, they presuppose the growth of transnational loyalties in the scientific community to promote the well-being of humanity rather than the interests of nation-states (Evan, 1997: 987–1003).

Rotblat, upon leaving the Manhattan Project, made a fundamental career decision based on ethical considerations:

> Work on the atom bomb convinced me that even pure research soon finds applications of one kind or another. If so, I wanted to decide myself how my work should be applied. I chose an aspect of nuclear physics that would definitely be beneficial to humanity: the applications to medicine. Thus I completely changed the direction of my research and spent the rest of my academic career working in a medical college and hospital. (Rotblat, 1986: 21)

A Nobel Laureate in physics, I.I. Rabi has articulated a similar conception: "Wisdom is the application of knowledge for the benefit of mankind" (Lederman, 1999: 15).

CONCLUSION

To ensure that the inventions of engineers and the discoveries of scientists promote the "benefit of mankind," both professions will have to abandon their stance of ethical neutrality. Through their role in institutions of higher learning, they can help inculcate future generations of engineers and scientists with a social responsibility perspective. Through their participation in professional societies, engineers and scientists can potentially implement the principles embodied in their codes of ethics. By translating codes of ethics principles into their everyday work experience, they can help significantly reduce the incidence of technological disasters.

References

Accreditation Board of Engineering Education and Technology (ABET). (1998). Engineering criteria 2000. Available on the World Wide Web at: www.abet.org/eac/EAC_99-00_Criteria.htm#EC2000

Akgerman, A., Anthony, R.G., and Darby, R. (1999). "Integrating process safety into ChE education and research," *Chemical Engineering Education* 33 (3): 198–204.

Beckman, Peter R., Campbell, Larry, Crumlish, Paul W., Dubkowski, Michael N., and Lee, Steven P. (1989). *The nuclear predicament.* Englewood Cliffs, NJ: Prentice Hall.

Bernstein, Jeremy. (2000). "Creators of the bomb," *The New York Review of Books,* May 11; 47: 19–21.

Bryan, Leslie. (1999). "Educating engineers on safety," *Journal of Management in Engineering* 15 (2): 30–34.

Carpenter, Stanley. (1984). "Redrawing the bottom line: The optimal character of technical design norms," *Technology in Society* 6: 329–340.

Durbin, P. (1989). "The challenge of the future for engineering educators." In A. Flores (Ed.), *Ethics and risk management in engineering.* Washington, DC: University Press of America.

Evan, William M. (1975). "The International Sociological Association and the internationalization of sociology," *International Social Science Journal* XXVII: 385–393.

Evan, William M. (1981). "Some dilemmas of knowledge and power." In William M. Evan (Ed.), *Knowledge and power in a global society.* Beverly Hills, CA: Sage Publications.

Evan, William M. (1997). "Identification with the human species: A challenge for the twenty-first century," *Human Relations* 50 (8): 987–1003.

Flores, Albert. (1980). "Problems with technology: Social responsibility and the public's trust." In Albert Flores (Ed.), *Ethical problems in engineering.* Troy, NY: Center for the Study of the Human Dimensions of Science and Technology: 211–214.

Flores, Albert. (1982). "Designing for safety: Organizational influences on professional ethics." In A. Flores (Ed.), *Designing for safety: Engineering ethics in an organizational context.* Troy, NY: Rensselaer Polytechnic Institute.

Glagola, C., Kam, M., Loui, M., and Whitbeck, C. (1997). "Teaching ethics in engineering and computer science: A panel discussion," *Science and Engineering Ethics* 3 (4): 463–480.

Grundfest, Warren, and Scott, Andrea. (1998). "Ethical issues in the development of medical devices: The role of education and training in the evaluation of safety and efficacy," *Critical Reviews in Biomedical Engineering* 26 (5): 378–383.

Harris, C., Pritchard, M., and Rabins, M. (1995). *Engineering ethics: Concepts and cases.* San Francisco, CA: Wadsworth Publishing Co.

King, Julia. (1998). "Incorporating safety into a unit operations laboratory course," *Chemical Engineering Education* 32 (3): 178–184.

Lakoff, Sanford. (1979)."Ethical responsibility and the scientific vocation." In Sanford A. Lakoff (Ed.), *Science and ethical responsibility.* Reading, MA: Addison-Wesley Publishing Company: 19–31.

Lau, Andrew. (1998). "Broadening student perspectives in engineering design courses," *IEEE Technology and Society Magazine* 17 (3): 18–25.

Lederman, Leon M. (1999). "The responsibility of the scientist," *The New York Times,* July 24: A15.

Leifer, Jeffrey R. (1979)."The origins and objectives of Student Pugwash." In Sanford A. Lakoff (Ed.), *Science and ethical responsibility.* Reading, MA: Addison-Wesley Publishing Company: 3–11.

Lichter, B. (1989). "Safety and the culture of engineering." In A. Flores (Ed.), *Ethics and risk management in engineering.* Washington, DC: University Press of America: 211–221.

Long, Thomas. (1983). "Informed consent in engineering: An essay review," *Business and Professional Ethics Journal* 3: 59–66.

Lowrance, William. (1976). *Of acceptable risk: Science and the determination of safety.* Los Altos, CA: William Kaufmann.

Manion, Mark, and Kam, Moshe. (2000). "Engineering ethics at Drexel University." *Proceedings of the American Society of Engineering Education (ASEE) 2000 Conference,* March 24: 45–63.

Martin, W., Mike and Schinzinger, Roland. (1996). *Ethics in engineering.* New York: McGraw-Hill.

McLean, G.F. (1993). "Integrating ethics and design," *IEEE Technology and Society Magazine* 12 (1): 19–30.

Merton, Robert K. (1973). *The sociology of science: Theoretical and empirical investigations.* Chicago: University of Chicago Press.

Moonashingha, Ananda. (1996). "Design—cornerstone of your career: Advice for young engineers," *Journal of Professional Issues in Engineering Education and Practice* 122 (3): 135–138.

Nandagopal, N.S. (1990). "Why and how should students in technical programs be exposed to ethical and international issues." In *A delicate balance: Technics, culture, and consequences.* The Institute of Electrical and Electronics Engineers. Catalog Number 89CH2931-4. Los Angeles: California State University: 87–93.

National Society of Professional Engineers. "NSPE code of ethics." Available on the World Wide Web in September 2000 at: www.nspe.org

Orr, R., and Pang, N. (1993). "The use of the Hippocratic oath: A review of 20th century practice and content analysis of oaths administered in medical schools in the U.S. and Canada." Available on the World Wide Web June 6, 2001 at: www.imagerynet.com/hippo.ama.html

Pentz, Michael J., and Slovo, Gillian (1981). "The political significance of Pugwash." In William M. Evan (Ed.), *Knowledge and power in a global society.* Beverly Hills, CA: Sage Publications: 175–203.

Petroski, Henry. (1985). *To engineer is human: The role of failure in successful design.* New York: St. Martin's Press.

Petroski, Henry. (1994). *Design paradigms: Case histories in error and judgment in engineering.* Cambridge, England: Cambridge University Press.

Pletta, Dan, and Hon, M. (1987). " 'Uninvolved professionals' and technical disasters," *Journal of Professional Issues in Engineering* 113 (1): 23–32.

Pugwash Conferences on Science and World Affairs. (1978). *The Pugwash movement at twenty-one.* Basingstoke, Hampshire: Taylor and Francis Ltd.

Rabins, Michael. (1998). "Teaching engineering ethics to undergraduates: Why? what? how?" *Science and Engineering Ethics* 4: 291–302.

Rhodes, Richard. (1988). *The making of the atomic bomb.* New York: Simon & Schuster.

Rogers, Carol L. (1981). "Science information for the public: The role of scientific societies," *Science, Technology and Human Values* 6 (36), Summer: 36–39.

Rotblat, Sir Joseph. (1986). "Leaving the bomb project." In Len Ackland and Stephen McGuire (Eds.), *Assessing the nuclear age.* Chicago, IL: 15–22.

Rotblat, Sir Joseph. (1999, July). "Science and humanity at the turn of the millennium," *Student Pugwash USA's 20th Anniversary*

International Reunion Conference, held at the University of California, San Diego.

Schinzinger, Roland. (1986). "Technological hazards and the engineer," *IEEE Technology and Society Magazine* 11 (1): 12–19.

Sherr, Sava. (1982). "Societal aspects of engineering standards," *IEEE Technology and Society Magazine* 1 (3): 13–16.

Slaymaker, Robert, and Bates, Christine. (1993). "Today's challenge in training new engineers," *IEEE Transactions on Industry Applications* 29 (1): 26–34.

Susskind, Charles. (1973). *Understanding technology.* Baltimore, MD: The Johns Hopkins University Press.

Tribe, Lawrence. (1971a)."Towards a new technological ethic: The role of legal liability," *Impact of Science on Society* 21 (3): 215–222.

Tribe, Lawrence. (1971b). "Legal frameworks for the assessment and control of technology," *Minerva,* October: 243–255.

Unger, Stephen. (1982). "The role of engineering schools in promoting design safety," *IEEE Technology and Society Magazine* 7 (1): 9–12.

Unger, Stephen. (1987). "Would helping ethical professionals get professional societies into trouble?" *IEEE Technology and Society Magazine* 6 (3): 19–21.

Unger, Stephen. (1994). *Controlling technology: Ethics and the responsible engineer* (2nd ed.). New York: John Wiley & Sons.

Unger, Stephen. (1995). "Engineering ethics: What's new?" *IEEE Technology and Society Magazine* (Fall): 4–5.

Vanderburg, Willem. (1995). "Preventive engineering: Strategy for dealing with negative social and environmental implications of technology," *Journal of Professional Issues in Engineering Education and Practice* 121 (1): 155–160.

Vanderburg, Willem. (1997). "Rethinking end-of-pipe engineering and business ethics," *Bulletin of Science, Technology & Society* 17 (93): 141–153.

Vesilind, Aarne. (1992). "Guidance for engineering-design-class lectures on ethics," *Journal of Professional Issues in Engineering Education and Practice* 118 (2): 214–219.

Wacker, G. (1990). "Teaching impacts of technology/professional practice in an engineering program." In *A delicate balance: Technics, culture, and consequences.* The Institute of Electrical and Electronics Engineers. Catalog Number 89CH2931-4. Los Angeles: California State University: 79–83.

Whitbeck. C. (1995). "Teaching ethics to scientists and engineers: Moral agents and moral problems," *Science and Engineering Ethics* 1: 397–408.

Willey, Ronald, and Piece, John. (1998). "Freshman design projects in the environmental health and safety department," *Chemical Engineering Education* 33 (3): 198–204.

Wilson, G. (1990). "Education for the social practice of engineering," *IEEE Spectrum* 25 (4): 117–122.

Zuckerman, Harriet. (1977). *Scientific elite: Nobel laureates in the United States*. New York: Free Press.

The Role of Corporations in the Management of Technological Disasters

*"The distinguishing mark of the executive
responsibility is that it requires not
merely conformance to a complex code
of morals but also the creation of moral
codes for others."*

—Chester J. Bernard

*I*f one wishes to identify the social, political, and moral dimensions of technological disasters, one must focus on the organizational context in which the failed technology is embedded. An analysis of the corporate environment surrounding many technological disasters often reveals a structure and pattern of corporate misconduct in the management and control of potentially hazardous technologies.

CORPORATE MANAGEMENT VERSUS MISMANAGEMENT

Research on "white-collar crime" and "organizational deviance" (Ermann and Lundman, 1992) has identified at least three hypothetical explanations, inferred from the investigation of dozens of cases, that help account for how and why corporations "socialize individuals into evildoing" (Darley, 1991). From this research we can gain some insight into the dynamics of corporate mismanagement and, hence, corporate responsibility and accountability for technological disaster. In the course of analyzing factors conducive to corporate *mismanagement* of hazardous technologies, we also uncover factors concerning the proper corporate *management* of technological crises.

One factor that can result in corporate mismanagement of technology is the diffusion and fragmentation of information in large organizations, as illustrated in the Challenger case (Vaughn, 1996). When it is discovered that a product that has been designed, manufactured, advertised, and sold is harmful to consumers, we find that knowledge of the product's potential for harm tends to be dispersed in various divisions of a corporation. This is because corporate divisions are often not in adequate communication with each other on issues of risk and safety.

The Challenger case provides a clear illustration of the compartmentalization of information and the failure of communication within a large corporation. Morton Thiokol engineers wrote a series of memoranda, even presenting data about potential failures of the O-rings the night before the fateful flight, but the information was never passed up from Level III to Level II or to decision makers at Level I during the all-important preflight readiness review process.

A second factor that can lead to corporate mismanagement of technology is the prior commitment to courses of action by corporate elites who establish norms and rewards that may encourage conduct leading to technological disaster. The Ford Pinto case, as we shall see, is a telling example of this. A third factor is that technological disaster may be the result of deviant actions by corporate top management, as illustrated in the Dalkon Shield case and in the Johns-Manville case. These cases of technological disaster were, for the most part, the result of management misconduct—which Lerbinger calls "crises of malevolence" or "crises of deception"— clearly the result of distorted management values (Lerbinger, 1997).

These three particular disasters, the Ford Pinto, Dalkon Shield, and Johns-Manville cases, are prime examples of a failure that spiraled into a "crisis," a situation that can threaten the very existence of a corporation. Such behavior by top management can have untold negative effects on its environment as well as on multiple stakeholders or constituencies. In the case of the Exxon *Valdez,* a major technological disaster, the spillage of millions of gallons of oil into the pristine waters off the coast of Alaska, turned into a crisis because the Exxon executives, by their misperception of the spills' consequence, mishandled the problem. In contrast to the three previous cases cited, the Exxon crisis was not the result of any deliberate action by top management. However, top management's handling of the Exxon *Valdez* oil spill was severely criticized for its failure to take responsibility for the spill, for its negligence in its cleanup efforts, and for its insensitivity in the management of the crisis. The Exxon *Valdez* crisis was more the result of corporate inaction and inappropriate action than intentional wrongdoing. In stark contrast to the behavior of top management in the Ford Pinto, Dalkon Shield, Johns-Manville, and Exxon *Valdez* cases, however, the reaction of top management of Johnson and Johnson to the news that its product, Tylenol, had been laced with cyanide and had caused the death of five people is indeed exemplary.

Given the prevalence of technological disasters that are the result of corporate mismanagement and given the growing number of failures that have turned into crises, it would be helpful if one could identify certain characteristics of organizations that make them crisis-prone. The identification of such factors would provide the framework for a theory of crisis management in which implementation could mitigate the incidence of technological disasters, especially when corporate mismanagement is at issue.

CASE STUDIES IN CRISIS MANAGEMENT

Ford Pinto Case

Top management can indirectly initiate deviant actions by establishing particular norms, rewards, and punishments for people occupying lower-level positions in a corporation. These top management decisions can take on the power of an unbreakable "law" of corporate decision making that no one within the organization is likely to question. Take, for example, the now-famous "Rule of 2,000," issued as a firm directive by Ford Motor Company's then Chief Executive Officer Lee Iacocca. He directed that the entire manufacturing process of Ford's new Pinto—from conception, testing, and production to the final product—should not weigh an ounce over 2,000 lbs. and cost not a penny over $2,000.

During the 1970s, the American automobile industry was in fierce competition with the small "economy" German and Japanese models. Iacocca identified the problem, devised a plan, and executed it with a firm hand using the strategy of concurrent engineering. As a result, the tooling of Ford's Pinto had to be done simultaneously with the design and testing. Consequently, by the time the fuel tank safety problem was discovered, the tooling and prototyping was well under way. No one raised the safety problem with management because, in the words of one engineer, "Safety wasn't a popular subject around Ford in those days" (Dowie, 1974: 21). No one intentionally ordered that a Ford car be built with a serious design defect and then be sold to unsuspecting customers. But ironically this was the unanticipated consequence of Iacocca's directive. Company employees, in the course of fulfilling the objectives of a mammoth corporation, may lose sight of the fact that, either by their action or inaction, other people will be put in harm's way. The cost-benefit calculations by Ford management are a case in point.

When management became aware of the problem with the rear-end gas tank design as a result of litigation over harm and death caused by a series of rear-end collisions involving Ford Pintos, they conducted a cost-benefit analysis and decided it was to their benefit *not* to install an $11.00 modification of the gas tank, because the costs of recalling and/or installing the $11.00 item would be greater than the company's projected costs and damage awards in wrongful death and injury lawsuits. The managers reasoned thus: projected sales of Ford Pinto sedans and trucks would be 12.5 million that year. The unit cost of repair would be $11.00. So, total costs to fix the problem would be 12,500,000 × $11 per car/truck = $137 million. On the other hand, highway statistics indicated that the maximum number of burn deaths would be 180, number of injuries 180, and number of burned vehicles 2,100. Assigning a value to an individual human life at $200,000 and values of $67,000 per injury and $700 per vehicle, the total "benefit" to the company would only be 180 × ($200,000) + 180 × ($67,000) + 2,100 × ($700) = $49.5 million. Any manager, thinking only of profit, would tend to depersonalize the 360 victims and rationalize his or her decision by claiming it was a sound "business" decision. This is exactly what Ford management did.

The national attention paid to the media treatment of various lawsuits surrounding Ford Pinto rear-end explosions caused a public relations nightmare for Ford Motor Company and tested the public's confidence in Ford vehicles. The overall result was a tarnished image and a reduction in brand name status for many Ford-made automobiles.

Ford-Firestone Conflict

In the years 2000–2001, the Ford Motor Company was embroiled in one of the largest product liability cases and product recalls in automotive history. On August 9, 2000, Firestone, Inc., voluntarily recalled 6.5 million of its 15-inch ATX and Wilderness AT tires installed on many popular SUVs (sports-utility vehicles) because of alleged defective tire treads. As of June 19, 2001, the defective Firestone tires had been implicated in the deaths of 203 people and the injuries of 700 motorists (White, Power, and Aeppel, 2001: A3). Almost all of these deaths involved Ford Explorers, which were equipped with either Firestone ATX or Firestone Wilderness AT tires. The initial cause of the accidents was traced to the tread of the tire, which had a tendency to separate from the tire itself, causing numerous fatal rollover accidents.

Unlike the other case studies of crisis management in this chapter that focus on the mismanagement of a single corporation, the Ford-Firestone clash is instructive because it involves two giant corporations, each of which blames the other for the disastrous effects of its products.

As more and more liability lawsuits were filed in the courts, Ford Motor Company went on the offensive, charging its longtime business partner Firestone with supplying Ford SUVs with defective tires. Ford announced it would sever all future business ties and no longer install Firestone tires on new Ford vehicles. Although eventually footing the bill for the 6.5 million tire recall, at a cost to the company of approximately $500 million, Firestone management also went on the offensive, charging that the fatal blowouts and rollovers were due, in fact, to the way the tires were installed on the Ford Explorer, citing evidence showing that Firestone ATX and Wilderness AT tires installed on other SUVs, similar to the Ford Explorer, were *not* prone to tread separation and consequent tire blowouts and vehicle rollovers. Firestone management blamed the deaths and injuries on Ford, claiming that the auto company informed consumers to underinflate their tires in order to create a smoother ride on the Explorer (Anonymous [a], 2000: 50). Bridgestone/Firestone also claims that the Explorer is too heavy for the 15-inch tires and that Ford should have installed 16-inch tires on its Explorer SUVs. The excess weight on the 15-inch tires is said to have contributed considerably to tire tread separation, especially when the tires were intentionally underinflated and the ambient average temperature was consistently high (Anonymous [a], 2000: 50).

On May 23, 2001, Ford Motor Company announced that it would spend an estimated $3 billion to replace 13 million tires on its SUVs because of perceived "unacceptable risks to our customers" (Kiley, 2001: 1B). Firestone manufactures many of the tires found on Ford Explorer and Expedition SUVs and Ford Ranger pickup trucks. Whereas Ford claims it was motivated to issue the recall in order to protect consumer safety, Firestone claims that Ford management is only attempting to disguise serious problems associated with its best-selling Ford Explorer SUV.

The controversy between the two giant corporations has turned into "rancorous finger-pointing." As one lawyer put it, "Ford and Firestone have chosen to point their guns at each other, and now they are confirming what we thought all along— that there are problems with both the tires and the Explorer" (Anonymous [b], 2001: C2). The result is that the 95-year Ford-Firestone relationship, one of the longest in American industry, was destroyed.

Rather than unite to deal with the impending crisis, Ford management and Firestone management have taken the unexpected route of creating enemies of one another. "Normally, crisis management dictates that companies in difficulties should present a unified front and act quickly. But this case was never normal" (Anonymous [c], 2001: 4). According to one analyst:

> The biggest product recall since Tylenol is spiraling into a fiasco for all involved—Ford Motor Co., Firestone Corp., regulators, Congress, even the U.S. judiciary system. Everyone is blaming someone else, no one can yet explain the true cause of the tire problem, and the reported death toll keeps rising around the world. The lesson of what may be the first global product recall

is that the mechanisms for ensuring consumer safety need fixing. (Anonymous [d], 2000: 178)

Such mechanisms include better safety standards at the major automotive manufacturers, more effective crisis management at the corporate level, and more attention and resources from Congress to help regulatory agencies such as the National Highway Traffic Safety Administration (NHTSA) to do their job effectively:

> [In] July 1998, State Farm Mutual Automobile Insurance Co. says it noticed a pattern of problems with Firestone ATX and Wilderness tires in the U.S. and e-mailed the NHTSA. But the agency took no action. It started scrambling to get a handle on the problem only four months ago. Congress shares much of the blame for this regulatory failure . . . Congress has consistently refused time and time again to grant the NHTSA money and authority, often under heavy lobbying pressure from Detroit. (Anonymous [d], 2000: 178)

When a bitter conflict breaks out between two corporations that affects the welfare of consumers, management would be well advised to resort to either an arbitration process or a regulatory process to resolve the conflict as expeditiously as possible.

Eastern Air Lines: Whistleblowing Case

Employees who turn to whistleblowing against an employer embark on a courageous but potentially hazardous journey (Alford, 2001). Convinced that their employer is "engaged in illegal, dangerous or unethical conduct . . . they try to have said conduct corrected through inside complaint" (Westin, 1981: 1). Voicing their concerns within the company's own channels, whistleblowers almost always meet with resistance and rejection; they are also frequently stigmatized as disloyal, harassed into resignation or retirement, or fired. After their appeals within the firm prove fruitless, whistleblowers often make their protests public, "submitting their complaints to various government boards or agencies, or bringing suits for wrongful discharge to the courts" (Westin, 1981: 2).

It should be obvious that the interests and therefore judgments of whistleblowers and those of top executives diverge. In the competitive arena in which most technological developments are exploited, the interests of executives are directed to maximizing production or profit, while the engineers and others who operate the technology are likely to place more emphasis on safety, avoiding exposure to workplace hazards and diminishing stress in the working environment. Since top executives enjoy the prerogative of making the decisions, other interested parties, including potential whistleblowers, can argue their case only by persuasion.

The experience of Dan Gellert (1981), a pilot employed by Eastern Air Lines who decided to blow the whistle on a serious defect in the Lockheed 1011 wide-bodied aircraft, is a telling account of the price a whistleblower pays for speaking out. A pilot with 25 years experience, the last 10 with Eastern Air Lines, Gellert became aware of the autopilot system problem in September 1972.

> I was flying an L-1011 with 230 passengers aboard, cruising on autopilot at 10,000 feet, when I accidentally dropped my flight plan. As I bent down to pick it up, my elbow bumped the control stick in front of me. Suddenly, the plane went into a steep dive—something that shouldn't have occurred. Fortunately, I was able to grab the control stick and ease the plane back on course.
>
> What had happened, I realized, was that in bumping the stick, I had tripped off the autopilot. Instead of holding the plane at 10,000 feet, it had switched from its "command mode" to "control wheel steering." As a result, when the stick moved forward, causing the plane to dive, the autopilot, rather than holding the aircraft on course, held it in the dive.
>
> There was no alarm system to warn me that the plane was off course. If I hadn't retrieved my flight plan quickly, observing the dive visually when I sat up in my seat, the plane could have easily crashed. Even more alarming, as I eased the plane back onto course, I noticed that the autopilot's altimeter indicated that the plane was flying at 10,000 feet—the instrumentation was giving me a dangerously false reading. (Gellert, 1981: 18)

Gellert subsequently made a verbal report of the near disaster to an Eastern management official. Three months later, on December 29, 1972, an Eastern Air Line L-1011, approaching the Miami International Airport, crashed into the everglades, killing 103 people. Convinced that the defect with the autopilot system contributed to the disaster, Gellert outlined the problem with the L-1011 autopilot system in a two-page memorandum addressed to the three top executives of Eastern Air Lines. Two months later, he received a reply from Frank Borman, then Vice President of Operations, stating that it was "pure folly" to say that one procedure could cause an accident (Gellert, 1981: 19). Faced with this off-hand dismissal of his assessment of the cause of the disaster, Gellert decided to send his memorandum to his union, the Airline Pilots Association (ALPA), and the National Transportation Safety Board (NTSB), which was about to conduct a hearing into the causes of the Miami crash.

A year after Gellert filed his initial complaint, he was flying an L-1011 to Atlanta when the autopilot disengaged twice when it shouldn't have. The crew had to engage full takeoff power in order to make the runway and land safely. Shaken up by this experience, Gellert wrote a 12-page memorandum to the NTSB explaining the continuing problem with the L-1011 autopilot; he also sent a copy to Frank Borman.

Shortly after sending this memorandum to NTSB, Gellert was demoted to copilot. Realizing that he was being penalized by the company for sending a memo to the NTSB, he wrote a letter of protest to Frank Borman. When Borman replied to Gellert's letter of protest, he grounded him for medical reasons. "Eastern questioned my ability to fly an aircraft since, they said, I had written so many letters concerning flight safety. Obviously, they maintained that there was something wrong with me mentally" (Gellert, 1981: 22).

Outraged by being medically grounded, he submitted a grievance to the Eastern Airlines Pilots System Board of Adjustment. After several months of deliberation, the System Board found no basis on which to ground Gellert. Once he was reinstated, Eastern began a campaign of disciplinary harassment, culminating in an indefinite grounding. On the advice of his attorney, he initiated a lawsuit in 1977, challenging Eastern's decision to ground him indefinitely and to discredit him professionally because of his concern about the safety of the L-1011 aircraft. To the astonishment of Eastern, a Florida jury awarded Gellert $1.5 million in damages and an extra $100,000 to penalize Eastern for its transparent attempts to discredit him. "The trial judge called Eastern's action against me 'outrageous and mischievous' " (Gellert, 1981: 25). Gellert's legal victory, however, was short-lived. Eastern appealed the judgment and a Florida circuit court set aside the $1.5 million verdict on a technical issue of tort law.

After a seven-year struggle with Eastern, Gellert found his situation ironic. While his associates told him, "You've won. You're a winner," he is not at all sure.

> What chance does an individual have against a $2 billion company that is willing to spend any amount to beat one of its employees . . . Airline safety involves more than just correcting specific defects like the L-1011 autopilot. Safety is an attitude which, ideally, should be shared throughout a company from top to bottom. But if Frank Borman and his legal department are consumed with pursuing a vendetta against me, then they aren't dealing with safety problems. (Gellert, 1981: 29)

Dalkon Shield Case

Corporate elites sometimes knowingly undertake actions that produce antisocial results and use the corporate hierarchy and policies to achieve their goals, even if it means suppressing incriminating evidence. This most likely occurs when corporate elites, in order to avoid litigation and incrimination, develop campaigns of cover-up that greatly exacerbate the individual and social harms done. This is because, while all of the attention is focused on the denials from the company, the harms continues to be perpetrated.

On August 21, 1985, A.H. Robins Company filed for bankruptcy following years of litigation concerning one of its products, the Dalkon Shield Intra-Uterine Device (IUD). At the time it filed for bankruptcy, Robins had been deluged with more than 12,000 personal injury lawsuits charging that the Dalkon Shield was responsible for countless serious illnesses and at least 20 deaths among the women who used it. Moreover, critics claimed that Robins had been involved in a prolonged institutional cover-up of the short-term and long-term effects of the use of the Shield (Gini and Sullivan, 1996). In addition, not only did the critics accuse Robins of stonewalling about the possible negative side effects of the Shield, Robins' management was also, as Gini and Sullivan state, ". . . guilty of marketing a product they knew to be relatively untested, undependable, and therefore potentially dangerous. Robins is accused of having deceived doctors, lied to women, perjured itself to federal judges, and falsified documentation to the FDA" (1996: 217). As a result, dozens of women suffered chronic infections, loss of childbearing capacity, birth of children with multiple birth defects, unwanted abortions, recurring health problems, and chronic pain (Hartley, 1993).

The primary cause for concern for the Shield users was the nylon tail of the Shield itself. The tail runs between the vagina, where bacteria are always present, and the uterus, which is germ-free. The string is not a single monofilament but a cylindrical sheath encasing 200 to 450 round monofilaments separated by spaces (Gini and Sullivan, 1996). Any bacteria that found its way between the spaces could be drawn into the uterus by the process of "wicking," a phenomenon similar to that by which a string draws the melting wax of a candle to the flame. The wicking of the Shield's tail could cause pelvic inflammatory disease (PID), a highly virulent and painful, difficult-to-cure, life-threatening infection. It may destroy a women's ability to bear children and may cause bacterially induced miscarriages called septic spontaneous abortions (Hartley, 1993).

During the entire controversy, company management embarked on a campaign of denial, all the while participating in a conspiracy to conceal information from the public, the court system, and the FDA. First, Robins was accused of quashing all documentation and denying the validity of the results of Dr. Hugh Davis, one of the original inventors of the Dalkon Shield (Hartley, 1993). Second, not only did the company know that the nylon sheath could degenerate and cause infection, but it received a warning from one of its consultants, Dr. Thad Earl, that the Shield should be removed from women to avoid "abortion and septic infection" (Hartley, 1993: 120). Third, it was reported that on at least three separate occasions executives and officials of Robins "lost" or destroyed company files and records specifically requested by the federal appellate court and the FDA.

At the time of filing for bankruptcy, Robins had already paid more than $500 million in settlements, litigation costs, and legal fees. This case became one of the biggest business blunders of all time, made so much worse by a firm that at first blinded itself to any danger, then tried to cover it up. As Hartley put it, "How could a respected management, one with the reputation of a multi-generational family firm at stake, have accepted such risks with a relatively untested new product in the crass pursuit of short-term profits?" (Hartley, 1993: 121). Answers to this question haunt every analysis of corporate misconduct leading to technological disaster.

Johns-Manville Asbestos Case

An especially tragic example of mismanagement resulting from conscious deviant action by corporate top management is the case of Johns-Manville and its manufacture and sale of asbestos over a 50-year period. Although banned in the 1970s, "asbestos has spawned more lawsuits than any other product in the history of personal-injury litigation" (Schmitt, 2001: A1).

Asbestos has been primarily associated with three forms of respiratory illnesses: asbestosis, mesothelioma, and lung cancer (Gini, 1996). Asbestosis is a chronic and sometimes fatal lung disease characterized by extensive scarring of the lung tissue and progressive shortness of breath, symptoms much like emphysema. Asbestosis has a latency period of 10 to 20 years. Mesothelioma is a fatal cancer of the chest or abdomen lining. Its average latency period is 25 to 40 years. Asbestos-related lung cancer is highly virulent and often fatal. Moreover, modern research has also shown a link between asbestos fibers and cancer of the gastrointestinal tract, larynx, and kidney (Gini, 1996).

When it filed for bankruptcy on August 26, 1982, approximately 17,000 lawsuits had been filed against the Johns-Manville Corporation in regard to its asbestos pipe insulation product line (Coplon, 1996). Experts claim that, during the last 40 years, 9 to 20 million Americans were exposed to large amounts of asbestos in the workplace. According to medical reports, at least 5,000 cancer-related deaths linked to asbestos occur annually. The courts have found that Johns-Manville, a leading manufacturer of asbestos, knew about the dangers of asbestos for decades and deliberately suppressed medical information from its workers (Trevino and Nelson, 1995). During court proceedings, documents surfaced that disclosed a conspiracy between Johns-Manville and Raybestos-Manhattan, another asbestos materials manufacturer. Correspondence between the President of Raybestos and the General Counsel of Manville indicated that not only did both companies know of the hazardous nature of asbestos but that they conspired to cover up that knowledge. This became the basis for punitive damage awards against Johns-Manville (Coplon, 1996).

When Johns-Manville and other asbestos manufacturers discontinued their operations when it became public knowledge that asbestos fibers could lodge in the lungs and cause cancer, W.R. Grace, Inc., reported a "research breakthrough" when it claimed that it had devised a "completely asbestos-free fireproofing spray." For approximately two decades the new formula was sprayed onto the steel skeletons of office buildings, schools, and hotels across the United States.

> Notwithstanding its disclaimers, Grace's new product was not completely asbestos-free. A little-known kind of asbestos, tremolite, laced the ore used in the spray. And while Grace knew this, for years it kept that knowledge largely hidden from workers who applied the fireproofing and clients who wanted their buildings asbestos-free. (Moss and Appel, 2001: 1)

Marketing its fireproofing spray under the brand-name Monokote, Grace claimed that the asbestos contamination was insignificant, like "gold in seawater." Government scientists, however, feared that asbestos might be unsafe at any level. In 1977, Grace was on the verge of disclosing the asbestos content of Monokote, when company executives met secretly at corporate headquarters in Cambridge, Mass., where they weighed the risks and benefits of disclosure. "While silence increased the danger of being sued, they calculated, disclosure could have meant the end of Monokote. So, they decided, customers who inquired if Monokote contained asbestos were to be told that it did not" (Moss and Appel, 2001:1).

In April 2001, W.R. Grace filed for Chapter 11 bankruptcy protection. The company said it had been named as the defendant in more than 325,000 personal injury cases and had spent $1.9 billion defending itself (Brick and Milford, 2001: C2).

Exxon *Valdez* Oil Spill

On March 23, 1989, the Exxon *Valdez*, the largest and best-equipped tanker in Exxon's fleet, left the Port of Valdez, Alaska, loaded with 52 million gallons of crude oil. As the *Valdez* found its way through the relatively narrow waters of the Valdez area of Prince William Sound, Captain Hazelwood retired to his cabin, leaving Third Mate Gregory Cousins at the helm. Hazelwood's actions violated company policy that the captain remain on the bridge until the ship reaches open water. In addition, Cousins was not licensed to pilot the vessel through the potentially treacherous waters of the Sound. In order to avoid icebergs, Cousins performed what officials later classified as an "unusual series of right turns," eventually causing the huge tanker to run aground on Bligh Reef just past midnight on March 24, spilling 11 million gallons of crude oil into the pristine waters of the Gulf of Alaska. The spill covered 2,600 miles in and around Prince William Sound. It eventually killed thousands of seafoul, otters, fish, and other aquatic life. In addition, the spill virtually destroyed the annual $100 million seafood harvest, greatly diminished Alaska's annual lucrative tourist industry, and ultimately threatened the entire ecosystem in and around the Sound (Ferrell and Fraedrich, 1997).

Captain Hazelwood was known to have a drinking problem and was twice convicted of drunk driving. He was tested nine hours after the tragic collision, and his blood alcohol level was 0.06, higher than the 0.04 permitted by the Coast Guard. However, if one assumes a normal metabolism rate, Hazelwood's blood alcohol rate would have been closer to 0.19 at the time of the collision, almost double the "legally drunk" criterion used by most states (Hartley, 1993: 221). Hazelwood was eventually found guilty of "negligent discharge of oil" due to his violation of company policy of leaving the bridge, which is a misdemeanor, and was acquitted by a jury of the more serious charges of criminal mischief, reckless endangerment, and operating a vessel while intoxicated (Hartley, 1993: 221).

The Exxon company and the Alyeska Pipeline Service, a consortium of seven oil companies formed to manage the Trans-Alaska oil project, were both severely criticized for the way they responded to the disaster, including how they handled the cleanup and managed the crisis in general. According to the comprehensive contingency plan filed by Alyeska, emergency crews were supposed to be able to fully encircle any spill within five hours. In the case of the Exxon *Valdez*, it took crews more than 36 hours to surround the spill with containment booms. As Ferrell and Fraedrich point out:

> A state audit of Alyeska's equipment demonstrated that the company was unprepared for the spill. It was supposed to have three tugboats and thirteen oil skimmers available but in fact it had only two and seven, respectively. The company also had only fourteen thousand feet of boom for containing spills; the contingency plan specified twenty-one thousand feet. The barge that carried the booms and stored skimmed oil was also out of

service because it had been damaged in a storm. In any case, the required equipment would not have been enough, because a tanker like the Exxon *Valdez* is almost one thousand feet long and holds 1.2 million barrels of oil. The booms available could barely encircle the giant ship, much less a sizable slick. (1997: 284)

Exxon's woefully inadequate response to the disaster and its consequent controversial cleanup efforts led an Alaska judge to impose a $5 billion fine for punitive damages on the Exxon company and to order Exxon to pay local fishermen $286 million in compensatory damages (Ferrell and Fraedrich, 1997: 286). In addition to these costs, Exxon was reported to have spent more than $2 billion to cover the costs of the cleanup.

Perhaps more damaging to Exxon in the long run was the tarnished public image it had acquired from its faulty crisis management strategies. The results of a survey of 200 American and Canadian business executives showed that "Exxon was slow to react, attempted to shift blame onto others, lacked preparation, seemed arrogant, was negligent, lost control of information processes, and ignored opportunities to build public support" (cited in Hartley, 1993: 228). As Ferrell and Fraedrich put it:

> Exxon's response to the crisis certainly hurt its reputation and credibility with the public. National consumer groups have urged the public to boycott all Exxon products, and nearly twenty thousand Exxon credit card holders cut up their cards and returned them to the company to express their dissatisfaction with its cleanup efforts. (1997: 285)

Johnson & Johnson

In recurrent reputational surveys of the 30 most admired companies in the United States, Johnson & Johnson (J&J) is always ranked first (Alsop, 1999). Benefiting from its 100-year history as manufacturer of various health care and pharmaceutical products, J&J has also been guided by core values embodied in its mission statement called "Credo." Its Credo sets forth J&J's responsibilities to its five stakeholders or constituencies in the following rank order: doctors and nurses; patients and other customers; employees; communities in which J&J operates; suppliers and distributors; and stockholders (Reidenbach and Robin, 1991: 280).

Its Credo was put to a test in September 1982 when three people in a Chicago suburb died from ingesting extra-strength Tylenol capsules, one of J&J's most popular products. A day later two other people died in another Chicago suburb. Investigators soon determined that somebody had laced the capsules with cyanide by tampering with the capsules. This prompted James Burke, CEO of J&J, to order a recall of all Tylenol capsules, regular and extra-strength, from all stores throughout the United States—35 million bottles of its Tylenol product at a cost to the company of more than $100 million.

In February 1986, another case of Tylenol poisoning occurred, causing the death of a woman in Bronxville, New York. The company responded by initiating another national recall of capsules and announcing plans to abandon manufacturing Tylenol in capsule form and substituting capsules with a caplet form of medication. "Recalls and production retooling were estimated to cost the company over $150 million" (Shrivastava, Mitroff, Miller, and Migliani, 1998: 295).

The Tylenol poisonings posed an immediate threat to the well-being of millions of Americans who take Tylenol. It also threatened the public image of J&J and its products. Its stock price dropped by 20 percent, a $2 billion decline in total value (Useem, 1998: 36), and its market share plummeted from 35 percent to 8 percent (Shrivastava *et al.*, 1998: 295).

J&J in general, and James Burke in particular, responded to the crisis with alacrity, candor, and openness. As CEO, Burke was obsessed with preserving the high level of trust in J&J products. He felt an obligation to maintain the history of trust developed over the years by his predecessors, which they codified in their corporate Credo. As Burke said: "All of the previous managements who built this corporation handed us on a silver platter the most powerful tool you could possibly have—institutional trust" (Smith and Tedlow, 1989: 6). In fact, Burke attributed the overall success in J&J's handling of the Tylenol crisis to the company's strict allegiance to the Credo and its principal value of institutional trust, which the Credo helped establish between the company and its stakeholders. J&J's response to the Tylenol poisonings is an exemplary study in crisis management.[1]

CRISIS MANAGEMENT THEORY

When technology fails—as it does more often than we think—it may endanger the lives of many people and destroy vast amounts of property. The seven case studies we just reviewed attest to this lamentable fact, too often ignored. What is the response of the corporation in which the failed technology is embedded? It can define the technological disaster as an accident or as a crisis. An accident, as Perrow points out, is any untoward failure of a subsystem component; a crisis, on the other hand, involves a threat to the system as a whole with dire consequences for the well-being of one or more suprasystems as well (Perrow, 1984). We speak of an automobile accident but refer to Chernobyl, for example, as a crisis, with systemic and suprasystemic consequences.

For about two decades, Mitroff and his colleagues have explored the dynamics of crisis management, setting forth a variety of frameworks (Mitroff and Kilmann, 1984; Mitroff and Pauchant, 1990; Pauchant and Mitroff, 1992; Pearson and Mitroff, 1993; Shrivastava et al., 1998). In a recent summary article, Mitroff identifies four components of his framework: crisis types, crisis mechanisms, crisis systems, and crisis stakeholders (Mitroff, 2000: 6–7).

Among the types of crises Mitroff enumerates are "economic, informational, physical, human resources, reputational, psychopathic and natural (disaster)" (Mitroff, 2000: 6). It is rare indeed that corporations in the face of crises are proactive in their management style and plan for all possible crises. Most corporations are likely to plan for natural disasters. Those that do are likely also to prepare for "core crises" common to a particular industry. In the chemical industry, for example, firms are likely to prepare for toxic spills and fires.

As for crisis mechanisms, a proactive as opposed to a reactive management style is more likely to be effective. It enables companies to "anticipate, sense, react to, contain, learn from and redesign effective organizational procedures for handling calamities" (Mitroff, 2000: 7).

Mitroff has identified five phases or stages through which virtually all crises pass—from signal detection to preparation/prevention, through containment/damage limitation and recovery, to organizational learning (see Figure 11–1).

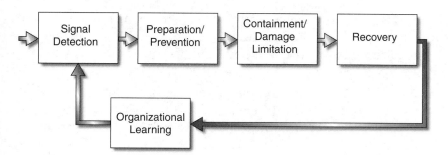

Figure 11–1
The five phases of crisis management.
Source: Academy of Management Executive by Christine Pearson and Ian Mitroff.
Copyright 1993 by Academy of Management via Copyright Clearence Center.

Of particular importance are the first phase, signal detection, and the last phase, lessons learned, for reflection on the crisis with a view to enhancing future organizational capabilities.

Crisis systems include technology, organizational structures, human factors, culture, and the psychology of senior management. Technological systems are not automatic in their operations; fallible human beings run them. Crisis management teams function best if they can avoid succumbing to the defense mechanisms of denial and grandiosity. If these teams fail to mitigate the costs of a crisis, they may inadvertently prolong it.

Crisis stakeholders include internal as well as external constituencies. They can be called upon to grapple with the issues that create crises. Stakeholders include not only employees but also nonorganizational elements in the environment whose assistance can be elicited, such as police and fire departments, hospitals, the Red Cross, and the media. To elicit the cooperation of key stakeholders for crisis management requires the development of training programs and the sharing of crisis management plans and experiences.

As one can see from the analysis of our seven cases, they all embody the variety of "crisis types" identified by Mitroff. For example, the huge numbers of fines and penalties levied against Ford, Exxon, Johns-Manville, and A.H. Robins were instrumental in creating an economic crisis for these corporations. Corporate reputation, status, public image, and brand name were all seriously threatened. Corporate moral character was questioned as consumers and the general public protested against what they regarded as callous mistreatment at the hands of the makers of the Ford Pinto and the Dalkon Shield, various asbestos manufacturers, and Exxon's management.

If one includes environmental elements as part of "physical crises," certainly the *Valdez* oil spill and the asbestos cases prove to be substantial environmental crises. These two cases were management failures that indeed resulted in substantial environmental harm. In fact, the asbestos case may very well be, in many ways, the worst human-made environmental crisis in recent times. As Barton (1993) points out:

> The mismanagement of the product [asbestos] from the time it was suspected to be injurious to human life remains one of the most outrageous environmental episodes of the century. At fault were a variety of manufacturers, distributors, and health officials . . . who failed to act with a precision and determination necessary to resolve an evolving catastrophe. (p. 154)

Despite its efforts to take responsibility for the massive oil spill, including full page statements of formal apology in major newspapers and Chairman Lawrence Rawl's personal apologies in numerous interviews, Exxon's crisis management and crisis communication strategies are treated in crisis management literature as a paradigmatic case of information mismanagement during a crisis (Lukaszewski, 1993; Small, 1991; Tyler, 1997; Williams and Olaniran, 1994). Corporate executives from other companies who were polled, as well as the public at large, felt that Exxon's top management was arrogant and negligent in dealing with public opinion concerning the disaster. This may very well be attributed to the manner in which it acknowledged the crisis. Although justifiably criticized for its lack of adequate preparation and slow reaction to the spill, eventually Exxon sent hundreds of people to the region to deal with the problem, including cleanup crews, claims adjustors, technical and scientific experts, and management specialists. However, by ignoring other stakeholders involved, especially the local people immediately affected by the disaster, it disregarded crucial psychological and sociological factors in its management of the crisis.

> By focusing on their tasks and outcomes instead of on the feelings, needs, fears, and apprehensions of the people involved, Exxon as a company may have driven an ever-widening wedge between it-

> self and the people it most needed on its side. And, since the peo-
> ple of Alaska already had a love-hate relationship with Big Oil—
> i.e., love the money, hate oil rigs and pipelines—the perception
> that a big company could just come in and throw lots of money
> at a problem that was more than just a money issue may have in-
> creased the growing resentment. (Albrecht, 1996: 65)

Exxon's exclusively technical approach to handling the crisis contributed to public cynicism and distrust.

No doubt, the insensitive response of Exxon's corporate executives ran deep in the individual and collective sentiments of
anger in the community and in the general public. Largely as a
consequence of the growing environmental consciousness of the
times, American citizens quickly lost faith in the capacity of corporations to act effectively in the public interest with a sense of
social responsibility that would have sensitized them to minimize
damage or harm citizens.

Obvious informational and communication factors that exacerbated an impending crisis are also evident in the misrepre-
sentation of facts, denial, deception, falsification and destruction
of subpoenaed documents at the hands of both A.H. Robins' and
Johns-Manville's executives. By contrast, the open, direct, and
honest crisis communication strategies exhibited by top executives of Johnson & Johnson were indeed exemplary. Chairman
Burke immediately formed a seven-member task force, which focused its attention on damage control and effective public communication about the events, immediately informing the press of
any and all information they received. This close working relationship with the press early on in the crisis greatly confirms
Johnson & Johnson as a credible, honest company genuinely concerned about the victims and all other stakeholders involved.

In short order, J&J management established its credibility and
concern, the hallmarks of good crisis management, through multiple channels of communication and action, including: setting up
toll-free consumer hot lines; putting out full-page news advertisements in major newspapers; offering free exchanges; appearing in
numerous interviews with top management in both print and on
television; and, finally, openly corresponding with all employees
asking for their support and help.

In retrospect, one of the most effective responses by Johnson
& Johnson was its immediate recall of 35 million bottles of

Tylenol, all of which were eventually destroyed. All in all, 8 million capsules were tested, 75 of which were found to be contaminated (Barton, 1993). This massive effort greatly helped convince the public of J&J's concern for product purity, safety, and integrity.

With government agencies, the media, and corporate elites all cooperating, the crisis was ultimately managed successfully. In turn, the public responded in a positive way to the efforts of Johnson & Johnson. This was reflected in Tylenol's quick recapture of its entire previous market share, which was a full 75 percent of the over-the-counter analgesic market (Pinsdorf, 1987: 51). In addition, the Tylenol case illustrates well how a corporation can respond to a crisis generated by a psychopath who sabotages its product.

The important lesson to learn from crisis management failures, as illustrated in the Exxon *Valdez* case, and from successes, as illustrated in the Johnson & Johnson case, is the need for corporations to demonstrate proper respect and care for all stakeholders with whom they interact and whose lives they affect.

In the Johnson & Johnson and Exxon *Valdez* cases, management quickly acknowledged that it had a crisis on its hands and made immediate efforts to forestall potential problems. In the case of J&J, crisis management was successful. By contrast, when the Ford-Firestone conflict surfaced, instead of seeking to discover the sources of the tire blowouts and fatal vehicle rollovers, the two companies blamed each other for the crisis in an effort to avoid responsibility and liability for the loss of life and property. In the Exxon *Valdez* case, managerial response was inadequate, leading to crisis mismanagement.

In the cases involving the Ford Pinto, the asbestos products, and the Dalkon Shield, corporate management ignored warnings and was in denial about the events of impending crises. This led to deficiencies in handling the crises and resulted in serious crisis mismanagement. In the whisleblowing case involving Eastern Air Lines, top executives dismissed the repeated efforts of an experienced pilot to correct a life-threatening autopilot system problem. Only after the National Transportation Safety Board, investigating the causes of an Eastern Air Lines disaster—in the course of which it identified the autopilot safety problem—did Eastern Air Lines finally modify the autopilot system. Such managerial responses to the various types of crises are summarized in Figure 11–2.

	Crisis Management	Crisis Mismanagement
Acknowledgement of Technological Disaster	Johnson & Johnson	Ford-Firestone Conflict Exxon *Valdez* Oil Spill
Denial of Technological Disaster	— —	Ford Pinto Case Johns-Manville Asbestos Case Dalkon Shield Case Eastern Air Lines Whistleblowing Case

Figure 11–2
A classification of corporate responses to crises.

Ignoring warnings from Ford engineers that the design of the rear-end gas tank was prone to fires from rear impacts, Ford management forged ahead with the "tooling" stage simultaneously with the design and testing stage. This strategy put a serious strain on safety issues, which were ultimately left out of the equation.

In the asbestos case, executives at Johns-Manville were found guilty of foreknowledge of the dangers of asbestos, which they deliberately suppressed for decades. Not only were they found guilty of suppressing research findings that demonstrated a direct linkage between asbestos and asbestosis, corporate executives at Johns-Manville developed an outrageous campaign of denial and denunciation that was shameless in its tactics to blame everyone *but* the Johns-Manville Company. In a full page statement, published in the nation's leading newspapers in 1982, John A. McKinney, Manville's chairman and chief executive officer, denied any responsibility and proceeded to blame everyone else except Johns-Manville's executives for the quandary in which the company found itself after being deluged with more than 16,500 lawsuits related to health problems associated with asbestos. As Brodeur (1985) reports:

> He [McKinney] began by blaming the federal government for
> refusing to admit responsibility for asbestos disease that had

developed among . . . shipyard workers. . . . He criticized Congress for failing to enact a statutory compensation program for the victims of asbestos disease. . . . He castigated the insurance companies . . . for refusing to pay claims against product-liability policies. (p. 4)

As far as Johns-Manville was concerned, McKinney brazenly denied any wrongdoing. McKinney's only defense of his actions was the claim that "not until 1964 was it known that excessive exposure to asbestos fiber released from asbestos-containing insulation products can sometimes cause lung diseases" (Brodeur, 1985: 4). What McKinney failed to mention, however, was that this claim, which had been the company's main legal defense for years, was rejected by jury after jury and had recently been struck down by the New Jersey Supreme Court (Brodeur, 1985: 4).

The deception and denial inherent in McKinney's statement becomes all the more evident in light of sworn testimony made by Charles Roemer, an attorney investigating reports of lung diseases in asbestos workers, which was given in 1942! The testimony maintains that:

Vandivier Brown [a Johns-Manville attorney at the time] stated that Johns-Manville's physical examination program had . . . produced findings of X-ray evidence of asbestos disease among workers exposed to asbestos and that it was Johns-Manville's policy not to do anything, nor to tell the employees of the X-ray findings. . . . If Johns-Manville's workers were told, they would stop working and file claims against the company, and that it was Johns-Manville's policy to let them work until they quit because of asbestosis or died as a result of asbestos-related diseases. (Coplon, 1996: 67–68)

In the Dalkon Shield case, denial of problems by top management at A.H. Robins created an atmosphere of complicity that ultimately led to intimidating tactics against women who filed lawsuits. In the course of various trials, lawyers for Robins began blaming the users of the product: they inquired into the sexual activities of the women, subpoenaed past and present sexual partners, and ruthlessly grilled them about the intimate details of their sexual life.

A.H. Robins' crisis management strategy of "Deny everything at all costs" is a paradigmatic case of crisis mismanagement and stands as a striking lesson of what corporations should *never* do. The problem is that too many corporations focus on a "do nothing until something happens" mentality and hence are often caught off-guard, lose control of the problem, and end up on the defensive. The result is an all too-narrow focus on the potential crisis effects on profits and stockholder interests, while ignoring other important stakeholders, including consumers.

Federal District Judge Miles Lord's reproach of the corporate executives of A.H. Robins expresses well the dire consequences that are the result of corporate denial and mismanagement of potentially dangerous technologies:

> . . . And when the time came for these women to make their claims against your company, you attacked their character. You inquired into their sexual practices and into the identity of their sex partners. You ruined families and reputations and careers in order to intimidate those who would raise their voices against you. You introduced issues that had no relationship to the fact that you had planted in the bodies of these women instruments of death, of mutilation, of diseases. . . . Another of your callous legal tactics is to force women of little means to withstand the onslaughts of your well-financed team of attorneys. . . . You have taken the bottom-line as your guiding beacon and the low road your route. (cited in Barton, 1993: 48)

The various "crisis systems"—technical, organizational, human resources, corporate culture, and the psychology of top management—are also evident in these seven case studies. According to Shrivastava and his colleagues (1998), technical factors include: "faulty design, defective equipment, contaminated or defective materials and supplies, and faulty technical procedures" (p. 290). Of course, one of the central factors involved in the Ford Pinto case was the defectively designed gas tank. Certainly, defective equipment and materials were involved in the making and marketing of the Dalkon Shield. In both cases, as was the case in the Exxon *Valdez* and in the asbestos cases, faulty technical procedures were followed.

Corporate factors that may precipitate a crisis include "policy failures, inadequate resource allocations for safety, strategic pressures that allow managers to overlook hazardous practices and conditions, communications failures, misperceptions of the extent and nature of hazards, inadequate emergency plans, and cost pressures which curtail safety" (Shrivastava *et al.*, 1998: 290). Aberrant corporate culture and the psychology of top management should also be added to the list. All of these factors can be identified in six of our seven case studies. In the Exxon and Johnson & Johnson cases, we have seen how crucial effective crisis communication is as a strategy in successfully managing corporate crises.

Looking at the Ford Pinto case, we see that cost pressures led Ford management to "curtail safety" in a quest for the ultimate bottom line. This is nowhere more evident than in the technical rationality of relying exclusively on cost-benefit analysis as the sole decision-making strategy, which ultimately led to the depersonalization of potential victims. This also created the preconditions for a general neutralization of obligations and responsibilities owed to consumers.

Inadequate resources for safety and inadequate emergency plans greatly exacerbated an already monumental environmental disaster in the case of the Exxon *Valdez*. The same can be said for the asbestos problem. Asbestos manufactures were lethargic in the face of extensive use of its product in literally tens of thousands of ships, factories, and buildings. To make matters worse, asbestos is costly and time-consuming to remove. During removal, fibers can become airborne and can actually do more damage than if left behind walls and around piping. Because of this, removal crews need to have special training and equipment. As a result, private companies and governmental agencies were totally unprepared to deal with the complex set of problems surrounding asbestos removal.

Corporate inattention to the possibility of crises—that is, the signal detection phase of effective crisis management (see Figure 11–1)—and top management's refusal to deal with crises once recognized, is all too often the result of normal corporate policy and embodied in the belief systems and attitudes of corporate top management. Such policies cause management to overlook, even deny, hazardous practices and conditions. These factors are especially evident in the corporate decisions made by top executives of A.H. Robins to develop, aggressively market, and sell a product they knew to be "relatively untested, unpredictable, and potentially dangerous." This is also evident in various asbestos manufacturing firms such as Johns-Manville, which produced and sold asbestos many years *after* they became aware of its dangerous properties.

The various factors involved in these crisis systems often culminate in *triggering events* that have the potential to cause a crisis. Triggering events often lead to physical, economic, social, or

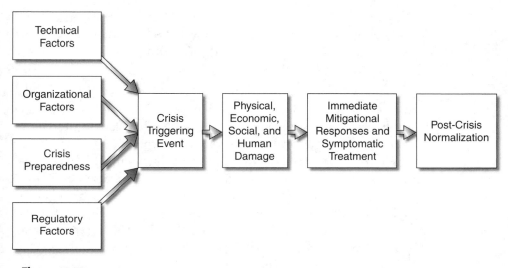

Figure 11–3
A model of crisis development.
Source: Adapted from Shrivastava, Paul; Mitroff, Ian I.; Miller, Danny; and Migliani, Anil (1998). "Understanding industrial crises," *Journal of Management Studies,* 25 July: 293.

individual harm, which forces the companies to respond in an attempt to halt the crisis. Finally, there is a period of *post-crisis normalization,* when the corporation involved assesses how the crisis has affected the status of the corporation. Figure 11–3 provides a model for how a crisis develops and how it can be managed. As the model indicates, technical factors, organizational factors, crisis preparedness factors, and regulatory factors often work in conjunction with precipitating crises. This process is exemplified in most of our case studies. In fact, regulatory factors were evident in each case study with the exception of the Tylenol case.

Part of the controversy surrounding the Ford Pinto case relates to the responsibility of Congress to devise adequate automobile safety regulations and to have its administrative agencies enforce the laws governing the manufacturing of safe and reliable automobiles. Consumer advocates have chastised the U.S. government for failing to enact adequate safety standards, charging that government officials fear stepping too forcefully on the toes of giant automakers such as the Ford Motor Company. In the asbestos case, the U.S. government was criticized for ineffectively dealing with the problem of its own use of the substance, which was installed on dozens of naval vessels and at various other military installations. Government officials were also criticized for the way they initially responded to the challenges of asbestos removal from military facilities, public schools, libraries, and other public buildings.

In light of the asbestos and Love Canal cases (see Chapter 8), Congress enacted the 1988 Community Right to Know Act amendment to the Superfund Law of 1986. Because of these new laws, corporations were, for the very first time, forced to disclose to the public the massive amounts of toxic chemicals they actually release into the environment each year. For example, Rohm and Haas, a major manufacturer of industrial and other chemicals, had to admit that it pumped 1 million tons of chemicals into the local environment, from its plant in Bristol, Pennsylvania (Barton, 1993: 156).

In the case of the Dalkon Shield, the Food and Drug Administration was criticized for allowing A.H. Robins' executives to convince the agency to classify the product as a "device" rather than as a "drug." This classification allowed Robins to market and sell the product without going through the normally rigorous efficacy- and safety-testing required by the FDA for drugs. Moreover, even after learning of a septic abortion death of an Arizona woman in June 1973, it took a full two years for FDA Commissioner Alexander Schmidt to "request" that A.H. Robins suspend marketing the Dalkon Shield until its "questionable safety" could be reviewed (Perry and Dawson, 1985: 346). Robins eventually complied with the FDA request in the United States, but continued to market the product in numerous foreign countries. For their part, FDA officials blamed their ineffectiveness on the lack of a strong law granting them the power to act. It wasn't until the FDA confirmed that 11 women had died and hundreds of others had fallen seriously ill that a law was finally passed by Congress in the form of the 1976 Medical Device Amendment to the Food, Drug, and Cosmetic Act. Even with the passage of this law, however, which required companies to demonstrate the safety and efficacy of an Intra-Uterine Device (IUD) before being marketed, the FDA was powerless to require that A.H. Robins recall all Dalkon Shields sold, which numbered in the millions. Hence, regulatory factors often interact with other factors to give rise to crises. As Barton (1993) points out:

As in most visible crises, lapse was laid upon lapse. Robins was not the sole responsible player. The media was lethargic in picking up the story and the federal government waited four years after the first reports of trouble to investigate the shield and then, after the product was removed from the market, did little to ensure the device was removed from women still using it. (p. 49)

Perry and Dawson (1985) interpret the Dalkon Shield case as an outrageous story of corporate greed, unwarranted consumer trust, government ineffectiveness, and medical indifference. With this interpretation, Perry and Dawson identify the complex web of corporate, legal, and medical institutions that interact with one another to bring about technological hazards and technological disasters. The function of regulatory factors in precipitating technological crises points to the need to analyze the role of the legal system in the management of technological disasters.

The lessons Mintz (1985) draws from his analysis of the Dalkon Shield can be expanded as a set of general lessons to learn from all of these cases:

First, the corporate structure itself—oriented as it is toward profit and away from liability [or ethics]—is a standing invitation to such conduct; secondly, the global scale of contemporary marketing has made hazardous corporate activities more perilous to ever larger numbers of people; and, third, all the deterrents and restraints that normally govern our lives—religion, conscience, criminal codes, economic competition, press exposure, social ostracism—have been overwhelmed. (p. 248)

By way of concluding this chapter, we turn to a discussion of the role of corporate social audits and corporate codes of conduct in helping corporations meet their social obligations and ethical responsibilities.

CONCLUSION

How can we explain such deliberate, illegal, and unethical behavior on the part of corporate management of the Ford Motor Company, A.H. Robins, Johns-Manville, Exxon Corporation, Eastern Air Lines, and Ford-Firestone? According to management theorist and consultant Peter Drucker, the three major unquestioned presuppositions of the classical theory and practice of management have to do with its "scope," "role," and "task" (Drucker, 1970). For Drucker, the scope of traditional management theory is "severely restricted." This is because economic activity becomes an end in itself, totally divorced from social and ethical concerns. Second is a minimum concern with social responsibility in many corporate boardrooms. Concerns that cannot be encompassed within an economic calculus are often seen as constraints and limitations rather than as management objectives. Third, the primary, perhaps only task of management is to mobilize the energies of the organization for the accomplishment of well-known and defined goals such as profit maximization, constant productivity, and growth.

For several decades, organization theorists and philosophers have focused on a "paradigm shift" in corporate management philosophy, from a "stockholder" model of the corporation to a "stakeholder" model of the corporation (e.g., see Ackoff, 1981; Ansoff, 1965; Eells and Walton, 1961; Evan, 1993; and Freeman, 1984). The stakeholder model of the corporation acknowledges that the corporation has an obligation to fulfill the legitimate expectations of constituencies other than stockholders; namely, employees, customers, suppliers, the community surrounding the corporation, and the public at large. Taking the claims and interests of these constituencies into account would deter management from producing potentially harmful products or services. In other words, by incorporating the stakeholder model of decision making, management could thereby minimize corporate deviance and would, in the long run, reduce the probability of generating technological disasters.

In addition to strategic stakeholder management theory, other policies such as corporate social audits and corporate codes of conduct may be effective ways to infuse humanistic and social values into the corporate culture.

Just as business executives use their annual financial audits to gauge the success of their profit making, some business ethicists have suggested that it behooves corporations to perform a "corporate-social audit" that would inform all stakeholders of the corporation's "success" in corporate social responsibility and ethical practices. Although the details of what an adequate corporate-social audit would look like are difficult to articulate, the philosophy of such a proposition rests on the implicit "social contract" that exists between corporations and society. As two of the first developers of such a corporate-social audit have argued:

> Society in America and elsewhere grants the corporation the right to exist. By issuing a corporate charter, it endows the corporation with certain privileges . . . and many unspecified, but valuable rights . . . In return for these and other legal privileges and rights society expects certain standards of behavior. . . . In the intervening years society has vastly increased its expectations. . . . It follows from this that the corporation will be called upon, formally or by the subtle pressures of public opinion, to make it known how it is measuring up to its responsibilities. Thus the social audit flows logically from the social contract and from the expectation rooted in the contract. (Corson and Steiner, 1974: 43–44; cited in Bowie, 1982: 107)

Even if for no other reason, some kind of corporate-social audit could play an important role in motivating management to articulate and disseminate a sound crisis preparedness strategy. Such a social audit could, in addition, provide an input for organizational learning, the crucial, but all-too-often neglected, last phase of effective crisis management (see Figure 11–1).

In addition to social audits, corporations have a variety of other mechanisms at their disposal through which they can integrate social and ethical values into their corporate culture. They include: formal codes of ethics, mission statements, credos, policies, and general codes of conduct. Just as important is their proper implementation through ethics training courses, ethics officers, and ombudspersons to deal with complaints of employees. Such programs promote corporate responsibility and help the corporation build and sustain trusting relationships with all of its stakeholders (Weiss, 1998). Such programs could guide CEOs and Boards of Directors to work more cooperatively and responsibly as partners with all stakeholders of the firm (Evan, 1998: 249–272; Weiss, 1998: 269–270).

An outstanding example of the potential effectiveness of a corporate code is the Credo of Johnson & Johnson. As Chairman James Burke explains:

> All of the previous managements who built this corporation handed us on a silver platter the most powerful tool you could

possibly have—institutional trust. Everybody who puts something into this organization that builds trust is enhancing the long-term value of the business. I think that these values were here. We traded off them. We articulated them through the Credo. We spent a lot of time getting people to understand what we meant in the Credo. A very dramatic event of unparalleled proportions challenged us. Not only did we face up to the challenge, but we also demonstrated that for all those . . . years of trust, worked to help solve problems no matter how serious they may be. (Smith and Tedlow, 1989: 6)

Regrettably, such cases as Johnson & Johnson are not as common as we would like to think. We thus recognize, unfortunately, that stakeholder management, corporate social audits, and corporate codes of conduct are not enough to mitigate the incidence of technological disasters that result from corporate mismanagement of technological innovations. As we have seen, legal, economic, and political factors also have important roles to play. Hence management practices alone are not enough. However, such enlightened, proactive management strategies may provide necessary conditions for the emergence of a commitment to corporate social responsibility that, if implemented, would greatly reduce the prevalence of technological disasters.

Judge Lord's harsh reprimand of A.H. Robins' management provides an authoritative argument for the necessity of corporate social responsibility to mitigate the harms of technological disasters:

Under your direction, your company has in fact continued to allow women, tens of thousands of them, to wear this device—a deadly depth charge in their wombs ready to explode at any time. . . . The only conceivable reasons you have not recalled this product are that it would hurt your balance sheet and alert women who have already been harmed that you may be liable for their injuries. . . . If this were a case of equity, I would order your company to make an effort to locate each and every women who still wears this device and recall your product. But this court does not have the power to do so. I must therefore resort to moral persuasion and a personal appeal to each of you. . . . Please, in the name of humanity, lift your eyes from the bottom-line. . . . Please, gentlemen, give consideration to tracking down the victims and sparing them the agony that will surely be theirs. (Sobol, 1991: 20)

References

Ackoff, R.L. (1981). *Creating the corporate future.* New York: John Wiley & Sons.

Albrecht, Steve. (1996). *Crisis management for corporate self-defense.* New York: American Management Association.

Alford, Fred. (2001). *Whistleblowers: Broken lives and organizational power.* Ithaca, NY: Cornell University Press.

Alsop, Ronald. (1999). "The best corporate reputations in America," *The Wall Street Journal,* September 23: B1.

Anonymous (a). (2000). "An unwieldy recall: Firestone and Ford can't stay clear of the fallout," *Time,* September 4; 156 (10): 50.

Anonymous (b). (2001). "Recall of 4 million Explorers sought," *The New York Times,* June 4: C2.

Anonymous (c). (2001). "Screeching to a halt: Last year's tire crisis has returned to haunt Ford and Firestone," *The Economist,* May 26: 4–5.

Anonymous (d). (2000). "Lessons from the tire fiasco," *Business Week,* September 18: 178.

Ansoff, H.L. (1965). *Corporate strategy.* New York: McGraw-Hill.

Barton, Lawrence. (1993). *Crisis in organizations: Managing and communicating in the heat of chaos.* Cincinnati, OH: South-Western Publishing Co.

Bowie, Norman. (1982). *Business ethics.* Englewood Cliffs, NJ: Prentice Hall.

Brick, Michael, and Milford, Maureen. (2001). "Grace files for Chapter 11, citing cost of asbestos suits," *The New York Times,* April 3: C2.

Brodeur, Paul. (1985). *Outrageous misconduct: The asbestos industry on trial.* New York: Pantheon Books.

Coplon, J. (1996). "When did Johns-Manville know?" In T. Donaldson and A. Gini (Eds.), *Case studies in business ethics* (4th ed.). Upper Saddle River, NJ: Prentice Hall: 67–69.

Corson, John, and Steiner, George. (1974). *Measuring business's social performance: The corporate social audit.* New York: Committee for Economic Development.

Darley, J. (1991). "How organizations socialize individuals into evildoing." In D. Messick and A. Tenbrunsei (Eds.), *Codes of conduct: Behavioral research into business ethics.* New York: Russell Sage Foundation: 13–43.

Dowie, M. (1974). "Pinto madness." In J. Fielder and D. Birsch (Eds.), (1994). *The Ford Pinto case: A study in applied ethics, business, and society.* Albany, NY: State University of New York Press: 54–70.

Drucker, P.F. (1970). *Technology, management, and society.* New York: Harper and Row: 34–35.

Eells, R., and Walton, C. (1961). *Conceptual foundations of business.* Homewood, IL: Richard D. Erwin.

Ermann, M.D., and Lundman, Richard J. (Eds.). (1992). *Corporate and governmental deviance: Problems of organizational behavior in contemporary society* (4th ed.). New York: Oxford University Press.

Evan, W.M. (1993). *Organization theory: Research and design.* New York: Macmillan Publishing.

Evan, W.M. (1998). "International codes of conduct for multinational corporations." In Ronald F. Duska (Ed.), *Education, leadership, and business ethics: Essays on the work of Clarence Walton.* Boston: Kluwer Academic Publishers.

Ferrell, O.C., and Fraedrich, J. (1997). *Business ethics: Ethical decision-making and cases.* Boston: Houghton Mifflin.

Freeman, R.E. (1984). *Strategic management: A stakeholder approach.* Boston: Pitman Co.

Gellert, Dan. (1981). "Insisting on safety in the skies." In Alan Westin (Ed.), *Whistleblowing: Loyalty and dissent in the corporation.* New York: McGraw-Hill: 17–30.

Gini, A. (1996). "Manville: The ethics of economic efficiency." In T. Donaldson and A. Gini (Eds.), *Case studies in business ethics* (4th ed.). Upper Saddle River, NJ: Prentice Hall: 58–66

Gini, A., and Sullivan, Terry. (1996). "A.H. Robins: The Dalkon Shield." In T. Donaldson and A. Gini (Eds.), *Case studies in business ethics* (4th ed.). Upper Saddle River, NJ: Prentice Hall: 215–224.

Hartley, R. (1993). *Business ethics: Violations of the public trust.* New York: John Wiley & Sons.

Kiley, David. (2001). "Ford bites $3 billion bullet to replace tires Firestones viewed as risk," *USA Today,* May 23: B1.

Lerbinger, Otto. (1997). *The crisis manager: Facing risk and responsibility.* Mahwah, NJ: Lawrence Erlbaum Associates.

Lukaszewski, James E. (1993). "The Exxon *Valdez* paradox," In Jack Gottschalk (Ed.), *Crisis response: Inside stories on managing image under siege.* Detroit, MI: Gale Research: 185–211.

Mintz, Morton. (1985). *At any cost: Corporate greed, women, and the Dalkon Shield.* New York: Pantheon Books.

Mitroff, Ian. (2000). "The essentials of crisis management," *Financial Times,* June 20, Part 9: 6–7.

Mitroff, Ian, and Kilmann, Ralph. (1984). *Corporate tragedies: Product tampering, sabotage, and other catastrophes.* New York: Praeger Press.

Mitroff, Ian, and Pauchant, Thierry. (1990). *We're so big and powerful nothing bad can happen to us: An investigation of America's crisis-prone corporations.* New York: Carol Publishing.

Moss, Michael, and Appel, Adrianne. (2001). "Company's silence countered safety fears about asbestos," *The New York Times*, July 7: 1.

Pauchant, Thierry, and Mitroff, Ian. (1992). *Transforming the crisis-prone organization: Preventing individual, organizational, and environmental tragedies.* San Francisco, CA: Jossey-Bass Publishers.

Pearson, Christine, and Mitroff, Ian. (1993). "From crisis prone to crisis prepared: A framework for crisis management," *Academy of Management Executive* 7(1): 48–59.

Perrow, Charles. (1984). *Normal accidents: Living with high-risk technologies.* New York: Basic Books.

Perry, Susan, and Dawson, Jim. (1985). *Nightmare: Women and the Dalkon Shield.* New York: Macmillan Publishing.

Pinsdorf, Marion K. (1987). *Communicating when your company is under siege: Surviving public crisis.* Lexington, MA: Lexington Books.

Reidenbach, R. Eric, and Donald, P. Robin. (1991). "A conceptual model of corporate moral development," *Journal of Business Ethics* 10: 273–284.

Schmitt, Richard. (2001). "How plaintiffs' lawyers have turned asbestos into a court perennial," *The New York Times*, March 5: A1.

Shrivastava, Paul, Mitroff, Ian I., Miller, Danny, and Migliani, Anil. (1998). "Understanding industrial crises," *Journal of Management Studies*, July 25: 285–303.

Small, William J. (1991). "Exxon *Valdez*: How to spend billions and still get a black eye," *Public Relations Review* 17 (1): 9–25.

Smith, W.K., and Tedlow, R. (1989). "James Burke: A career in American business," *Harvard Business School Case*, 9-389-177.

Sobol, Richard B. (1991). *Bending the law: The story of the Dalkon Shield bankruptcy.* Chicago: The University of Chicago Press.

Trevino, L., and Nelson, K. (1995). *Managing business ethics.* New York: John Wiley & Sons.

Tyler, Lisa. (1997). "Liability means never being able to say you're sorry," *Management Communication Quarterly* 11 (1): 51–73.

Useem, Michael. (1998). *The leadership moment: Nine true stories of triumph and disaster and their lessons for us all.* New York: Times Business, Random House.

Vaughn, Diane. (1996). *The Challenger launch decision: Risky technology, culture, and deviance at NASA.* Chicago: University of Chicago Press.

Weiss, Joseph W. (1998). *Business ethics: A stakeholder and issues management approach.* Fort Worth, TX: The Dryden Press.

Westin, Alan. (1981). *Whistleblowing: Loyalty and dissent in the corporation.* New York: McGraw-Hill.

White, Joseph, Power, Steven, and Aeppel, Timothy. (2001). "Death count linked to failure of Firestone tires risen to 203," *The Wall Street Journal,* June 19: A3.

Williams, David E., and Olaniran, Bolane A. (1994). "Exxon's decision-making flaws: The hypervigilant response to the *Valdez* grounding," *Public Relations Review* 20 (1): 5–18.

Endnote

1. A kindred managerial response occurred in the Merck pharmaceutical company in the 1980s. Following the discovery of Mectizen, a drug effective in a killing parasite that causes river blindness—a disease afflicting millions of people in sub-Saharan Africa and Latin America—scientists at Merck discovered that one could treat this disease by giving victims one tablet of Mectizen once a year and thus completely preventing the progression of the disease. However, this discovery posed a quandary for Merck because the victims who could benefit from the drug were completely unable to pay for it. Hence, a debate raged within the ranks of management as to whether to shelve this miracle drug or to embark on a philanthropic venture to produce and distribute the drug free of charge. Under the imaginative leadership of Dr. Roy Vagelos, the then CEO of Merck, he embarked on an effort to elicit the cooperation of the World Health Organization, various foundations, and other organizations to distribute the drug free of charge—at considerable cost to Merck—in the interest of curing approximately 20 million people suffering from the river blindness disease.

   In 1987, Merck announced that it would donate Mectizen for the treatment of river blindness as long as needed (www.merck.com/overview/philanthropy/mectizan/p12.htm).

   Merck's decision was based on a simple, yet profound Credo, articulated by George W. Merck, the company's President from 1945–1950. In a 1950 address at the Medical College of Virginia, George Merck stated, "Medicine is for the people. It is not for the profits. The profits follow, and if we have remembered that, they have never failed to appear."

CHAPTER *12*

The Role of the Legal System in Technology Policy Decisions

"Justice can be attained only by the careful regard for fundamental facts, since justice is but truth in action."
—Justice Louis D. Brandeis

*T*he case studies of technological disasters presented in Chapters 5, 8, and 11 raise questions about the ability of our legal system to cope successfully with the negative consequences of technology. The most general point to be made is that lawyers, judges, and politicians are rarely in a position to assess the risks of complex technology about which they make decisions. Technology policy is made by politicians who—although a large proportion may have a legal background—are nevertheless mostly innocent of technical matters. It is because of these inadequacies that the need for independent, objective, and expert advice arises at every level of the legal system.

The legal system in the United States consists of four subsystems, each of which performs a lawmaking function that bears on technology policy decisions: the executive, legislative, administrative, and judiciary branches.

THE EXECUTIVE BRANCH

The executive branch consists of a complex of organizations reporting to the president and responsible for assisting him in dealing with

Congress (Congressional Quarterly, 1989; Edwards and Wayne, 1994; Ragsdale, 1998). Three types of organizations can be distinguished: the Executive Office of the President, the cabinet and executive departments, and agencies and corporations (Thomas *et al.*, 1994).

Twelve separate divisions comprise the executive office: the White House Office, Office of Management and Budget (OMB), Council of Economic Advisers, National Security Council (NSC), Office of Policy Development, Office of the U.S. Trade Representative, Council on Environmental Quality, Office of Science and Technology Policy, Office of Administration, Office of the Vice President, Office of National Drug Policy, and Executive Residence of the White House. Figure 12–1 diagrams the complex organizational relationships of the president.

The cabinet consists of the president, the vice president, the heads of the 14 executive departments, and any other officials the president wishes to include such as the directors of the OMB and NSC. In recent years cabinets have served as a forum for the president to discuss policy proposals.

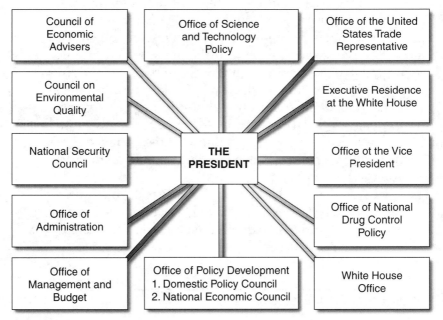

Figure 12–1
Executive office of the president.
Source: *U.S. Government Manual* (Washington, DC: Government Printing Office, 1997–1998: 90–108).

The executive branch includes many agencies for which the president is responsible, such as the National Aeronautics and Space Administration (NASA). It also encompasses government corporations, such as the U.S. Postal Service and Amtrak, which operate like business enterprises to provide a necessary public service though, they are unprofitable.

In 1950, William Golden, then serving as a special consultant to the White House, drafted a memorandum to President Truman recommending ". . . the prompt and immediate appointment of an outstanding scientific leader as Scientific Advisor to the President" (Rosenzweig, 1993: 307). He reasoned that the proliferation of uncoordinated scientific programs in individual federal agencies dictated the need for ". . . centralization of knowledge of all scientific programs in one independent and technically competent individual to whom the President can turn for advice" (Rosenzweig, 1993: 307). The scale and complexity of the government's stake in science and technology has grown since Golden's recommendation was accepted by President Eisenhower in 1957, when he created a President's Science Advisory Committee (PSAC). Ever since the PSAC was abolished by President Nixon in 1973, the executive branch has struggled with maintaining "independent and comprehensive advice." However, in 1989, President George Bush established the President's Council of Advisors on Science and Technology (PCAST) to ensure that the president would receive advice from the private sector and the academic community on technology and scientific research priorities.

One recurring proposal for solving the related problems of improved coordination and oversight of federal science activities has been to create a Department of Science (Brooks, 1986; Golden, 1986). There are difficulties with this proposal, but the most significant aspect of the idea of a Department of Science is that its secretary, as a cabinet member, would have regular access to the president and to the heads of all other federal departments and agencies. One compromise is to grant the president's principal science advisor cabinet-level status without creating a Department of Science (Wiesner, 1989). By appointing John H. Gibbons Assistant to the President for Science and Technology in 1993, President Clinton availed himself of a science advisor although the position lacks cabinet-level status. As it happens, John Gibbons was formerly director of the Office of Technology Assessment for 13 years, before Congress abolished it in 1995.

When a natural disaster devastates a geographic area, the president has the authority to designate it as a disaster area and thus make it eligible for federal relief funds. The president can also direct the Federal Emergency Management Agency (FEMA), an independent agency reporting to the president, to investigate the disaster and coordinate relief operations.

In the case of a technological disaster, the president is likely to consult the Office of Science and Technology Policy for advice as to how to respond. For example, when the Y2K problem surfaced, it was assessed as a major looming problem; the president issued an executive order providing for the establishment of a President's Council on Year 2000 Conversion Information Center to coordinate all the remedial programs for the federal government (see Chapter 3).

If a technological disaster occurs that affects the health and well-being of the citizenry, the president has the authority to issue an executive order to mitigate the effects of the problem—unless he defines the problem as falling under the province of one or another of the federal regulatory agencies that do not report to the president.

During President Clinton's administration, 227 executive orders were issued (http://www.pub.whitehouse.gov/retrieve-documents. html). The subjects varied greatly: employment discrimination, drug use, computer software privacy, illegal immigration, and child support for pregnant women. The following 17 executive orders deal with a range of technological problems:

1993-04-21

Executive Order 12843 on Ozone Depletion

1993-03-09

Executive Order 12840 on Nuclear Cooperation

1994-02-11

Executive Order 12898 on Federal Actions to Address Environmental Justice in Minority Populations and Low Income Populations

1994-11-14

Executive Order 12938 on Mass Destruction Weapons

1995-01-20

Executive Order 12946 on Arms Proliferation Policy Board

1996-04-17

Executive Order 12999 on Computer Technology for Education

1996-07-16

Executive Order 13011 on Federal Information Technology

1997-06-11

Executive Order 13049 on Prohibition of Chemical Weapons

1998-02-04

Executive Order 13073 on Year 2000 Conversion

1998-06-29

Executive Order 13091 on Arms Export Controls

1998-07-28

Executive Order 13092 on Information Technology Advisory

Executive Order 13094 on Weapons of Mass Destruction

1998-09-30

Executive Order 13103 on Computer Software Piracy

1998-08-25

Executive Order 13100 on Food Safety

1999-06-25

Executive Order 13128 on Chemical Weapons

2000-04-22

Executive Order 13148 on Environmental Management

2000-04-27

Executive Order 13151 on Global Disaster Information Network

2000-05-26

Executive Order 13158 on Marine Protected Areas (Executive Orders)

Without the guidance of his Assistant to the President for Science and Technology, it is unlikely that President Clinton would have issued these 17 executive orders.

THE LEGISLATIVE BRANCH

In 1957, Congress passed the Price-Anderson Act. This act contained a number of provisions, one of which guarantees that the law will "hold harmless the (nuclear) licensee and other persons indemnified" from public liability claims arising from nuclear power accidents over a certain predetermined dollar limit (Atomic Energy Commission, 1974: 220). This liability limit was seen as a necessary condition for private industry's interest in nuclear power; fears of public liability had previously blocked corporate investment in nuclear power. Given the possibility of a core meltdown, corporate executives said unequivocally that they would not invest in nuclear power because "the legal consequences of litigation would destroy a company" (Shrader-Frechette, 1980: 12). The passing of the Price-Anderson Act created the incentive needed for capital investment, but at the same time Congress found itself being solicitous to the interests of private companies in order to induce them to invest in commercial forms of nuclear power.

Ironically, Congress's actions to limit the liability of companies operating nuclear power plants ran counter to the rise of strict liability in the law of torts. One problem was that the Atomic Energy Commission (AEC), founded in 1946, operated with a dual goal of nuclear promotion and regulatory control. The Energy Reorganization Act of 1974 separated the promotional and regulatory functions by creating the Energy Research and Development Agency (to promote nuclear energy) and the Nuclear Regulatory Commission (to assess the safety and security of nuclear power plants). The Energy Reorganization Act became necessary because of numerous lawsuits filed against the AEC. The public repeatedly sued the AEC for failing to regulate safety hazards of nuclear power plants and for catering to industry's demands rather than to the public interest (Shrader-Frechette, 1980).

The AEC's *Reactor Safety Study* of 1974 (WASH-1400), also known as the Rasmussen Report, was expressly published to persuade the public about nuclear reactor safety. The deficiencies and inadequacies of this report are, however, well documented. Shrader-Frechette (1980) discusses: (1) various faulty mathematical assumptions underlying accident probabilities, (2) the suppression of data regarding nuclear hazards, (3) the reliability of the emergency core cooling system, and (4) faulty assumptions concerning the evacuation of people—in the event of a nuclear power plant accident—before being exposed to dangerous levels of radiation.

The technological and policy problems posed by the operations of the AEC increasingly sensitized members of Congress to the need for a technically reliable source of information on the effects of advancing technologies. In establishing the Office of Technology Assessment (OTA), Congress, in the Technology Assessment Act of 1972, specified that, "The basic function of the office shall be to provide early indications of the probable beneficial and adverse impacts of technology which may assist Congress" (Whiteman, 1982: 52). During its operation, the OTA produced dozens of technology assessment reports on a broad range of topics. The philosophy of OTA was to produce studies in policy analysis and decision making; it was an attempt to build into national political institutions a mechanism for identifying the unexpected, unplanned, and undesirable effects of advancing technologies. In many respects, the OTA served an invaluable function to the understanding and assessment of technology. It provided independent and competent technological advice to aid politicians, who are generally uninformed about technical matters. OTA was established to provide Congress with the capacity to judge technology proposals independent of the executive branch. Its 202-member staff drew on outside experts for reports on topics ranging from science education to remote-satellite sensing. Sometimes referred to as Congress's own think tank, OTA was a pioneering example of scholarship embedded in a legislative body.

During OTA's 23-year history, staff analysts had issued some 750 reports in response to requests from Congressional committees. These ran the gamut from investigations into unconventional cancer therapies, the reliability and ethicality of polygraph tests, and the impacts of antibiotic-resistant bacteria to ways of reducing urban ozone, designing less polluting products and processes, and simulated combat. All of the analyses were grounded in the highest of academic standards, and they often laid the groundwork for legislation. But, in September 1995, Congress—in an effort to trim the federal budget—abolished this low-budget, prestigious research center, renowned for its thorough analyses of technology and its impartial assessments of the public policy options stemming from its reports.

One researcher lamented that now "Congress will have to increase its reliance on people with a stake in the outcome. And that's bad news" (Bimber, 1996: 23). As Bimber argues, even analyses done by the National Academy of Sciences (NAS) may reflect biases. This is because NAS panels typically enlist academic scientists "who are supported heavily by the federal government, and obviously have an interest in those programs, whereas people at OTA weren't financed by anybody but the U.S. Congress" (Bimber, 1996: 26). Perhaps Congress will yet see the wisdom of reviving the OTA. In the meantime, we must be satisfied in knowing that its former director served as the Science and Technology Advisor to the President.

Since the late 1960s, Congress has embarked on a legislative program that gave rise to a new field of environmental law (Percival and Alevizatos, 1997). Prior to the emergence of this body of law, environmental litigation revolved around such common law tort doctrines as nuisance, trespass, negligence, and strict liability. In contrast to statutory law, which can be rapidly altered by new statutes, common law is built on a case-by-case basis, evolving slowly in response to lawsuits filed between disputing parties (Meiners and Morris, 2000).

In addition to the shortcomings of common law, some states enacted pollution control statutes; many did not, however. Moreover, many state courts were reluctant to enjoin corporations from polluting for fear that the polluters could possibly retaliate by harming the local economy by increasing prices or by laying off employees. Clearly, federal environmental laws were necessary. Congress thus enacted air, water, and soil pollution standards and natural wilderness and wildlife preservation regulations.

The pivotal environmental legislation was the National Environmental Policy Act, passed in 1969, requiring the federal government to consider the environmental impacts of major decisions. This legislation was followed in 1970 by the Clean Air

Act Amendments and the Federal Water Pollution Control Act in 1972. In 1976, the Toxic Substance Control Act was enacted to regulate toxic chemicals, and in 1980, the Comprehensive Environmental Response, Compensation and Liability Act (CERCLA), also known as the Superfund, was passed.

Since all of these environmental laws involved promulgating technical standards, Congress delegated the task to an administrative agency, the Environmental Protection Agency (EPA). In 1970, President Richard Nixon issued an executive order, Reorganization Plan No. 3, which created the EPA. This executive order transferred to EPA the powers of 15 agencies and parts of other agencies. The EPA started with a staff of 6,000 employees and a budget of $455 million (Patton-Hulce, 1995: 134). As a result of the wave of environmental legislation, the EPA became the primary agency for overseeing compliance to environmental laws in the United States.

THE ADMINISTRATIVE BRANCH

For over a century, Congress, in the course of enacting laws on technically complex subjects, has seen fit to delegate their implementation to administrative agencies. Beginning with the Internal Revenue Service (1862) and the Interstate Commerce Commission (1887), followed by the Food and Drug Administration (1907), the Federal Bureau of Investigation (1908), the Federal Reserve System (1913), and the Federal Trade Commission (1914), administrative agencies have greatly proliferated. In the 1930s, as the country was struggling with the Great Depression, President Roosevelt and Congress created many agencies such as the Securities and Exchange Commission, the National Labor Relations Board, the Federal Communications Commission, the Social Security Administration, the Federal Savings and Loan Insurance Corporation, and the Federal Aviation Administration.

One of the main reasons for establishing these administrative agencies is the need for expertise in translating statutes into rules and regulations (Bowers, 1990; Kerwin, 1999). Since members of Congress are generally not technically qualified to determine whether a chemical compound should be used as a pesticide, whether a drug is safe and effective, or whether an airplane is airworthy, they delegate, in enabling legislation, to administrative agencies the responsibility for rule making, enforcement, and adjudication. Administrative agencies thus defy the doctrine of separation of powers by performing simultaneously legislative, executive, and judicial functions.

In discharging its legislative function, an administrative agency promulgates rules and regulations, which it must publish in the *Federal Register*, to afford the public an opportunity to comment. When the final regulation is published, at which point it becomes law, the agency must provide a summary of the comments and its responses to them. The EPA, for example, has identified toxic air, water pollutants, and hazardous substances; testing criteria have been established; and cleanup criteria for hazardous waste sites have been detailed.

In performing its executive function, an administrative agency maintains records, issues permits, investigates allegations of unlawful activity, and enforces the law.

In performing its judicial function, an administrative agency has the power to judge whether a law has been violated. The judicial function is exercised both informally and formally. In informal hearings, a regulated corporate official meets informally with an agency official and has an opportunity to explain the finding of a violation. A regional administrator determines what action should be taken based on the regulated corporate official's evidence. In a formal hearing, an administrative law judge (ALJ) listens to the evidence, reviews the record, and reaches a decision. The administrator of the agency has the authority to reject the ALJ's decision.

Because of the combined legislative, executive, and judicial functions of administrative agencies, their powers tend to be limited in the enabling statutes. In addition, to safeguard the rights of citizens to due process of law, Congress enacted the Administrative Procedure Act in 1946 to guard against arbitrary and capricious actions by administrative agencies.

Yet another mechanism to guard against agency arbitrariness is the requirement of judicial review of administrative action. For many years "a series of judicial decisions articulated the 'nondelegation doctrine,' i.e., the notion that it is illegitimate for elected legislators to delegate their policymaking responsibilities to appointed officials. The cases implied that the only power legislatures could delegate is the power to apply policies previously set by those who are responsible to the electorate" (Merrill, 1984: S421). Congress, however, recognized that some policymaking responsibilities would have to be delegated to administrative agencies. To resolve the conflict between the presumption that policy is to be made by those who are responsible to the electorate and the reality that many important policy decisions have to be made by appointed expert officials, the doctrine of judicial review evolved.

An example of judicial review involves the case of *Gulf South Insulation Co. v. Consumer Product Safety Commission* (1983). The plaintiff challenged the commission's ban of urea formaldehyde foam insulation on the grounds that it is carcinogenic and can affect adversely the residents of homes in which the foam is installed. The federal appellate court reviewing this case explicitly criticized the commission for relying on animal data instead of making use of available human epidemiological data. On judicial review, the U.S. Court of Appeals for the Fifth Circuit overturned the commission's decision.

How do administrative agencies interpret and implement their legal mandates? Research on several agencies, such as the Securities Exchange Commission (SEC), the Interstate Commerce Commission (ICC), the Food and Drug Administration (FDA), and the Federal Trade Commission (FTC), supports a theory that regulators tend to be co-opted by regulatees (Evan, 1993: 164–170). Mintz (1965), who has conducted an extensive study of the FDA's regulation of the drug industry, points to a problem of the interchange of personnel between the FDA and its client organizations:

> Men from the drug industry have gone on to the FDA jobs and—more important—FDA specialists have gone on to lucrative executive jobs in industry. . . . It does not seem desirable to have in decision-making positions, scientists who are consciously or unconsciously always contemplating the possibility that their futures may be determined by their rapport with industry. . . . There has also been a rather more complicated series of movements among the three power centers, the Food and Drug Administration, the Pharmaceutical Manufacturers Association and its member firms, and the American Medical Association—the three components of the FDA-PMA-AMA molecule. (pp. 175–178)

In view of the flow of personnel among the three interacting organizations—FDA, PMA, and AMA—it is not surprising that the FDA has opted for a program of voluntary compliance rather than one of strict enforcement of the law (Evan, 1993: 168).

Research on the performance of the Federal Trade Commission reveals a similar pattern. Like the FDA, the FTC has de-emphasized formal complaint procedures and concentrated on a program of voluntary compliance. Among the informal devices it employs, each of which lacks the force of law, are: (1) industry guides, (2) advisory opinions, (3) trade regulation rules, and (4) assurances of voluntary compliance and informal corrective actions (Kirkpatrick, 1969: 27).

To the extent that these regulatory agencies allow themselves to be co-opted by private corporations, they fail to promote the public interest; and to the extent that experts, upon whom regulatory agencies depend for information and advice, are also co-opted by vested interests, they fail in their purported role as disinterested purveyors of scientific truth. Thus, when a technological disaster occurs, as happened in the marketing of the Dalkon Shields, the relevant regulatory agency was slow to respond to mitigate the harm to the public.

THE JUDICIAL BRANCH

Unlike the executive and legislative branches of the legal system, which are proactive in discharging their functions, the judiciary is reactive (Boggs, 1995; O'Brien, 1982). In performing its function of adjudicating disputes, the judiciary is governed by common law, statutory law, and procedural law.

The court system in the United States is a complex structure consisting of 50 state judicial systems plus local courts of the District of Columbia and Puerto Rico, federal district and appellate courts and, finally, the U.S. Supreme Court. Regardless of jurisdiction, courts seek to reconcile a problem presented by conflicting parties with a relevant body of law. Our adversarial system of justice relies on counsel for plaintiffs and defendants to present the facts of a case and the legal rationale for deciding a case.

Rapid developments of science and technology have presented the judiciary with a multiplicity of cases for adjudication.

> Courts are emerging as forums for technology assessment partially because of the pressures of litigation brought by individuals and special-interest groups seeking either compensation for personal injuries or the prevention of ostensibly detrimental health, safety, and environmental consequences of developing chemical and industrial processes. Judicial intervention has also been invited by the expansion of congressional legislation and administrative regulations on health, safety, and environmental matters. Health, safety, and environmental regulations precipitate litigation. (O'Brien, 1982: 80)

The publication of the *Restatement of Torts* in 1965, imposing strict liability upon sellers of defective products, stimulated a wave of tort litigation. The strict liability provision reads as follows:

402A Special Liability of Seller of Product for Physical Harm
to User or Consumer

(1) One who sells any product in a defective condition unreasonably dangerous to the user or consumer or to his property is subject to liability for physical harm thereby caused to the ultimate user or consumer, or to his property, if

  (a) the seller is engaged in the business of selling such a product, and
  (b) it is expected to and does reach the user or consumer without substantial change in the condition in which it is sold.

(2) The rule stated in Subsection (1) applies although

  (a) the seller exercised all possible care in the preparation and sale of his product, and
  (b) the user or consumer has not bought the product from or entered into any contractual relation with the seller (*Restatement of Torts,* second, vol. 2: 347–348).

Several tort cases will now be considered to illustrate the role of the courts in handling policy disputes involving science and technology issues.

Diethylstilbestrol (DES), a toxic synthetic estrogen used to prevent miscarriages, has caused harm to two generations of women. In *Sindell v. Abbott Laboratories* (1980), the plaintiff and other women with vaginal disorders filed suit against manufacturers of DES, claiming that their disorders resulted from their mothers' use of DES. Approximately 200 companies, with FDA approval, manufactured this prescription drug. According to traditional tort law, plaintiffs must demonstrate that the defendant negligently caused the injuries they suffered. Mrs. Sindell and other women were unable to identify conclusively which of the 200 companies had manufactured the DES taken by their respective mothers. The California State Supreme Court, in deciding the case, advanced a novel theory for recovery: a plaintiff who cannot identify the wrongdoer may sue the manufacturers who collectively account for a substantial market share. If the plaintiff demonstrates that the defendant acted negligently, the burden shifts to each defendant to prove that its product could not have caused the injury. The court's market-share theory of recovery and imposition of joint liability expands the liability of the pharmaceutical industry. Abbott Laboratories appealed the decision to the Supreme Court, which affirmed the decision of the California State Supreme Court.

Two other toxic substances, benzene and cotton dust, were the subject of permissible exposure standards promulgated by Occupational Safety and Health Administration (OSHA). In *Industrial Union Department, AFL-CIO v. American Petroleum Institute* (1980), the Supreme Court, in a 5–4 decision, struck down OSHA's standard of reducing the permissible exposure limit on airborne concentration of benzene from 10 parts per million (ppm) of air to 1 part per million (ppm). OSHA maintained that benzene is carcinogenic and that there is no safe threshold level of exposure; moreover, the present state of science makes it impossible to calculate the number of lives that would be saved by a 1 ppm benzene standard. Justice John Paul Stevens, writing for the plurality, rejected lowering the standard, implying that OSHA should undertake a cost-benefit analysis.

In *American Textile Manufacturers Institute Inc. et al. v. Donovan, Secretary of Labor et al.* (1981), the Supreme Court upheld OSHA's cotton dust standard. Justice William Brennan, writing for the court, ruled that a cost-benefit analysis was not required under the Occupational Safety and Health Act, unlike other statutes.

One other toxic tort case warrants attention because of the gravity of the injury and the failure of the court's response. In *Tacoma v. Michigan Chemical Corporation and Michigan Farm Bureau Services et al.* (1977), the plaintiffs charged the companies with negligently contaminating their cattle feed with polybrominated biphenyl (PBB). The Michigan Chemical Company allegedly shipped to the Farm Bureau bags of a fire retardant, Firemaster, containing PBB, instead of a cattle-feed additive, Nutrimaster. The fire retardant was subsequently fed to the cattle, resulting in widespread injuries and contamination to the cattle as well as to the general population of Michigan. During the trial court proceeding, the following evidence was produced:

> That both the feed and chemical companies knew—long before the PBB incident—that their plants had a potential for cross-contamination. Michigan Chemical knew that the natural brine pumped into the plant became contaminated with toxic residues, yet it continued to use this to wash magnesium oxide and salt which went to Farm Bureau Services. Farm Bureau Services knew that its mixing machines still had traces of PBB months after they had been cleaned, yet it went on using them to turn out cattle feed. (Egginton, 1980: 302)

Judge Peterson's verdict in the case came as a surprise to both sides:

> He dismissed all the Tacomas' claims against Michigan Chemical and Farm Bureau Services, and ordered the plaintiffs to pay these companies' court costs. The Tacomas had been unable to prove, he wrote, that the low levels of PBB in their herd had been harmful or that they were justified in shooting many of their cattle. On the contrary, Judge Peterson concluded that "in small amounts PBB is not toxic," and that the greatest tragedy of the contamination had been the "needless destruction of animals exposed to low levels of polybrominated biphenyl, and even of animals that never received PBB. (Egginton, 1980: 309–310)

Finally, Judge Peterson took the plaintiffs to task when he wrote:

> The claims of plaintiffs that defendants have been guilty of criminal concealment, gross neglect, misrepresentation and of working a cover-up with governmental authorities have not only been unproved but appear to be flagrantly irresponsible in view of proofs to the contrary. (Egginton, 1980: 310)

In short, the judge found the case for the feed and chemical companies more persuasive than testimony presented by the counsel for the Tacomas. This case is a startling example of the failure of a trial court to handle—with comprehension and an appreciation of its technical ramifications—a technology-based lawsuit involving a toxic tort.

In addition to corporate torts that elicit civil lawsuits from adversely affected plaintiffs, some cases of illegal corporate behavior evoke criminal liability proceedings (Sethi, 1982: 99–132; Koenig and Rustad, 1998). Federal and state agencies have the authority to initiate liability proceedings against corporations, their officers, and employees who are charged with criminal violations such as polluting the water supply, selling contaminated food, or endangering the health or safety of employees.

Such proceedings are especially important in view of the magnitude of the costs of corporate illegality; estimated costs from white-collar crimes are $200 billion a year as compared with $3 billion to $4 billion for blue-collar crimes (Clinard, 1990). When technological disasters occur that involve corporate illegality, criminal proceedings are particularly appropriate to dramatize the outrage of society. One of our cases illustrates this point. In the Exxon *Valdez* oil spill, the Exxon Corporation was convicted of violations of federal and environmental laws and fined $5 billion in a class-action suit for environmental damages (Koenig and Rustad, 1998). Unlike civil liability, criminal liability conveys the retributive concerns felt by society (Friedman, 2000). As Kahan puts it:

> Punishing corporations, just like punishing natural persons, is also understood to be the right way for society to repudiate the false valuations that their crimes express. Criminal liability "sends the message" that people matter more than profits and reaffirms the value of those who were sacrificed to "corporate greed." (Quoted in Friedman, 2000: 855)

THE LEGAL PROFESSION

Undergirding the four legal subsystems is the legal profession. It is a source of personnel for each of the subsystems. Lawyers advise, guide, and not infrequently control decision making in each of the four subsystems. The executive branch is heavily staffed by lawyers, and many presidents have started their political careers by becoming lawyers. Members of the legislative branch, with relatively few exceptions, are members of the bar. In administrative agencies, a high proportion of technical personnel are lawyers. And members of the judiciary are, of course, all recruited from the bar.

Unlike some other highly industrialized countries such as Great Britain, Germany, and Japan, the American bar is large, highly differentiated, highly organized, and highly influential in the public as well as the private sector. In the past half-century, the legal profession has quadrupled in size, currently numbering about 923,000 members (U.S. Census Bureau, 1999). With the growth of the U.S. population and the increasing industrialization of the country, conflicts have likewise increased, thereby increasing the demand for legal services.

The profession is both highly differentiated by type of specialization and by degree of stratification, namely, level of prestige, income, and power. At the top of the professional pyramid are lawyers working in large law firms specializing in corporate law, antitrust, contracts, mergers and acquisitions, and taxation. A high proportion of such lawyers have graduated from elite law schools such as Harvard, Yale, or Columbia University, graduating in the top 10 percent of their classes (Heinz and Laumann, 1982: 353–365; Kidder, 1983: 216–219). In the United States, the field of specialization correlates highly with type and status of client. Those who work for corporate clients are invariably higher in prestige, income, and power. An especially important category of lawyers is known as corporate house counsel. They provide legal advice to corporate executives and, on occasion, will outsource corporate legal problems to outside law firms (Heinz and Laumann, 1982: 365–368). A comparable group of in-house counsel works for government agencies and performs key regulatory functions.

By contrast, lawyers whose clients are individuals, whether criminal defendants or parties in divorce proceedings or personal injury cases, enjoy a lower level of income, prestige, and power in the professional pyramid—except for a few criminal defense attorneys with "star" status. At the base of the pyramid are the overwhelming majority of lawyers who are solo practitioners. Most such lawyers attended second- or third-rate law schools and were not among the top 10 percent of their classes. Their clients are, by and large, middle-class or lower-income people, and there tends to be a similarity in the level of earnings of the lawyer and the client, as is the case for client and attorney at the apex of the pyramid (Kidder, 1983: 216–218).

Regardless of their position in the professional pyramid, lawyers historically are a learned profession, which requires the acquisition of a large body of technical knowledge. The other requirement of the profession is an ethic of service, which is regulated by a code of professional responsibility (American Bar Association, 1999). Since 1908, the American Bar Association (ABA), has promulgated canons of professional ethics. Standing committees of the ABA have periodically amended these canons to adapt them to the needs of a changing society.

For example, in 1983 a new rule was introduced in ABA's Code of Professional Conduct permitting lawyers to advertise. Once prohibited, lawyer advertising now became permissible. Rule 7.2 of the 1983 ABA code states:

> Subject to the requirements of rule 7.1 ('a lawyer shall not make a false or misleading communication about a lawyer or the lawyer's services') a lawyer may advertise services through public media, such as a telephone directory, legal directory, newspaper or other periodical, outdoor, radio or television, or through written communication . . . .(Rotunda, 1988: 352–353)

In 1977, the ABA created the Commission on the Evaluation of Professional Standards to undertake a comprehensive analysis of the ethical premises and problems of the legal profession. In its review of rules of professional conduct, the ABA Commission considers no less than 54 rules, beginning with Rule 1.1, which deals with competence, and ending with Rule 8.5, which deals with disciplinary authority (American Bar Association, 1999: v–vi). The overall objective of these rules is to ensure that lawyers, while applying their specialized talents to further the interests of their clients—thereby earning a living for themselves—must serve their clients within the bounds of ethical and legal limits. As professionals, they are expected to fulfill their social duty by providing legal services to all who need them, if necessary on a pro bono basis. Thus, lawyers have multiple demands and multiple loyalties. As officers of the court, they have a duty to uphold the integrity of the legal process and the administration of justice. In serving their clients, they are expected to be loyal to their clients' interests and needs and to represent them in a fiduciary capacity. Finally, lawyers fulfill their duty to colleagues and to the professional community by maintaining the dignity and honor of the profession (Rueschemeyer, 1973: 123–145).

One of the 54 rules of professional conduct is of particular concern to our interest in mitigating technological hazards. Rule 1.6

deals with the confidentiality of information: a lawyer should preserve the confidences and secrets of a client. For many years, the privilege governing lawyer-client communication was sacrosanct and absolute. Under no circumstances could a lawyer divulge any information obtained from the client. In recent years this has been amended to provide for exceptions such as if a client reveals an intent to commit a crime and the attorney is unable to dissuade him or her from the plan. Likewise, if the client, in the course of the representation, perpetuated a fraud upon a person or a tribunal, and if the client refuses to rectify his or her behavior, the lawyer is no longer bound to protect the confidentiality of the relationship.

At the 2001 meeting of the ABA convention, lawyers considered recommendations by ABA's commission that has studied several ethical-legal rules for four years. "The commission has proposed a rule letting lawyers reveal client confidences 'to the extent the lawyer reasonably believes necessary to prevent reasonably certain death or substantial bodily harm.' The main effect of the proposal would be to allow lawyers to warn about dangerous products" (Gillers, 2001: A25). The change in Rule 1.6, recommended by the commission, was adopted in a close vote by the House of Delegates at the 2001 ABA convention.

The new Rule of 1.6(b1) reads as follows: "A lawyer may reveal information relating to the representation of a client to the extent the lawyer reasonably believes necessary to prevent reasonably certain death or substantial bodily harm." The decision of the House of Delegates, however, has no standing as far as the Model Rules of Professional Conduct are concerned until the House of Delegates votes on all rule changes recommended by the commission, scheduled to occur in 2002.

Gillers, vice dean and professor at New York University School of Law, contends that the commission did not go far enough in protecting human life and welfare, and he is arguing for *mandatory* disclosure, not *permissive* disclosure under Rule 1.6(b1). Other proponents say that "the proposed rule might have allowed corporate lawyers to sound an alarm about such dangers as tobacco, asbestos or defective tires" (Glaberson, 2001: A1).

It should be clear that the legal profession, working within a highly technological society for a variety of industrial and advocacy groups holding differing levels of countervailing political power, must periodically amend its code of professional conduct and struggle to make its rules compatible with the perceptions of the times. To permit a lawyer "to blow the whistle on a dangerous product may seem obvious, but it is not obvious to lawyers. They view any exception to the profession's confidentiality rules with great suspicion" (Gillers, 2001: A25). Thus we see that if the commission's recommendation pertaining to Rule 1.6(b1)—though approved by the House of Delegates—is eventually incorporated in ABA's Model Rules of Professional Conduct, it may have a positive affect on public health and safety for the foreseeable future.

RELATIVE EFFECTIVENESS OF U.S. LEGAL SUBSYSTEMS IN TECHNOLOGY POLICY DECISIONS

In a pioneering article on "The Limits of Effective Legal Action," the renowned legal scholar Roscoe Pound (1917) traced the evolution of law through four stages: primitive law, strict law, equity courts, and maturity of law. He suggested a fifth stage, legislative lawmaking, which was just emerging when he wrote his article. Although Pound's focus is on the "intrinsic limitations of effective legal action" of the courts, he notes that "experiments of all sorts are in the air, and all manner of administrative tribunals, proceeding summarily upon principles yet to be defined are acquiring jurisdiction at the expense of the courts" (Pound, 1917: 64).

We will apply Pound's concept of "limits of effective legal action" to all four legal subsystems of the United States. It is reasonable to assume that the legal subsystems differ along at least four dimensions, as presented in Figure 12–2: (1) decisional time horizon, (2) level of resources (informational and financial), (3) level of expertise, and (4) perception of legitimacy for resolving science and technology policy issues.

With reference to the first dimension, decisional time horizon, the differences among some of the subsystems are substantial. The executive branch is confronted with a multitude of domestic and foreign events, some in the nature of crises that require a prompt response. In addition, virtually all of the organizational components comprising the Executive Office of the President,

Legal Subsystem	Decisional Time Horizon	Level of Resources (Informational and Financial)	Level of Expertise	Perception of Legitimacy for Resolving Science and Technology Policy Issues
Executive	Proactive	High	High	High
Legislative	Proactive	High	Medium	High
Administrative	Active	Medium	High	Medium
Judiciary	Reactive	Low	Low	Medium

Figure 12–2
Comparison of the relative effectiveness of U.S. legal subsystems in technology policy decisions.

enumerated in Figure 12–1, are engaged in policy planning and development. Hence a proactive decisional orientation is the order of the day. The legislative branch likewise exhibits a proactive decisional orientation with respect to a great variety of pressing policy issues, be it social security, Medicare, missile defense, or tax reduction.

Administrative agencies, on the other hand, since they are constrained by the parameters of enabling laws, are not proactive in their decisional orientation. In developing rules and regulations to implement the will of Congress, they may be considered *active* rather than proactive in their decisional time horizon.

The judiciary differs fundamentally from the other three subsystems in being *reactive* in its decisional orientation. Courts are passive institutions that "must await litigation that focuses retroactively on disputes" (O'Brien, 1982: 101).

The legal subsystems also differ in a marked manner as regards the second dimension, informational and financial resources. The executive branch has at its disposal a high level of resources, not only from the organizations comprising the Executive Office of the President itself, but also from the 14 cabinet departments, the independent executive agencies, and the presidential commissions.

The legislative subsystem is likewise relatively high in its level of resources, informational and financial, especially since Congress controls the purse strings for various bodies. Three of its agencies are a source of a wealth of information: the General Accounting

Office, the Congressional Research Service, and the Congressional Budget Office.

Relative to the executive and legislative Branches, the administrative branch is only medium in its level of resources. Dependent on annual appropriations from Congress, administrative agencies frequently find themselves with insufficient resources to fulfill their mandate. Although staffed by technically sophisticated personnel, they are often lacking sufficient personnel to discharge their responsibilities.

The judiciary differs from the other three legal subsystems in having a relatively low level of resources. "The judiciary possesses neither the informational nor the financial resources of Congress or the Executive Branch" (O'Brien, 1982: 91).

Turning to the third dimension of our comparison of legal subsystems, the level of expertise, some noticeable differences are also observable. Much has been written about the president's need for science and technology advice (Branscomb, 1995; Garwin, 1995; Gibbons, 1995; Robinson, 1995; Tape, 1995). In 1958, the executive branch took the initiative in establishing the President's Science Advisory Committee (PSAC). For some years, members of PSAC had access to the President. In 1973, however, this committee was abolished. Congress, recognizing the importance of a science advisory system, passed the National Science and Technology Policy and Priorities Act of 1976. It established the Office of Science and Technology Policy (OSTP) and designated its director as the president's science advisor. It also assigned to OSTP specific coordination, evaluation, analysis, and planning functions.

Although Congress never allocated sufficient resources to OSTP, the executive branch, depending on the policy orientation of the president, has at its disposal considerable scientific resources, including the capability of its mission-oriented cabinet departments and the National Academies of Science and Engineering. "The President has available an awesome armada of potential sources of science advice if he feels the need for advices" (Lederman, 1995: 224–225). For this reason, the current level of expertise available to the executive branch may be ranked as high.

By comparison, the level of expertise of Congress has been relatively low, especially since very few members of Congress have backgrounds in science and engineering. To make up for this deficiency, congressional committees have added staff trained in science and engineering. "Faced with a deluge of complex technological proposals sent to it by an Executive branch resplendent in scientific resources, Congress felt increasingly ill-equipped to evaluate those proposals" (Gibbons, 1995: 416). Consequently, in 1972 Congress passed the Office of Technology Assessment Act to provide itself with an in-house capability of examining emerging technologies and their direct and indirect implications. The Office of Technology Assessment (OTA) was designed to critically review new technologies, develop findings, describe options, and communicate the results to Congress. By all accounts, the OTA undertook numerous technological assessment studies in a nonpartisan fashion. Nevertheless, in 1995, in an effort to trim the budget, Congress voted to abolish the OTA. With the abolition of the OTA, Congress now has a relatively medium level of expertise.

Administrative agencies in the federal government, by definition, are repositories of technical experts to translate enabling laws into rules and regulations, to undertake adjudication of disputes, and to enforce its law-making function. They are staffed with scientists and engineers in diverse specialties, along with the usual complement of lawyers knowledgeable about rule making, adjudication, and enforcement.

The caseloads of these agencies frequently revolve around issues based on scientific and technical data. For example, the Federal Trade Commission makes use of statistical data in deciding deceptive advertising and antitrust law cases; the Federal Aviation Administration must rely on complex technical data in making decisions regarding aviation safety; the Food and Drug Administration must assess complex biological data in ruling upon new drug applications; and the Nuclear Regulatory Commission must apply scientific data in determining nuclear reactor safety. In short, most federal administrative agencies enjoy a relatively high level of expertise.

Notwithstanding that "our society is becoming increasingly technological and scientific and that it is becoming increasingly litigious" (Boggs, 1995: 484), the judiciary, with the possible exception of the federal judiciary, has a relatively low level of expertise. The dramatic growth in toxic torts and environmental litigation has put new pressures on the judiciary. Almost all cases involving scientific and technological issues are decided by judges who derive their understanding of the technical issues from the parties' presentation at trial. Expert testimony by plaintiffs and defense counsel is a frequent feature of civil cases. Nevertheless, it is a "regrettable but indisputable fact that most lawyers and judges are scientific illiterates" (McKay, 1995: 476). To cope with this problem, the National Conference of Lawyers and Scientists seeks to familiarize lawyers with science (Gerjuoy, 1995). This, however, is a tall order:

> Science and technology have invaded our courts and are here to stay. If, as a society, we want better judicial resolutions of cases that involve complex scientific and technological matters, then we must figure out some way to make our judges more adept at handling scientific evidence and the policy implications of new technology. (Wald, 1995: 488)

A case in point is the antitrust suit against Microsoft (*United States v. Microsoft Corporation,* 2000). In its appeal of the federal district court decision to split the company in two, Microsoft contended that Judge Thomas Penfield Jackson, who presided over the case, was "unable to comprehend the facts" (Brick, 2000: 4). Judge Jackson admitted that he supported the position of the Justice Department and the State Attorneys General to break up Microsoft because "there is no way I can equip myself to do a better job than they have done."

In a subsequent appellate procedure, the U.S. Court of Appeals for the District of Columbia Circuit, on June 28, 2001, unanimously reversed the lower court's order that the Microsoft Corporation should be broken up. However, it upheld the District Court's finding that the company repeatedly abused its monopoly power in the software industry. This ruling all but ends the risk of a court-ordered breakup of Microsoft (*United States v. Microsoft Corporation,* 2001).

The question of the technological complexity of lawsuits has prompted the legislature of Maryland to convene a task force to study whether to establish a specialized court to hear high-technology cases (Stern, 2000). James L. Thompson, former President of the Maryland State Bar Association who appointed 12 of the 18 task force members, favors the creation of a technology court. There is already a precedent for a specialized court to hear technology cases. In 1982, the United States Court of Appeals for the Federal Circuit in Washington was created to hear, among other cases, appeals of final decisions of the federal district court in patent cases (Brick, 2000: 4).

With respect to the fourth dimension of our comparison of legal subsystems, perception of legitimacy in resolving science-technology policy issues, this is a difficult question to answer in the absence of empirical data. Hence our conclusions are, in part, hypothetical in nature.

The Constitution accords the President a panoply of powers in domestic and foreign arenas. These include appointing judges and ambassadors, signing treaties, issuing executive orders, submitting budgets to Congress, committing troops to combat, and so on. These powers presuppose a high level of perceived legitimacy, especially since the president is answerable to the electorate for all presidential decisions. As regards science-technology policy issues, the president, as we have seen, has access to extensive technical advice.

Congress, like the executive branch, is also responsible to the electorate and thus enjoys a high level of perceived legitimacy.

> For, in the minds of the average citizens, legislatures—whether at the local, state or federal level—are probably perceived to be the proper and legitimate forums for the enactment of new laws. This perception may be due to the fact that legislatures are more sensitive to public pressures and sentiments than are the other sources of lawmaking. (Evan, 1980: 557)

O'Brien, a legal scholar, who has undertaken a systematic analysis of the role of courts in technology assessment and science-policy disputes, concludes as follows:

> Within the framework of the Constitution and a pluralistic political process, the Congress and the executive branch, not the judiciary, are the legitimate and primary institutions for assessing, structuring, and resolving the vexations science-policy issues posed by industrial and technological developments. (O'Brien, 1982: 103)

Federal judges of district courts and appellate courts, including members of the Supreme Court, are all appointed by the president and hence are not responsible to the electorate. In view of their limited expertise in science and technology issues, we hypothesize that they have a medium level of perceived legitimacy.

As for administrative agency personnel, they, too, are appointed by the executive branch and hence are not answerable to the electorate. However, because of their widely acknowledged level of expertise in handling science-technology policy issues, we hypothesize that their level of legitimacy is medium, as high as judges but lower than that of the executive branch and of the Congress.

CONCLUSION

In our comparison of the four subsystems of the American legal system, we have assessed their differential performance—or their "limits of effective legal action"—in handling science-technology policy issues. To the extent that the capabilities of each of the subsystems are upgraded with respect to our four dimensions of analysis in Figure 12–2, their capacity to manage technological disasters will be enhanced.

In discussing the unique role of the legal profession in undergirding the four legal subsystems, we focused on the potential significance of changes in the confidentiality rules of professional conduct for managing technological hazards. If the profession's confidentiality rules, such as Rule 1.6(b1) are changed—as already voted upon by the House of Delegates—it may have a profound effect in promoting health and safety in the United States.

References

American Bar Association. (1999). *Model rules of professional conduct.* Chicago: American Bar Association

American Textile Manufacturers Institute, Inc. et al. v. Donovan, Secretary of Labor, et al., 542 U.S. 490 (1981).

Atomic Energy Commission. (1974). *Reactor safety study: An assessment of accident risks in U.S. commercial nuclear power plants (WASH-1400).* Washington, DC: U.S. Government Printing Office.

Bimber, Bruce. (1996). *The politics of expertise in Congress: The rise and fall of the Office of Technology Assessment.* Albany, NY: State University of New York Press.

Boggs, Danny S. (1995). "Science and technology advice in the Judiciary." In William T. Golden (Ed.), *Science and technology advice to the president, Congress, and judiciary.* New Brunswick, NJ: Transaction Publishers: 455–460.

Bowers, James R. (1990). *Regulating the regulators: An introduction to the legislative oversight of administrative rulemaking.* New York: Praeger.

Branscomb, Lewis M. (1995). "Supporting the president's need for technical advice." In William T. Golden (Ed.), *Science and technology advice to the president, Congress, and judiciary.* New Brunswick, NJ: Transaction Publishers: 44–50.

Brick, Michael. (2000). "When the judge can't really judge," *The New York Times,* September 11: C4.

Brooks, Harvey. (1986). "An analysis of proposals for a Department of Science," *Technology in Society* 8: 19–31.

Clinard, Marshal. (1990). *Corporate corruption: The abuse of power.* New York: Praeger.

Congressional Quarterly. (1989). *Cabinets and counselors: The President and the Executive Branch.* Washington, DC: Congressional Quarterly Press.

Edwards, George C. III, and Wayne, Stephen J. (1994). *Presidential leadership* (3rd ed.). New York: St. Martin's Press.

Egginton, Joyce. (1980). *The poisoning of Michigan.* New York: W. W. Norton & Co.

Evan, William M. (1980). "Law as an instrument of social change." In William M. Evan (Ed.), *The sociology of law.* New York: The Free Press: 554–562.

Evan, William M. (1993). *Organization theory: Research and design.* New York: Macmillan Publishing Co.

Executive Orders. http://www.pub.whitehouse.gov/retrieve-documents.html

Friedman, Lawrence. (2000). "In defense of corporate criminal liability," *Harvard Journal of Law and Public Policy* 23: 833–858.

Garwin, Richard L. (1995). "Toward a nation healthy, wealthy and wise." In William T. Golden (Ed.), *Science and technology advice to the president, Congress, and judiciary.* New Brunswick, NJ: Transaction Publishers: 144–147.

Gerjuoy, Edward. (1995). "Judicial understanding of science." In William T. Golden (Ed.), *Science and technology advice to the president, Congress, and judiciary.* New Brunswick, NJ: Transaction Publishers: 470–474.

Gibbons, John H. (1995). "Science, technology and law in the third century of the Constitution." In William T. Golden (Ed.), *Science and technology advice to the president, Congress, and judiciary.* New Brunswick, NJ: Transaction Publishers: 415–419.

Gillers, Stephen. (2001). "A duty to warn," *The New York Times,* July 26: A25.

Glaberson, William. (2001). "Lawyers consider easing restriction on client secrecy," *The New York Times,* July 31: A1.

Golden, William. (1986). "A Department of Science: An illusion— PSAC: A prescription," *Technology in Society* 8: 107–110.

Gulf South Insulation Co. v. Consumer Product Safety Commission, 701 F.2d 1137 (5th Cor. 1983).

Heinz, John, and Laumann, Edward. (1982). *Chicago lawyers: The social structure of the bar.* New York: Russell Sage Foundation.

Industrial Union Department, AFL-CIO v. American Petroleum Institute, 100 S. Ct. 2844 (1980).

Kerwin, Cornelius M. (1999). *Rulemaking: How government agencies write law and make policy* (2nd ed.). Washington, DC: Congressional Quarterly Press.

Kidder, Robert L. (1983). *Connecting law and society.* Englewood Cliffs, NJ: Prentice Hall.

Kirkpatrick, Miles W. (1969). *Report of the ABA Commission to study the Federal Trade Commission.* Chicago: American Bar Association.

Koenig, Thomas, and Rustad, Michael. (1998). "Crimtorts as corporate just desserts," *University of Michigan Journal of Law and Reform* 31: 289–352.

Lederman, Leon M. (1995). "Science advising." In William T. Golden (Ed.), *Science and technology advice to the president, Congress, and judiciary.* New Brunswick, NJ: Transaction Publishers: 224–225.

McKay, Robert B. (1995). "Science and technology advice to the Judiciary." In William T. Golden (Ed.), *Science and technology advice to the president, Congress, and judiciary.* New Brunswick, NJ: Transaction Publishers: 475–479.

Meiners, Roger E., and Morris, Andrea P. (Eds.). (2000). *The common law and the environment: Rethinking the statutory basis for modern environmental law.* New York: Rowman & Littlefield Publishers.

Merrill, Richard A. (1984). "The legal system's response to scientific uncertainty: The role of judicial review," *Fundamental and Applied Toxicology* 4: S418–S425.

Mintz, Morton. (1965). *The therapeutic nightmare.* Boston: Houghton Mifflin.

O'Brien, David M. (1982). "The courts, technology assessment, and science-policy disputes." In David M. O'Brien and Donald A.

Marchand (Eds.), *The politics of technology assessment*. Lexington, MA: Lexington Books: 79–115.

Patton-Hulce, Vicki R. (1995). *Environment and the law: A dictionary*. Santa Barbara, CA: ABC-Clio.

Percival, Robert V., and Alevizatos, Dorothy C. (Eds.). (1997). *Law and the environment*. Philadelphia, PA: Temple University Press.

Pound, Roscoe. (1917). "The limits of effective legal action," *American Bar Association Journal* 3: 55–70.

Ragsdale, Lyn. (1998). *Vital statistics on the Presidency* (rev. ed.). Washington, DC: Congressional Quarterly Press.

Restatement (Second) of Torts, Vol. 2. (1965). St. Paul, MN: American Law Institute.

Robinson, David Z. (1995). "Science and technology advice to government." In William T. Golden (Ed.), *Science and technology advice to the president, Congress, and judiciary*. New Brunswick, NJ: Transaction Publishers: 298–301.

Rosenzweig, R. (1993). "Advising the president on science and technology: Why can't we get it right?" In William T. Golden (Ed.), *Science and technology advice to the president, Congress, and judiciary*. Washington, DC: AAAS Press.

Rotunda, Ronald. (1988). *Professional responsibility* (2d ed.). St. Paul, MN: West Publishing Co.

Rueschemeyer, Dietrich. (1973). *Lawyers and their society*. Cambridge, MA: Harvard University Press.

Sethi, Praskash S. (1982). *Up against the corporate wall* (4th ed.). Englewood Cliffs, NJ: Prentice Hall.

Shrader-Frechette K. (1980). *Nuclear power and public policy: The social and ethical problems of fission technology*. Boston, MA: D. Reidel Pub. Co.

Sindell v. Abbott Laboratories, 26 Cal. 3d 588, 607, p. 2d 924 (1980), cert. denied 101 S. Ct. 286 (1980).

Stern, Sethi. (2000). "Science-savvy judges in short supply," *The Christian Science Monitor*, December: 17.

Tacoma v. Michigan Chemical Corporation and Michigan Farm Bureau Services et al., Circuit Court for the County of Wexford, Michigan, May 1977.

Tape, Gerald F. (1995). "Science advice in the government: Formulation and utilization." In William T. Golden (Ed.), *Science and technology advice to the president, Congress, and judiciary*. New Brunswick, NJ: Transaction Publishers: 336–343.

Thomas, Norman C., Pika, Joseph, and Watson, Richard A. (1994). *The politics of the presidency* (3rd ed.). Washington, DC: Congressional Quarterly Press.

United States of America v. Microsoft Corporation, State of New York, et al. v. Microsoft Corporation, 97 F. Supp. 2d59 (2000).

United States of America v. Microsoft Corporation, 2001 U.S. App. P73, 321 (June 28, 2001).

U.S. Census Bureau. (1999). *Statistical Abstract of the United States* (119th ed.). Washington, DC: Author.

U.S. Government Manual. (1997–1998). Washington, DC: Government Printing Office.

Wald, Patricia M. (1995). "Technology and the courts." In William T. Golden (Ed.), *Science and technology advice to the president, Congress, and judiciary.* New Brunswick, NJ: Transaction Publishers: 484–488.

Whiteman, D. (1982). "Congressional use of technology assessment." In D. O'Brien and D. Marchand (Eds.), *The Politics of technology management.* Toronto: Lexington Books.

Wiesner, J. (1989). "On science advice to the president," *Scientific American* 260: 34–40.

CHAPTER 13

Assessing the Risks of Technology

*"In all traditional cultures . . . human
beings worried about the risks coming
from external nature . . . very recently . . .
we started worrying less about what nature
can do to us, and more about what we
have done to nature. This marks the
transition from the predominance of
external risk to that of manufactured risk."*
—Anthony Giddens

The prevalence of technological disasters points to deficiencies in
the way technology is assessed—the way technologies are determined to be or not to be risky. In this chapter we focus on two prominent methods for assessing risk: probabilistic risk assessment (PRA)
and risk-cost-benefit analysis (RCBA). If we find the standard methods deficient, then it should not be surprising that we have more disasters than we can expect. The overall lesson is that, if we want to
reduce the number and magnitude of technological disasters, we
must reform our methods of evaluating the potential impacts of
technology—PRA and RCBA—and develop more effective methods.

Experts attempt to separate risk assessment techniques into two
independent procedures—the risk identification or risk estimation
level, which is supposedly factual, scientific, objective, and value-
neutral, and the risk assessment or risk management level—which
is supposedly normative, political, subjective, and value-laden
(Humphreys, 1987). This rigid demarcation into the "factual" or
scientific measurement of risks vs. the "normative" management of
the social acceptability of risks is thought to secure for risk assessors
a level of scientific objectivity and value neutrality.

We will argue, however, that the factual/normative split is no longer adequate for the proper identification, assessment, and management of technological risk. Consequently, the so-called objective activities of risk identification or risk estimation need to be integrated with the normative and evaluative aspects of risk evaluation and risk management.

Probabilistic Risk Assessment

Probabilistic risk assessment (PRA), or quantitative risk assessment (QRA), attempts to provide a model of the causal interactions of the technological system under study. The goal of PRA is to supply a mathematical technique for estimating the probability of events that cause physical damage or loss of life (Thompson, 1982: 114). A comprehensive probabilistic risk assessment involves three steps: (1) the identification of events that lead to, or initiate, unwanted consequences; (2) the modeling of identified event sequences with respect to probabilities and consequences; and (3) the determination of the magnitude of risks and harms involved (Bier, 1997).

PRA has become one of the standard methods used by engineers for determining the likelihood of an industrial accident or a technological disaster. This method makes use of technical procedures called *fault-tree analysis* and *event-tree analysis*. Fault-trees and event-trees generate diagrams that trace out the possible ways a malfunction can occur in complex technological systems. They enable design engineers to analyze in a systematic fashion the various failure modes associated with a potential engineering design (Henley and Kumamoto, 1981: 24–28). Failure modes are the ways in which a structure, mechanism, or process can malfunction.

In an event-tree analysis one begins with an initial event—such as a loss of electrical power to a nuclear power plant—and, using inductive logic, reasons *forward,* trying to determine the state of the system to which the event can lead (Henley and Kumamoto, 1981: 24–28). Figure 13–1 illustrates an event-tree analysis of the probability of radiation release from a standard nuclear power plant. The diagram is based on the authoritative reactor safety study, WASH 1400, the so-called "Rasmussen Report," commissioned by the U.S. government in 1975 (*Reactor Safety Study,* 1975).

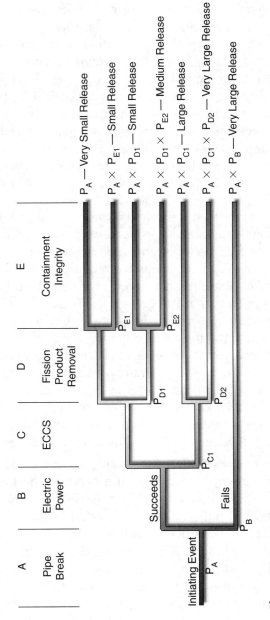

A	B	C	D	E
Pipe Break	Electric Power	ECCS	Fission Product Removal	Containment Integrity

Initiating Event

P_A

Succeeds

Fails

P_B

P_{C1}

P_{D1}

P_{D2}

P_{E1}

P_{E2}

P_A — Very Small Release

$P_A \times P_{E1}$ — Small Release

$P_A \times P_{D1}$ — Small Release

$P_A \times P_{D1} \times P_{E2}$ — Medium Release

$P_A \times P_{C1}$ — Large Release

$P_A \times P_{C1} \times P_{D2}$ — Very Large Release

$P_A \times P_B$ — Very Large Release

Figure 13-1

Event-tree analysis of a possible fission product release from a nuclear power plant.
Source: Henley, Ernest and Kumamoto, Hiromitsu (1981). *Reliability engineering and risk assessment.* Englewood Cliffs, NJ: Prentice Hall: 25.

The Reactor Safety Study was performed to determine the public risks associated with existing and planned nuclear power plants. The simplified event tree begins with a definite accident-initiating event and tries to identify all the safety systems that can be called upon to mitigate the consequences. The study determined that the failure of the reactor cooling system is the most critical component that could lead to a radiation release. The analysis therefore begins with the initiating event that a (coolant) pipe might break. If the pipe breaks, without any other system failing at the same time, then the probability that there would be a radiation release is P_A, a very small release, as shown in Figure 13–1. Possible failures are next defined for each system; and accident sequences are constructed, consisting of the initiating event with specific systems failing and specific systems succeeding. For example, the probability of both a pipe breaking and a corresponding loss of electrical power to the plant ($P_A \times P_B$), would lead to a very large release of radiation. Keep in mind that a probability of 1 means that an event is certain to happen. Therefore, the probability that P_A occurs is less than one. In fact, all of the assigned probabilities will be less than 1, and, as the probabilities are multiplied, the total probability will diminish. For example, if the probability of P_A occurring is 0.01 and the probability of P_B occurring is 0.001, then the overall probability that a pipe will break per day and the electrical power will fail is 0.01×0.001, which equals 0.00001 or one in 100,000 events. Even though the probability of a pipe breaking at the same time as the electrical system failing is rather low, the potential consequences are very high.

If electric power does not fail during the pipe break, the analysis moves to determining the relationship between the breaking pipe and the emergency core coolant system (ECCS). The ECCS could either succeed or fail. The probability that ECCS fails at the same time that a pipe breaks would lead to a large release, namely, ($P_A \times P_{C1}$). If the ECCS succeeds, then the probability that the fission product removal is inhibited and that a pipe breaks would lead to a small release, namely, ($P_A \times P_{D1}$). Likewise, the probability of a pipe breaking and a failure of the ECCS as well as the fission product removal being inhibited would lead to a very large release, namely, ($P_A \times P_{C2} \times P_{D2}$). In the end, each possible system state (failure or success) is connected through a branching logic to give all the specific accident sequences that can arise. The event tree is particularly useful when many individual systems and subsystems interact.

In a fault-tree study, the analyst begins with a hypothetical undesirable event, then, using deductive logic, reasons *backwards* to determine what might have led to the event. Fault-trees follow a cause-and-effect model and can be used whenever hypothetical events can be resolved into more basic, discrete units for which failure data exist or for which failure probabilities are generally easily calculable. Figure 13–2 illustrates a fault tree for analyzing the possible causes of why a car would not start. For example, a good mechanic would, more than likely, already be aware of all of the possible reasons of why a car will not start. The mechanic would then construct a fault tree as illustrated in Figure 13–2, checking each subsystem as a possible cause of why the car will not start. An insufficient battery charge is the most likely cause of a car not starting, so the mechanic would begin there and proceed to check whether there was a faulty ground connection, the battery terminals were loose or corroded, or the battery charge was weak, etc. If, for example, it is determined that the terminals are not loose or corroded or the battery charge was not weak, then the mechanic would check to see if there was rust on the ground connections, the connections were corroded or otherwise dirty, or the ground connections were loose. If the battery checks out, the mechanic would move to an analysis of the starter system, checking each related subsystem. If this checks out, then the mechanic would move to the fuel system and each of the other related subsystems. The mechanic reasons step-by-step through the fault tree until the cause of the car not starting is identified.

Since the first comprehensive application of probabilistic risk assessment in 1975—the U.S. Reactor Safety Study—more than 15 large-scale PRAs have been carried out for nuclear power plants in the United States. In addition, large-scale PRAs have been carried out in Sweden and West Germany and have also been used in determining levels of safety in such varied industries as chemical production, liquid gas transport and storage, oil-drilling rigs, transport of toxic chemicals, and the aerospace and nuclear industries (Linnerooth-Bayer and Wahlstrom, 1991: 240). Even though PRA has been used extensively, there are substantial methodological problems, some of which we will now consider.

The first set of methodological problems in PRA arises when experts attempt to determine which factors to include and which to exclude. The second set of problems arises when scientific uncertainties appear in the modeling process (Rowe, 1994). Thirdly, there are

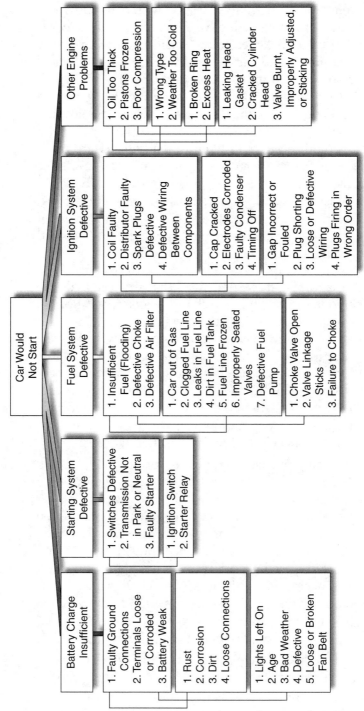

Figure 13–2

Fault-tree analysis of the failure of an automobile to start.

Source: Fischoff B., Slovick P., and Lichtenstein S.A., "Fault trees: Sensitivity and estimated failure problem representation," *Journal of Experimental Psychology: Human Perception and Performance*, 4 (1978). Copyright © 1978 by the American Psychological Association. Reprinted with permission.

reservations about the adequacy of the method due to uncertainties that arise in attempting to trace out unknown cause-and-effect relationships. A fourth deficiency of the method is its inability to account for uncertainties that inevitably arise due to operator error and other human factors. Fifth, the very complexity of many large-scale technologies renders the PRA method inadequate as the sole source of assessing risks and adequate safety levels. Figure 13–3 lists the five problems pertaining to PRA and the associated issues that arise with each problem.

The first set of problems with PRA methodology arises when experts are confronted with difficulties in taking into account all of the ways the components of a system are interrelated. Furthermore, risk or reliability analysts can fail to foresee all of the interactions among individually-separated problems. Methodological uncertainties abound, causing large knowledge gaps in how systems operate and how systems fail. Such uncertainties arise if fault trees or event trees are inaccurate, incomplete, or inappropriate (Bier, 1997: 72–73).

The second set of difficulties of PRA methodology arises when researchers inadvertently construct, select, and validate an

Problems	Issues
1. Problems of identifying all potential risk factors	1. Uncertainties arise when experts attempt to anticipate all of the mechanical, physical, electrical, and/or chemical factors to be included in a fault-tree or an event-tree analysis.
2. Problems with uncertainties in the modeling of systems	2. Uncertainties arise from the failure to incorporate in the model important characteristics of the process under investigation.
3. Problems associated with determining cause-and-effect relationships	3. Direct cause-and-effect relationships between potential hazards and consequent harms are often not demonstrable.
4. Uncertainties due to human factors	4. Potential errors are associated with human operators, which often cannot be "modeled" and hence are rarely anticipatable.
5. Problems of complexity and coupling	5. Tight coupling and interactive complexity between system components disallow any complete modeling of potential system failures.

Figure 13–3
Problems and issues with probabilistic risk assessment.

inaccurate or faulty modeling of the system under analysis. "Models are simplified representations of real world processes; as such, they make certain assumptions concerning the true state of nature" (Haimes, 1998: 240). Model uncertainty arises from the failure to incorporate important characteristics of the process under investigation. "If this uncertainty is improperly understood, it can be potentially the largest contributor of error, leading to significant misrepresentations of processes" (Haimes, 1998: 240).

The third set of methodological shortcomings of PRA arises because the probabilities assigned to various failure modes are, by and large, conjectural and based on analyses that often cannot be corroborated by experimental testing. This is especially true of uncertainties that arise due to hidden or unknown cause-and-effect relationships (Bougumil, 1986). For example, researchers often come upon uncertainties when inferring risks to humans from animal experimentation models.

The fourth set of problems associated with the PRA method is its inability to anticipate all of the opportunities for human error that could lead to failure. As we have pointed out in Chapter 5, such human factors include cognitive processing of information; insufficiencies in human perception, memory retention, or an individual operator's capacity for stress; cognitive and emotional overload; and worker burnout. All of these factors may lead to bad judgments and/or inaccurate perceptions. Simple human inattention, carelessness, or even negligence can lead to operational failure. Such factors were evident in the TMI, Bhopal, Chernobyl, USS *Vincennes*, and *Titanic* cases. In spite of decades of research on human-factors problems, we still have a long way to go to understand adequately human-machine interactions. Other human-factors problems that render PRA insufficient as a method for determining risks were operating in the Ford Pinto case and in the B.F. Goodrich brake scandal case. Risk assessors may, on occasion, underrate risks in a risk assessment report or fail to see the significance of risks that are present. The Challenger disaster is a case in point.

In fact, a research survey of about 1,500 members of the Society of Risk Analysis found that such human factors as data fabrication and bias in research design are more common in the risk assessment process than one would think. As the researchers put it:

Surveys of almost 1,500 members of . . . professional societies that do risk analysis (e.g., environmental economics, epidemiology, exposure assessment, industrial hygiene, toxicology) found that 3 in 10 respondents had observed a biased research design, 2 in 10 had observed plagiarism, and 1 in 10 observed data fabrication or falsification. Respondents with many years in risk analysis, business consultants, and industrial hygienists reported the greatest prevalence of misconduct. These respondents perceived poor science, economic implications of the research, and lack of training in ethics as causes of misconduct. (Greenberg and Goldberg, 1994: 223)

The fifth and final set of problems with PRA is the increasing complexity and interdependence of technological systems. Large-scale risk assessments, such as those in the nuclear industry, for example, encompass a myriad of different systems requiring the involvement of expertise from numerous areas. As Perrow points out, the description of possible chains of events is so complex and open to multiple interpretations that there is often room for dispute after an accident as to whether the destructive chain of events was even described in the initial PRA.

Given the methodological deficiencies of PRA discussed above, it is safe to assume that technical methods alone, no matter how sophisticated, cannot be the only way to assess the benefits and burdens of technology. Moreover, exclusive focus on probabilities leads analysts to ignore *low-probability but high-consequence events*. In other words, PRA often ignores the category of "catastrophe," because catastrophes entail low-probability, high-consequence events. Ignoring low-probability, high-consequence events is unwise, given the immense complexity of many technologies, especially large-scale sociotechnical systems. No matter how detailed a fault-tree or an event-tree analysis may be, the methodology simply cannot begin to capture all of the common mode failure events that are possible. The inadequacy of PRA in treating low-probability, high-consequence events is glaringly evident in the Three Mile Island, Chernobyl, Challenger, and Bhopal cases. As Lanthrop puts it, "deciding that, say, a nuclear power plant is safe because it is only expected to fail once in every 10,000 successful usages does not rule out that a catastrophe may happen tomorrow, or next year, or the next" (Lanthrop, 1982: 171). This is exactly the kind of assessment failure that happened in the Three Mile Island case.

Even if PRA were an effective method for determining the risks of technology, which it is not, it would not be enough in any event. As we have seen in numerous cases discussed previously, beyond technical factors, human, organizational, and socio-cultural factors are often at the root of technological disasters.

RISK-COST-BENEFIT ANALYSIS

Along with PRA, cost-benefit analysis (CBA) and risk-cost-benefit analysis (RCBA) arose as the preeminent methods of assessing the risks of technology during the late 1960s and early 1970s, as Congress began to enact legislation on the regulation and monitoring of technology and its social and environmental impacts. RCBA is a variant of CBA in which human health and welfare are brought into the equations, along with the material costs and benefits of a proposed technology.

Comprehensive statutes such as the National Environmental Policy Act of 1970 (NEPA), the Federal Water Pollution Control Act Amendments of 1972, the Consumer Protection Act, and the Clean Air and Clean Water Acts require a government agency to consider technical and economic feasibility characteristics and health and environmental effects when contemplating a technological intervention. In order to accomplish these goals, organizations turned, and continue to turn, to CBA and RCBA in an effort to comply with statutory and judicial requirements (Baram, 1977). The National Aeronautics and Space Administration (NASA) uses RCBA in its feasibility and safety studies. The Nuclear Regulatory Commission (NRC) has followed NASA's lead in employing RCBA almost exclusively in setting "acceptable" radiation standards and in decisions concerning the licensing of nuclear facility construction and operation (Kneese, Ben-David, and Schultze, 1983: 60–61).

RCBA has also been the leading method used by experts as a basis for policy choices concerning controversial problems surrounding the storage and disposal of nuclear waste (Grossman and Cassedy, 1985). In addition, RCBA is utilized frequently in medical economics for assessments of medical interventions and other health-care contexts (Gewirth, 1990: 222). RCBA is also used widely in analysis of and policy making concerning environmental toxins (Baram, 1976). Finally, RCBA is used frequently in large-scale water and waste management technologies.

In order to set up a risk-cost-benefit analysis, one begins by trying to enumerate all adverse consequences that might arise from the implementation of a given technology. Next, one attempts to estimate the probability that each of these adverse consequences will occur. The third step is to estimate the cost or loss to social and individual health and well-being should any or all of the projected adverse consequences come to pass. Fourth, one tries to calculate the expected loss from each possible consequence. Finally, one attempts to compute the total expected losses from the proposed project by summing the expected losses for each of the various possible consequences. One follows a similar procedure to calculate the benefits. In the end, one subtracts the overall costs from the overall benefits. If the benefits outweigh the costs, the project is generally described as feasible.

However, there are significant methodological deficiencies in the RCBA method, especially those that raise ethical problems. Our analysis has identified five methodological deficiencies. They are: (1) problems of identification, (2) the value-of-life problem, (3) the commensurability problem, (4) problems associated with values and market mechanisms, and (5) problems of social and ecological justice. These problems and associated issues are listed in Figure 13–4.

The first methodological problem associated with RCBA is the unquestioned assumption that *all* significant consequences can be enumerated in advance. The assumption is that all of the costs and benefits of a particular implementation of a new technology or extension of a "known" technology can be clearly identified and catalogued, that meaningful probability, cost, and benefit values can be obtained and assigned to them, and that often disparate costs and benefits can somehow be made comparable to one another. Such judgments are grounded in unrealistic assumptions about the availability of the data needed to complete the analysis.

As with probabilistic risk assessment, not all of the crucial questions regarding the nature, estimation, or acceptability of the risks, costs, and benefits can be answered with quantitative analysis alone. Conscious normative judgments arise in determining what will be included and what will be excluded. In other words, at least as far as the "problems of identification are concerned, the same problems associated with PRA also arise with RCBA (or CBA). As Martin (1982) points out:

Problems	Issues
1. Problems of identification	1. It is almost impossible to arrive at a complete enumeration of *all* risks and benefits because one can never know all of the variables that need be assigned diagnostic values, let alone be able to calculate all the costs and benefits.
2. The value-of-life problem	2. A fundamental moral problem arises in assigning a monetary value to human life, a necessary requirement of RCBA.
3. The commensurability problem	3. The erroneous assumption that disparate costs and benefits are quantifiable according to an identical metric leads analysts to believe that all values are commensurable with one another.
4. Human values and market mechanisms	4. Utility maximizations fail to provide satisfaction for all crucial human needs and values.
5. Problems of social and ecological justice	5. RCBA fails to take into account issues of fairness in the distribution of risks and harms across social groups, between different generations, and throughout the natural environment.

Figure 13–4
Problems and issues with risk-cost-benefit analysis.

PRA and CBA are different techniques, but they have important similarities. Both attempt to translate seemingly incomparable sorts of considerations into a quantifiable common denominator of some sort (whether dollars or mathematical formulae), then tallying up the results for various options, and finally presenting this information in a form that can be readily digested by decision makers . . . [However] . . . cost-benefit analyses, as well as probabilistic risk assessment are value-laden, both in what they count out (usually, for example, considerations of rights and justice) and in what they count in (for example, assumptions about what sorts of things constitute costs and benefits, whose costs and benefits are to be weighed, and how relative values are to be assigned to them). (p. 147)

Uncertainties as to how one should define "harm" or "risk" of a particular action force analysts to make judgments that are value-laden. For example, one contested assumption of both PRA and RCBA is that mortality rates—ignoring morbidity rates—are usually chosen as the focus of analysis.

The second methodological problem with RCBA is a hotly debated issue: the assignment of a monetary value to human life, a necessary requirement of a robust risk-cost-benefit analysis (Byrne, 1988; Kahn, 1986; MacKinnon, 1986; Rescher, 1987). As MacKinnon puts it, "of all the difficulties that surround the attempt to calculate the economic 'value of a life' one of the thorniest is a moral one, namely whether it is morally permissible to place any 'price' on a human life" (MacKinnon, 1986: 29).

Of course, certain practices are used by insurance companies, economists, and risk assessors that demonstrate that society does place some implicit monetary value on human lives (VOL). As one philosopher argues, "If it is permissible to forego life-saving treatment due to its cost, life has a monetary price" (Bayless, 1978: 29). On the other hand, there is a long and venerable tradition in our philosophical attitudes toward the VOL problem, perhaps best articulated by the Enlightenment philosopher Immanuel Kant (1785), when he wrote:

> In the realm of ends everything has either a price or a dignity. Whatever has a price can be replaced by something else as its equivalent; on the other hand, whatever is above all price, and therefore admits of no equivalent, has a dignity.

Of course for Kant, human persons are such creatures who exhibit "dignity." As Rescher puts it "How much is it worth to prevent the death of a person? . . . the question has no answer . . . it assumes that 'life' and 'risk to life' is some measurable quantity that actually exists in a stable and determinable way" (Rescher, 1987: 226). But, since this is false, Rescher concludes: "The question of value of life pushes beyond the proper limits of cost-benefit analysis in its insistence on quantifying something that is inherently unquantifiable" (Rescher, 1987: 226).

Byrne's analysis reveals three general methods to assess the value of life that are used in RCBAs: insurance-based, earnings-related, and willingness-to-pay (WTP) strategies (Byrne, 1988). Unsurprisingly, each one of these methods has serious limitations and deficiencies. Rescher (1987) points out the limitations of the earnings-related method. As he puts it:

> One study that examined salary as a function of occupational risk concluded that a premium of about $200 per year (1986) was sufficient to induce workers in risky occupations to accept an increase of 0.001 in their annual probability of accidental

death, a finding that was interpreted to indicate a life-valuation of around $200,000. . . . The linearity assumption involved in such calculations is questionable—the man who accepts a 1% chance of death for $10,000 may well balk at accepting $1,000,000 for certain death. (p. 227)

In other words, the supposedly higher or lower wages people accept for different types of hazardous jobs are interpreted as a valid measure of the cash value people are thought to place on their own lives. All too frequently, however, when lives are valued based on such criteria as economic worth or expected earnings, this turns into "life is cheap" in poorer neighborhoods or less developed nations. This issue is clearly illustrated in the Bhopal case. Life in Bhopal was implicitly valued less than life in the United States. Therefore, the safety equipment and emergency preparedness at the Bhopal plant in India were far less adequate than those at a similar plant operated by Union Carbide in Institute, West Virginia.

Barbour (1980) states the consequences of following the valuation of life principle to its logical conclusion:

If applied consistently, the method would require that the lives of the elderly would be valueless. If future earnings are discounted, a child's life would be worth much less than an adult's. . . . I would maintain that there are distinctive characteristics of human life that should make us hesitant to treat it as if it were a commodity on the market. Life cannot be transferred and its loss to a person is irreversible and irreplaceable. (p. 73)

The third methodological problem of RCBA is how to deduce the value attribution of all the identified risks, costs, and benefits. Analysts automatically assume that often disparate costs and benefits can somehow be compared with another—that is, that all values are commensurable and can be fully quantified to reasonably determine whether the benefits of the proposed technological intervention or policy do, in fact, outweigh the risks and costs. Such calculations are necessary for RCBA so that disparate values can be compared and traded off, one against the other. Money becomes the common metric so that "goods" and "bads" can be compared with one another, and price becomes the medium through which all alternatives are evaluated, even those that are not normally perceived to have a market value (Kelman, 1981). This is evident in the "willingness-to-pay" criterion of a free-market economy: what a willing buyer will pay a willing seller. Take, for example, our aesthetic relationship to nature. How much is a beautiful view worth in monetary terms? How much is a landscape worth? A sunset? How much would someone be willing to pay to avoid having a toxic waste dump, a power plant, or an oil refinery built in his or her community?

The fourth methodological deficiency of RCBA becomes visible when one begins to probe the unquestioned assumption that market values provide the best opportunities for human beings to advance their life goals (Kelman, 1981). In other words, an RCBA methodology makes the assumption that the decisions people make in the marketplace are rational with regard to price, needs, and wants. However, it must be admitted that even in the open market the notion of utility maximization does not fully satisfy the variety of human needs and purposes. Notions such as freedom, equality, justice, and aesthetics also matter (Hausman and McPhearson, 1996: 77). In other words, one cannot always trust the market to satisfy all of our preferences and sustain all of our values. This became all too evident in the case of the Ford Pinto (see Chapter 11). The public was outraged when they were informed, perhaps for the first time, as to how decisions like this are made. In the end, the problem is that:

> By regarding human happiness, human well-being, human life, and non-human life as mere commodities, cost-benefit analysis ignores the non-market value of these things and the central role they should play in public policy. (Anderson, 1993: 190)

These sentiments are reflected in a sign that Albert Einstein is reported to have had hanging in his office. The sign read: "Not everything that counts can be counted, and not everything that can be counted, counts" (Diwan, 2000). After everything is said, Einstein's aphorism perhaps best sums up the problems that beset using risk-cost-benefit analysis as the preeminent method for assessing the risks of technology. The aphorism also points to why risk-cost-benefit analysis fails as the sole method of determining the appropriate and equitable level of acceptability of those risks. This is no more evident than in RCBA's neglect of social values that contribute to our idea of justice, qualities that one can be sure Einstein would consider among those things that "count, but cannot be counted."

The fifth set of problems that beset the RCBA method are the well-known criticisms that RCBA fails to address adequately issues of fairness associated with the equitable distributions of risks and harms. For one thing, RCBA places exclusive focus on aggregate benefits and cannot address the ways in which those benefits are distributed. It is *not* designed to pay attention to the ethically crucial question: "Who pays the costs, and who gets the benefits?" Typically, such analysis reaches its "bottom line" by aggregating all costs, all risks, and all benefits. Its goal is to determine, within its limited definition of the goods and harms involved, the *net* good, or harm that a technological intervention will produce. In other words, risk-benefit analysis is concerned only with the amounts of "goods" and "bads" in society, not with their fair or equitable distribution. For example, if the oil refinery in a neighborhood can be calculated to allow millions of distant persons to benefit from the gasoline and other products of that refinery, this can be multiplied into a major benefit. On the other hand, if the refinery results in higher cancer rates, greater medical costs, and residential property devaluation in the immediate neighborhood, this can also be calculated as part of the net costs, or harms, and subtracted from the "greater good." Although the net benefit may greatly outweigh the overall costs, the distribution of goods and harms may not be fair, because as Ferre (1995) puts it: "the principle of beneficence, to create greater good, is satisfied, but the principle of justice has been overlooked" (p. 83).

Justice across geographical, economic, and social space is one crucial set of values that RCBA leaves out of its calculations and equations. In addition, justice across time is almost totally neglected. Since RCBA is geared toward favoring short-range exploitation of opportunities and resources, it tends to ignore what Barbour (1980) calls "intergenerational justice" (p. 173). In other words, RCBA fails to address questions about the duties, obligations, and responsibilities one generation has to the next. Given recent concern over questions of ecological sustainability, resource depletion, and harm to future generations, this constitutes a major ethical flaw in the RCBA method of risk assessment.

In addition to overlooking questions about our duties and obligations to future generations, economists and policy makers seem to either ignore or deny that the market process in general, and cost- and risk-benefit analysis in particular, systematically undervalue irreplaceable natural assets. RCBA tends to ignore considerations of what Ferre (1995) calls "ecological justice" (p. 84). The scarcity of nonrenewable resources, the irreversibility of habitat and land destruction, the extinction of endangered species, the depletion of the ozone layer, global warming, etc., are all pressing concerns that RCBA fails to address.

TECHNOLOGY ASSESSMENT

PRA and RCBA are not the only ways to assess the risks and harms of technology. Another approach is called *technology assessment* (TA). As originally conceived, TA was sensitive to the problems and issues previously discussed. Take, for example, the definition of TA given by one early theorist:

> Technology assessment is the process of taking a purposeful look at the consequences of technological change. It includes the primary cost-benefit balance of short-term localized market-place economics, but particularly goes beyond these to identify affected parties and unanticipated impacts in as broad and long-range fashion as is possible . . . both 'good' and 'bad' side-effects are investigated since a missed opportunity for benefit may be detrimental to society just as an unexpected hazard. (Coates, 1976: 141)

This definition introduces two ideas: the first points to a feasibility analysis performed so as to determine whether a proposed technology would maximize public utility. The second idea calls for mechanisms that focus on second- and higher-order (noneconomic) consequences, which are to be balanced against first-order (economic) benefits. Only with the aid of such an analysis is it possible to take account of unanticipated impacts of technology and also identify how they affect different stakeholders or constituencies. These two different but complementary concerns give voice to two general models, a "narrow" and a "broad" definition of technology assessment.

The narrow definition tends to restrict the meaning of TA to basically an operational analysis of particular technologies defined as concretely as possible (as in PRA):

> Technology assessment is viewed as a systematic planning
> and forecasting process which encompasses an analysis of
> a given production method or a line of products . . . it may
> be considered as a natural follow-up to systems engineer-
> ing . . . (Coates, 1976: 142)

The broad definition, on the other hand, tends to consider technology assessment as a framework for societal analysis. This requires a systematic and interdisciplinary analysis of the impacts
of technological innovation on the social, political, ethical, and
medical aspects of life.

CONCLUSION

To enhance our capacity to prevent technological disasters, a
broad concept of technology assessment is in order, the features
of which are as follows:

1. *Social impacts.* TA should be concerned with second-,
   third-, and higher-order impacts such as impacts on human health, society, and the environment, as distinguished
   from economic utility of exclusively first-order concerns.
2. *Multi-disciplinary analysis.* TA should require that all
   pertinent aspects—economic, social, ethical, cultural,
   environmental, and political—be taken into account.
   Diverse methodologies and inputs from all disciplines
   are to be employed.
3. *Multi-constituency impacts.* TA should consider the
   widest range of stakeholders that may be affected by
   the proposed technology. Comprehensive TAs should
   require the informed consent of all affected stakeholders, inviting their active participation in the decision-
   making process.
4. *Policy-making tool.* TA should not be concerned with just
   technical expertise but, more essentially, with the sociopolitical problems associated with the impacts and consequences of a proposed technological innovation.

Such principles for a broad technology assessment can only be realized if risk assessment becomes a democratic process rather than

one that is dominated by a technocratic and power elite. This critical issue will be the subject of our final chapter.

References

Anderson, Elizabeth. (1993). *Value in ethics and economics.* Cambridge, MA: Harvard University Press.

Baram, Michael. (1976). "Regulation of environmental carcinogens: Why cost–benefit analysis may be harmful to your health," *Technology Review,* July/August; 78: 40–42.

Baram, Michael. (1977). "An assessment of the use of cost-benefit analysis." In Joel Tarr (Ed.), *Retrospective technology assessment.* San Francisco, CA: San Francisco Press: 15–30.

Barbour, Ian. (1980). *Technology, environment, and human values.* New York: Praeger.

Bayless, Michael. (1978). "The price of life," *Ethics* 89 (1): 28–39.

Bier, Vicki. (1997). "An overview of probabilistic risk analysis for complex engineered systems."In Vlasta Molak (Ed.), *Fundamentals of risk analysis and risk management.* Boston, MA: Lewis Publishers: 67–85.

Bougumil, R.J. (1986). "Limitations of probabilistic assessment," *IEEE Technology and Society Magazine* 24 (8): 24–27.

Byrne, L.J. (1988). "The value of life: The state of the art." In Larry Martin (Ed.), *Risk assessment and management: Emergency planning perspectives.* Waterloo, Canada: University of Waterloo Press: 79–101.

Coates, Joseph. (1976). "The role of formal models in technology assessment," *Technological Forecasting and Social Change* 9: 140–146.

Diwan, Romesh. (2000). "Relational wealth and the quality of life," *The Journal of Social Economics* 29 (4): 305–322.

Ferre, Frederick. (1995). *Philosophy of technology.* Athens, GA: University of Georgia Press.

Gewirth, Alan. (1990). "Two types of cost-benefit analysis." In Donald Scherer (Ed.), *Upstream/downstream: Issues in environmental ethics.* Philadelphia: Temple University Press: 205–232.

Greenberg, Michael, and Goldberg, Laura. (1994). "Ethical challenges to risk scientist exploratory analysis of survey data," *Science, Technology, and Human Values* 19 (2): 223–241.

Grossman, P.Z., and Cassedy, E.S. (1985). "Cost-benefit analysis of nuclear waste disposal: Accounting for safeguards," *Science, Technology, and Human Values* 10 (4): 47–54.

Haimes, Yacov. (1998). *Risk modeling, assessment, and management.* New York: John Wiley & Sons.

Harris, Charles E., Pritchard, Michael S., and Rabins, Michael J. (2000). *Engineering ethics: Concepts and cases.* Belmont, CA: Wadsworth.

Hausman, Daniel, and McPhearson, Michael. (1996). *Economic analysis and moral philosophy.* London: Cambridge University Press.

Henley, Ernest, and Kumamoto, Hiromitsu. (1981). *Reliability engineering and risk assessment.* Englewood Cliffs, NJ: Prentice Hall.

Humphreys, Paul. (1987). "Philosophical issues in the scientific basis of quantitative risk analyses." In James Humber and Robert Almeder (Eds.), *Quantitative risk assessment: Biomedical ethics reviews.* Clifton, NJ: Humana Press: 205–223.

Kahn, Shulamit. (1986). "Economic estimates of the value of life," *IEEE Technology and Society Magazine,* June: 24–29.

Kant, Immanuel. (1785). *Groundwork for the metaphysics of morals.* Translated by James W. Ellington. (1981). Indianapolis, IN: Hackett Publishing Company.

Kelman, Stephen. (1981). "Cost benefit analysis: An ethical critique." Reprinted in Thomas Donaldson and Patricia Werhane (Eds.), *Ethical issues in business* (5th ed.). Upper Saddle River, NJ: Prentice Hall.

Kneese, Allen, Ben-David, Shaul, and Schultze, William. (1983). "The ethical foundations of cost-benefit analysis." In Douglas MacLean and Peter Brown (Eds.), *Energy and the future.* Totowa, NJ: Rowman and Littlefield: 59–74.

Lanthrop, John. (1982). "Evaluating technological risk: Prescriptive and descriptive perspectives." In Howard Kunreuther and Eryl Levy (Eds.), *The risk analysis controversy: An institutional perspective.* Heidelberg, Germany: Springer-Verlag: 165–180.

Linnerooth-Bayer, Joanne, and Wahlstrom, Bjorn. (1991). "Applications of probabilistic risk assessments: The selection of appropriate tools," *Risk Analysis* 11 (2): 239–248.

MacKinnon, Barbara. (1986). "Pricing human life," *Science, Technology, and Human Values* 11 (2): 29–39.

Martin, Mike. (1982). "Comments on Levy and Copp and Thompson." In Vivian Weil (Ed.), *Beyond whistleblowing: Defining engineers' responsibilities.* Chicago: Center for the Study of Ethics in the Profession, Illinois Institute of Technology: 146–152.

Reactor safety study—An assessment of accident risks in U.S. commercial nuclear power plants. WASH-1400, NUREG-75/014, October 1975. Washington, DC: U.S. Nuclear Regulatory Commission, 1974.

Rescher, Nicholas. (1987). "Risk and the social value of a life." In James Humber and Robert Almeder (Eds.), *Quantitative risk assessment: Biomedical ethics reviews.*Clifton, NJ: Humana Press: 225–237.

Rowe, William. (1994). "Understanding uncertainty," *Risk Analysis* 14 (5): 743–750.

Thompson, Paul. (1982). "Ethics and probabilistic risk assessment." In Vivian Weil (Ed.), *Beyond whistleblowing: Defining engineers' responsibilities.* Chicago: Center for the Study of Ethics in the Profession, Illinois Institute of Technology: 114–126.

14

Technology Decisions and the Democratic Process

> *"I know of no safe depositor of the ultimate powers of society but the people themselves; and if we think them not enlightened enough to exercise their control with a wholesome discretion, the remedy is not to take it away from them, but to inform their discretion."*
>
> —Thomas Jefferson

We now turn to the question of how technological decisions in a democracy are made and how they might be made, especially when there is a discrepancy between the experts' judgment of risk, policy-makers' judgments, and the public's perception of risk. The separation into "expert" and "lay" judgments has polarized society's understanding of risk and has created two schools of thought, which Fiorino and others call the "technocratic" and the "democratic" modes of risk interpretation (Funtowicz, 1983; Lathrop, 1982; Fiorino, 1989; see also Spangler, 1982).

TECHNOCRATIC VERSUS DEMOCRATIC ASSESSMENTS OF RISK

Mitcham (1997) expresses the difference between these two approaches to risk assessment as follows:

> . . . sociologists and political scientists have analyzed technology as harboring within itself fundamental anti-democratic

possibilities and tendencies. The general name for such tendencies is 'technocracy,' rule by technical elites or technical information rather than by the people (demos). . . . (p. 40)

The unexamined assumption of the technocratic approach is
that "scientific" risk analyses yield totally objective, factual, value-
free assessments of risk, whereas the lay public's assessments of
risk are thought to be subjective and value-laden. Because of this
bias, technical experts often view the judgments of laypeople as
capricious and unreasonable, sensitive to irrelevant factors and insensitive to relevant ones. This leads experts to assess the risks of
technologies completely divorced from the public's perception, assessment, and understanding of technologically induced risk and
harm. Technocrats argue, for example, that since public perceptions of risk are supposedly derived exclusively from their subjective opinions, which can differ substantially from person to
person, only objective risk assessments can protect an uniformed
public from the dangers of technology (Slovic, Fischoff, and Lichtenstein, 1980).

Traditionally, the risks and failures of technology have
mainly been perceived as a technical problem, a problem rele-
gated to experts, not to public debate. But controversies have
politicized the issue of risk. Risk assessment is no longer seen as
simply an exercise in the technical measurement of risk. Questions of risk can no longer be defined simply in technical terms;
they must also be defined in political and social terms, because
the real question is not how safe it is, but how safe is safe
enough for individuals and society? Moreover, since technical
risk assessors are no more qualified than the general public to
assess value judgments such as those involving welfare, fairness,
social justice, and informed consent, they should not be the only
ones who participate in the assessment of the risks of technology that affect us all.

It is often true that the public's interpretation and assessment of risk differs considerably from the interpretations and
assessments of technical experts. However, the view that the
public perception of risk is distorted by subjective biases and
that only experts can define the "real" risks is overly simplistic.
As we have argued in Chapter 13, experts are also subject to biases in interpreting quantitative data, especially when objective
uncertainties are present. Many so-called objective assessments

ultimately depend on the subjective interpretations and normative judgments of engineers and applied scientists. As Barus (1982) points out:

> Human value judgments—conscious or unconscious—cannot be escaped by resort to modeling and analysis. These judgments get built into the model itself. Seldom are they re-examined after the analysis, for by then the ethical ground-rules have become tacitly established. (p. 8)

This point is no more evident than when the experts themselves disagree on the identification and estimation of some particular risk, or when they disagree on the appropriate level of acceptability of that risk. As Shrader-Frechette (1985[a]) points out:

> Many controversies over technology- and environmental-related policy have been characterized by experts' disagreement of the relevant scientific facts. Prominent examples include conflict over nuclear fusion, fluoridation, food additives, depletion of the ozone layer, and high-voltage transmission lines. (p. 103)

Uncertainties surrounding the risks and benefits of the relevant technical facts result in recurrent heated debates among scientists and engineers over the objective determination of risk. Such technical controversies demonstrate that the experts' tendency to view their objective characterizations of risk as somehow more "real" or more valid than the perceptions of risk by the public at large is questionable. It is clear that decisions about science and technology, long considered the domain of experts, are increasingly subjects of political controversy. Dorothy Nelkin (1994), a renowned social scientist, has compiled a set of scientific and technological controversies that are especially instructive in this regard:

- Nuclear waste disposal
- Setting occupational health standards, as in the vinyl chloride case
- Holes in the ozone layer
- Genetic testing in the workplace
- Regulating recombinant DNA research and its applications

One might have expected TMI, DC-10, Challenger, Bhopal, Chernobyl, and countless other cases of technological disasters to have tempered the overconfidence of experts in their "objective" assessment strategies. Unfortunately, this generally has not been the case. The idea that people's fears should be viewed as any "less real" than the results of abstract, complex, arcane calculations certainly needs explaining and justification. Standard approaches to defining technological risk tend to reflect technocratic rather than democratic values (Fiorino, 1989). As Renn (1990) puts it:

> The artificially-constructed contrast between an allegedly rational assessment by professionals and an allegedly irrational perception by laypersons has not only disguised the limitations and values of both approaches, but has put considerable constraints on an effective and acceptable risk management approach. (p. 8)

The point is this: both the technocratic approach to risk as well as the democratic approach to risk have merits and limitations for designing sound and fair risk-assessment and management strategies. While experts may confine the concept of risk to a combination of magnitude and probability of adverse effects— defined, for example, in terms of lives lost and property damage— laypersons associate risk with a variety of qualitative criteria, such as familiarity with the technology and the extent of personal and political control over technology decisions.

The public assessment of technology incorporates a larger number of dimensions and concerns, such as society's ability to cope with relatively rare but catastrophic events, which are totally ignored or "averaged out" in professional risk estimates. For the public, "risks are interpreted as 'higher' if the activity is perceived as involuntary, catastrophic, not personally controllable, inequitable in the distribution of risks and benefits, unfamiliar, and highly complex" (Covello, 1983: 289). Other factors influencing the social interpretation of risk are: whether adverse effects are immediate or delayed, whether exposure to the hazard is continuous or occasional, whether the technology is perceived to be necessary or a luxury, whether the adverse effects are well known or uncertain, and to what degree protection from technologically induced harm is provided for by the organization that manages the potentially risky technology. The denial of such factors—which people who have to bear the risks consider to be violations of their values and rights concerning risks and interests—must be regarded as important determinants of any sound management approach in order to balance the risks and benefits of a proposed technology decision. Renn (1990) makes a telling point when he asserts that:

> It is not irrational to base one's policy on a concept that is different from the concept suggested by experts. To put extra weight on risks with high uncertainties, to avoid risks that have high catastrophic potential in spite of minute probabilities of such a catastrophe materializing, to adopt a more cautious strategy to cope with unfamiliar risks, and to assure a sufficient level of institutional control and monitoring before a risky technology is implemented are all valid and reasonable tools which can play an active role in technology assessment. (p. 3)

Fairness and control in the distribution of risks and benefits are also of central importance in determining a socially acceptable level of risk. By measure of control we mean the extent to which individuals participate in decisions about their exposure to risk, the decision to deploy the technology in the first place, and the decision as to how the risk and benefits are to be distributed. These two concerns are interrelated. People want greater control and more active participation to ensure consideration of their notions of fairness and justice in the distribution of risk and in the management of risk.

The democratic approaches to technology decisions do not deny the importance of technical and economic analyses in informing risk decisions; they welcome all important voices to the dialogue on risk. It is to be noted that this is contrary to technocratic approaches, which deny any role for nonscientific or societal input. Proponents of the democratic approach to technology assessment criticize the technocrats for insisting that quantitative models alone should govern technological decisionmaking (Shrader-Frechette, 1985 [b]).

As research has shown, laypersons tend to misunderstand statistical evidence (Kasper, 1980); hence, scientific expertise is needed to explain the scientific data to the public. As we have seen, equally compelling research has shown that experts tend to misunderstand the social factors of risk perception and assessment; hence, input from the social sciences and concerned citizen groups is needed. Scientific assessments are necessary for making prudent decisions, but they should not serve as the only criteria for evaluating the acceptability of a risk.

The professional calculation of risk should be an important and essential component of the decision-making process with respect to risk acceptance and risk management. This demand is hardly disputed by anyone. As Renn (1990) puts it, "Nobody wants to substitute scientific knowledge with intuition" (p. 7). Problems arise, however, when experts make professional assessments the sole criteria for judging the acceptability or desirability of a proposed technology, or risk management policy without taking the public's interest into account and without doing so in accordance with socially-shared values. Without including such factors, technology assessment may become socially irresponsible.

Given the inherent methodological and ethical limitations of technocratic approaches to risk assessment, merely improving upon technical analyses cannot be the key to an enhanced risk-assessment process. With the assessment of large-scale sociotechnical systems, risk-assessment problems require the development of institutional procedures for structuring a critical dialogue among different stakeholders. The normative approach provides an alternative policy method by "starting from the risk perceiver and bearer, the inherent value-laden nature of risk decisions, and by emphasizing the essential role of participation by a range of perspectives in societal decisions about risk" (Bradbury, 1989: 395).

In sum, there are numerous methodological and substantive differences between the technocratic and democratic assessments of risk. Each approach differs in its concept of risk, in its understanding of the nature of risk, in its assessment of risk, and in the subsequent application of risk assessment decision making. Figure 14–1 summarizes the major differences between the technocratic and democratic assessments of risk.

Taking into account the benefits of both approaches to risk assessment would accomplish two crucial tasks. First, the technocratic approach brings scientific methodology to bear on the risk assessment process. Second, the democratic approach, which includes the citizenry in the risk assessment process, reflects the normative values necessary for a fair analysis of the risks of technology and a more equitable distribution of the harms and benefits of technology.

	Technocratic	Democratic
Concept of Risk	Risk is defined as a quantitative concept.	Risk is defined as a qualitative concept.
Nature of Risk	Risk is an objective property of phenomena discovered by scientists.	Risk is a social construct.
Assessment of Risk	Probability of the event happening × projected magnitude of the event; e.g., mortality rates.	Potential for catastrophe; familiarity with the technology; degree of control over the technology; perceived fairness of distribution of benefits and harms of the technology in question.
Application to Risk Assessment Decisions	Fact-value dichotomy is presumed in risk assessment decisions.	Risk assessment decisions are value-laden.

Figure 14–1
Differences between technocratic and democratic assessments of risk.

PARTICIPATORY TECHNOLOGY

The political question concerning the acceptability of risk has led many researchers to develop democratic models of technology development. One influential model is called "participatory technology." James Carroll coined this term in 1971. Carroll defines *participatory technology* as:

> . . . [T]he inclusion of people in the social and technical processes of developing, implementing, and regulating a technology, directly and through agents under their control, when the people included assert that their interests will be substantially affected by the technology and when they advance a claim to a legitimate and substantial participatory role in its development or redevelopment and implementation. (p. 647)

The advocates of participatory technology seem to feel that if the public is given sufficient information it would arrive at a consensus on what technology it needs and wants, and what it would reject or forgo (Nelkin, 1975). The general public has some knowledge by virtue of its direct involvement or interest concerning such problems as toxic waste or nuclear-waste disposal, pesticide regulation, energy consumption, mining policy, and water treatment policy. A succession of technological failures and disasters—safety recalls of poorly-designed automobiles, contaminated water, poisonous food additives, medical and pharmaceutical mishaps, and toxic chemicals in the air—has heightened public awareness of the risks of modern technology.

Distrust of the judgments of technical elites comes from the public's perception that the elites assume there is something immutably wrong with laypeople's judgments. But, as Cerezo and Garcia (1997) and Fiorino (1990) argue, and as we have seen previously, "lay-knowledge" is in a position to aid the decision-making process rather than to hinder it. As Cerezo and Garcia point out:

> Local knowledge can provide useful information concerning known parameters (e.g., economic or biological variables) and their relative significance to the socio-system's equilibrium. Local knowledge can also point out [a] new perspective in the sense of showing the relevance of dimensions (e.g., culture and traditions, local economic practice) omitted by expert knowledge. In addition, the inclusion of lay knowledge into policy-making, including the consequent promotions of active public participation, can avoid negative public perception, consequent social resistance, and the political manipulation of public opinion. (p. 38)

Fiorino (1990) identifies three arguments against the technocratic orientation of elites. The first point is that lay judgments about risk are often as sound as expert judgments. Nonexperts see problems, issues, and solutions that experts may miss. Lay judgments about technological hazards are sensitive to social, political, and ethical issues that expert models ignore, either intentionally or unintentionally. The technical expert is constrained in ways that the lay public is not. For example, decision makers tend to co-opt experts in order to support policy decisions in the public and private sectors that, in actuality, reflect "vested interests" (Fiorino, 1990).

Moreover, there are intrinsic limitations of expert knowledge. Bias and prejudice are unavoidable in expert opinion because it is often based on uncertainties of risk measurement, including factual as well as conceptual indeterminacies in the assessment and forecasting of the impact of technological systems on society. As we have shown previously, expert appraisal of risks is not a purely objective process of fact determination but a complex web of facts, normative judgments and value-laden decisions.

Fiorino's second point is that the inclusion of citizen or lay participation in the technology decision process may even contribute to better decision making by incorporating a broader range of values into decisions, hence reducing the probability of error. Taking into account lay knowledge in the process of producing policy decisions about technology and its risks could optimize the decision-making process by providing useful information not included in experts' assessments. One reason is that "citizens can gather data otherwise unavailable to officials because of professional ignorance or the expense of collecting it" (Johnson, 1987: 107).

One instructive example of this is Wynne's research on the Chernobyl disaster. Wynne (1989) conducted a case study of experts' assessment of the risks and that of sheep farmers affected by the fallout from the Chernobyl reactor explosion and fire. He discovered that the officials' advice often missed the mark because they did not understand the practical realities of raising livestock. As Wynne points out "the officially claimed notice period did not consider the time needed to gather, sort, and get sheep to the market . . . the effect on prices of the rush to sell created by the deadline was not officially acknowledged. . ." (p. 34). The experts gave unrealistic instruction to farmers, such as feeding the sheep straw and delaying spring lamb sales; the delay in sales resulted in overgrazing and the inability of farmers to sell lambs in their prime. In addition, "many local practices and judgments important to hill-farming were unknown to the experts, who assumed that scientific knowledge could be applied without adjusting to local circumstances" (p. 34).

On the other hand, when experts take community wisdom into account, a successful decision can often result. Grabill and Simmons (1998) discuss a case in which information from drivers in Seattle helped address local traffic problems that transportation officials had not been able to solve. The problem was heavy traffic congestion during peak commuting hours. Local transportation officials tried to solve the problem by applying quantitative analysis—measuring traffic flows and applying statistical methods—to ascertain where to reroute traffic and which roads to widen. These methods did little, however, to solve the traffic gridlock difficulties. A team of researchers from the University of Washington decided to employ a different method. Rather than study traffic patterns abstractly, they solicited information from the local motorists—through surveys, interviews, and focus groups—and determined their driving preferences and habits. Such qualitative information, garnered from citizen input, eventually helped solve the traffic situation in downtown Seattle. "In terms of the user as citizen . . . the point is made most strongly: the users are represented as an important force in the design of the system . . . because they are asked to help determine the best solution to the problem" (Grabill and Simmons, 1998: 433).

Gary Severson of Waste Tech Services, Inc., attributed his company's success in siting a hazardous waste incinerator to community collaboration (Belsten, 1996: 39). According to Belsten, Waste Tech Services representatives "fully acknowledged that hazardous waste incinerators are a highly controversial and diverse issue. . . . They told community representatives that if they determined they did not want the hazardous waste incinerator in their community, Waste Tech would not remain in the community" (Belsten, 1996: 39). Severson offered the community access to the company's staff and the company's written information on incinerators, and he also "met one-on-one with literally hundreds of people" in the community (p. 40). To his surprise, he found that "individual members were less concerned about the technology and the quantitative risk-assessments [*per se*] . . . than they were about the extent to which they believed they could trust Waste Tech to operate the incinerator safely" (p. 40). Finally, Severson "listened to the community's concerns . . . [and] incorporated community suggestions into the final design" (p. 40).

Citizen participation renders expert advice sounder by going beyond limited technocratic standards such as the ratio of inputs and outputs typical of cost-benefit analysis. In addition, citizens' questions and observations can expose experts' indeterminacies and uncertainties and thus point to a more cautious and flexible way of decision making (Cerezo and Garcia, 1997). This, in turn, can contribute to preserving the trustworthiness of expert advice and avoid social mistrust in policy makers. The result would be that laypeople might become less cynical about experts' advice. Consequently, citizen input would help create a more positive attitude toward technology and encourage more active democratic participation.

Fiorino's third point is that the technocratic orientation is incompatible with democratic ideals. Overreliance on the views of elites tends to ignore the ethical and social dimensions of policy analysis and disenfranchises the public who ought to have some control over that policy. The very definition of a citizen includes the capacity to participate in decisions that affect one's self and one's community. In many ways, the problem of the control and management of science and technology in modern society is simply a special case of the general problems of democracy. The main lesson for our purposes is that public participation can and does reduce the probability of error, risk, and failure by incorporating a broader range of information and values into the decision-making process.

Recent debates as to what should be done with more than 28,000 tons of poisonous nerve gas—manufactured for use in biological warfare—will serve as an example of the possibility of active citizen involvement in technology policy making (Robbins, 1998). Currently, the U.S. Army stores tons of Sarin and mustard blister agents at nine sites nationwide. In accordance with the Chemical Weapons Treaty that the United States has signed, it is obligated to destroy the entire stockpile of nerve gas by 2007. In 1985, Congress ordered the Defense Department to start destroying its aging chemical weapons stocks. Three years later the Army announced, after little or no public consultation, that it would burn the stocks at each of the sites in "high-tech" incinerators. Experts from the National Academy of Science agreed that incineration was safer than continuing to store the deteriorating weapons, some more than 50 years old. The Army's original plan was to destroy the stockpile by 1994 at a cost of $1.7 billion, but protests by citizens living near the chemical munitions bases have stalled the Army's plan. Opponents of incineration have kept it tied up in lawsuits and permit fights. With the deadline drawing near, only two nerve gas sites have been destroyed so far. In 1968, shifts in the wind pattern during a nerve gas test at Utah's Dugway Proving Grounds had killed more than 6,400 sheep. This incident caused consternation among citizen activists.

To attempt to solve the problem, the usually tight-lipped Pentagon invited the public into its deliberations over the safe disposal of these deadly gases. It organized a 32-member "Dialogue" of local citizens, activists, state regulators, and Pentagon experts to monitor and shape the search for solutions. An example of one such "citizen dialogue" occurred in Anniston City, Alabama. The group met formally six times. Of the 15 community members, a third have their degrees in chemistry or engineering. None of the remaining members—including a librarian, a political aide, a clinical psychologist from Kentucky, a housewife, and a graduate student—had much technical training. The group was, however, well organized. Before each meeting, the technical team briefed the group on the technical and engineering problems. Group motivation was attested to by the fact that the librarian in the group prepared a resource guide on chemical weapons for members with limited knowledge.

The Pentagon even helped fill in gaps by hiring an independent engineering firm to represent citizen Dialogue members to make sure the technologies were evaluated against the citizen Dialogue's agreed-upon selection criteria.

Responses to these "Dialogue" meetings have been mixed. Critics claim that all the Army has done is to institutionalize a citizen-lobbying group against incineration. For the critics, these Dialogues will only slow down the stockpile destruction at rising costs to taxpayers. In addition, these Dialogues create fears that the Army would not meet the mandated deadline of the Chemical Weapons Treaty. The Citizen Dialogue was, in effect, an attempt to integrate both democratic and technocratic approaches to a technology policy decision.

Fiorino (1989) articulates three tasks that would have to be undertaken in order to integrate fully the technocratic and democratic models of risk assessment and management. The first includes the reevaluation "of our reliance on formal analytical models, some of which displace political judgment and further isolate decisions from democratic control" (p. 127). The second task would be to "adapt those analytic models we continue to use so that they incorporate lay values more effectively" (p. 127). Finally, risk assessors must reassess "current mechanisms for citizen participation, from the public hearing to the initiative, to the advisory panel, as well as designing and experimenting with new institutional forms" (p. 127).

Notwithstanding the arguments in favor of the potential contributions of laypersons to the quality and fairness of technological decision making, the emergence of grass-roots-community environmental activism poses a dilemma. In defense of environmental values, these protest groups have embraced the concept of "not-in-my-back-yard" (NIMBYism), refusing to allow the siting of nuclear waste disposal and toxic wastes in their communities (Piller, 1991). Some protest groups, rejecting the argument of protecting selfish interests, or NIMBYism, have advanced instead a concept of "not-in-anyone's-back-yard" (NIABYism). As difficult as these activist charges may be, they have a beneficial effect of compelling policy makers and experts to reassess their decisions from the vantage point of protecting the interests of diverse constituencies.

Though they may be a thorn in the side of policy makers, these environmental activist groups have exercised their right to have their voices heard. In the process, they have not only learned how political power is exercised but have also learned the limitations of pursuing exclusively parochial interests. By transcending their own activist interests, citizen groups can learn to enhance their effectiveness in influencing decisions concerning technological risks. As Hunold and Young (1998) put it:

> Through public discussions citizens often transform their understanding of the problem and proposed solutions, because public communication forces them to take account of the needs and interests of others and may also give them information that changes their perceptions of the problems and alternatives for solving it. (p. 87)

This is a lesson technocratic and political elites also need to learn. In fact, the preservation of the democratic process requires a commitment to learning, through debate, the needs and interests of all relevant parties: the lay public, technical experts, and policy makers.

MECHANISMS FOR CITIZEN PARTICIPATION

In the enabling legislation establishing the National Foundation for the Arts and Humanities, Congress declared:

> ... [T]hat democracy demands wisdom and vision in its citizens and that it must therefore foster and support a form of educa-

tion designed to make men masters of their technology and not its unthinking servants. (Public Law 89-209, 1965: 845)

This is indeed a prescient declaration. However, this inspiring statement does not spell out how our educational institutions can make citizens "masters of their technology and not its unthinking servants."

Since the mid-1960s, various mechanisms for instituting public participation in technology policy have been tried in numerous contexts and in countries around the world (Fiorino, 1990; Renn, Webler, and Wiedemann, 1995; Rowe and Frewer, 2000). Public participation is defined as "forums of exchange that are organized for the purpose of facilitating communication between government, citizens, stakeholders and interest groups, and businesses regarding a specific decision or problem (Renn, Webler, and Wiedemann, 1995: 2). Such mechanisms have the goal of enabling the lay public to get involved and influence decisions concerning technological risk. They permit those affected by a decision to have an input into that decision. Through such mechanisms, an ethic of informed consent is maintained between those that may pose a risk by way of a technological innovation and those that are likely to be adversely affected by the technology (Fischoff, 1979).

There are various levels and degrees of public involvement, however. At one level, informed public opinion is solicited and perhaps incorporated in the decision-making process (public hearings and public surveys). At another level, citizens are invited to become actively involved in discussions and decisions about the benefits and harms of a proposed technological intervention that will affect them (referenda, consensus conference, and citizen advisory groups). We will now briefly consider each of these mechanisms for citizen participation.

Public hearings serve as a forum for discussing specific subjects in depth and as a channel for the expression of a range of options. Take, for example, the Canadian Berger Commission, created to assess the Mackenzie Pipeline controversy in northwestern Canada (Fischer, 2000). The commission heard from a great variety of people—from fishermen to legislators—and considered political testimony as important as technical information.

From the outset, Berger introduced a boldly innovative strategy for conducting the inquiry. . . . The goal was not just to take testimony but rather to supply the communities with information and education. . . . In particular, the staff sought to draw out the communities' local knowledge concerning ways of life, economic and social needs, unique or peculiar local circumstances, and social values and beliefs. (Fischer, 2000: 232)

According to Nelkin and Pollak (1980), the Environmental Protection Agency convenes hundreds of public hearings each year. One example is the EPA's use of hearings as one element in a larger program to involve the community in setting standards for inorganic arsenic emissions from the ASARCO smelter in Tacoma, Washington (Ruckelshaus, 1985: 30–33).

Public opinion surveys and other techniques of social research provide advice to decision makers about public preferences regarding controversial technologies. The public may also be informed of these results. In 1970, the U.S. Forest Service employed "code-resolve," a computerized content analysis system, to obtain a better understanding of public attitudes about the use of DDT on Douglas fir trees (Nelkin and Pollak, 1980). Code-resolve is designed to transfer information from diverse sources into a condensed form useful for policy review. This resource can tap a wider spectrum of attitudes than public hearings. Nevertheless, as with traditional public opinion polls, it is only a surrogate for direct participation.

Referenda are often used to settle disputes about technology decisions. These allow the voting public to resolve a controversial question by direct vote. In many respects, the referendum is the prototype of the democratic process. Referenda allow citizens to initiate legislation or appeal existing laws. According to Nelkin and Pollak (1980), the United States, Austria, Switzerland, and Holland have all seen referenda on the ballot banning various nuclear technologies. Citizens in many cities and towns around the United States have passed referenda declaring their communities to be "nuclear-free zones." Two such communities are Oakland, California, and Takoma Park, Maryland. Oakland's ordinance prohibits the production and storage of nuclear weapons, weapon components, and radioactive materials within city limits and restricts the transportation of such materials through the city (Marcus, 1989: A2). Takoma Park's "nuclear-free" ordinance goes even further by prohibiting any businesses operating within its city limits from doing business with companies linked to the nuclear arms industry (Montgomery, 1995: B1).

In the year 2000, voters in 42 states faced 204 ballot initiatives or referenda on questions ranging from local property taxes to airport expansion or nuclear power plant development (Verhovek, 2000: A24). The growing popularity of the referendum demonstrates the public's interest in dealing with controversial technical questions that affect their lives.

The *consensus conference*, a recent Danish innovation, gives ordinary citizens a chance to make their voices heard in debates on technology policy. Pioneered during the late 1980s by the Danish Board of Technology, a parliamentary agency charged with assessing technologies, the consensus conference is intended to stimulate broad and intelligent social debate on technological issues. To accomplish this, a carefully planned program of reading and discussion before the scheduled conference ensures that the public is well informed prior to rendering its judgment. So far, 12 consensus conferences have been organized by the Board of Technology on topics ranging from genetic engineering to educational technology, food irradiation, air pollution, human infertility, sustainable agriculture, and the future of private automobiles (Sclove, 1996).

Citizen advisory groups or citizen review panels are often organized to allow a local community to influence decisions affecting local interests. Such groups generally consist of members of the technical elite and members of the lay citizenry interested in such matters. The entire process could take as little as a few days or could last over an extended period of time. The panel hears testimony, questions technical experts, and deliberates about the issues. In contrast to public hearings, participants in these review panels have the opportunity to ask questions, challenge experts, and explore issues in some depth. For example, a citizen review board was set up in Cambridge, Massachusetts, to advise the local city council on a policy being considered to allow recombinant DNA research within the city. The review board was organized on the following principle:

> Decisions regarding the appropriate course between the risks and benefits of potentially dangerous scientific inquiry must not be adjudicated within the inner circles of the scientific establishment. . . . A key citizens group can face a technical scientific matter of general and deep public concern, educate itself appropriately to the task and reach a fair decision. (Nelkin and Pollak, 1980: 240)

One of the more comprehensive and successful citizen advisory groups is the State of Oregon's Statewide Planning Program. With the force of administrative law, the Oregon Planning Program requires that all citizens of the state "be involved in all phases of the planning process that every city and county undertakes" (Renn, Webler, and Wiedemann, 1995: 91). Since its inception in 1975, dozens of state-sponsored and state-mandated citizen advisory groups have been convened, addressing a range of issues from highway planning to industrial waste disposal plant sitings.

TOWARD AN ALLIANCE OF CITIZENS' ORGANIZATIONS

Participatory technology and mechanisms for citizen participation presuppose that if citizens exercise their voices on technological decisions, they will succeed in exerting influence on policy makers. This democratic presupposition unfortunately runs into the hard realities of power politics. Political elites, on the one hand, and the experts whom they call on for corroboration, on the other, together wield substantially more power than the citizenry. Moreover, special interest groups, such as those representing trade associations, professional organizations and the like, lobby policy makers by means of the enormous resources at their disposal. Given these hard political realities, how can citizens effectively participate in technology decisions? The answer is they can do little as *individual* citizens. Only when citizens organize themselves into an *alliance of complementary interest groups* have they been able to achieve countervailing power to be marshaled against the actions of political elites, experts, and special interest groups.

At least four nongovernmental organizations (NGOs) at the community level have, on occasion, been welded together into an effective citizen-participatory alliance. The first consists of women's organizations. Potentially dangerous technologies are likely to elicit the attention and efforts of women's groups who are sensitive to adverse effects of new technologies on women's health and the changing structure of the family. A second group is comprised of civil rights organizations that reach out to a broad constituency of impoverished and discriminated citizens. In decades of struggle to achieve civil rights for their disadvantaged constituents, they have acquired considerable political clout. Environmental organizations represent the third group. Since technological disasters often have a devastating effect on the environment, environmentalists are ready allies of citizens in protesting dangerous technologies. Finally, consumers form the fourth group. Consumers, like citizens, are a global category that includes everybody. For several decades the consumer movement has alerted the citizenry to various product defects that adversely affect their lives.

Women's Organizations

While the "first wave" of the women's movement of the 19th Century and early 20th Century focused on women's legal rights, such as the right to vote, own property, and divorce their husbands, the second wave of the women's movement, which started in the 1960s and 1970s, has become increasingly concerned with issues of technology and its impact on women. Women's organizations now seek to establish equal rights and opportunities for women in the job market, equal participation in the political process, and political control over their bodies, particularly concerning the reproductive process.

Having gained important legal rights during the early part of the 20th Century, the second wave of the women's movement turned its attention to the ways in which technological developments were affecting basic issues of reproduction: abortion, birth control, in vitro fertilization, genetic engineering, and the medicalization of the birth process, including an increasing use of high-tech diagnosis and treatment techniques such as electronic fetal monitoring, fetal imaging and, most recently, ultrasound three-dimensional imaging in utero. A dramatic success of the women's movement in the reproductive area was the legalization of abortion in 1973 with the U.S. Supreme Court's ruling in *Roe v. Wade*. The pressing concerns expressed by various women's organizations about technologies related to women's health care is best understood, at least partly, as a reaction to several technological disasters involving various forms of reproductive technology. One reproductive technology disaster involved DES (diethylstilbestrol), a synthetic estrogen administered to women during the 1950s to 1970s to counteract the possibility of miscarriage. A second reproductive technology disaster involved the drug thalidomide, prescribed to pregnant women during the 1960s and 1970s to fight off "morning sickness" and anxiety.

- DES . . . is now known to cause serious health problems in children of women who took the drug. . . . In addition to their higher incidence of vaginal cancer, DES daughters were subject to a host of anatomical and functional irregularities of the reproductive tract. . . . When these young women reached childbearing age, they encountered high rates of infertility, tubal pregnancies, and premature births. . . . Sons of DES mothers have a higher than average rate of infertility, testicular cancer, and other non-cancerous anatomical abnormalities. The DES mothers themselves, decades after taking the pills, may be especially prone to vaginal adenocarcinoma, breast cancer, and cancer of the uterus. (Williams, 1994: 465)

- In the early 1960s, the over-the-counter (nonprescription) pharmaceutical thalidomide, marketed as a sedative and antinausea drug in more than forty countries, was found to have caused ghastly birth defects in thousands of babies whose mothers had taken the drug to prevent morning sickness. Some of the babies were born . . . without arms or without legs. Others were born blind and deaf. Some were born with heart defects or intestinal abnormalities. Some of the babies were nothing more than a trunk with an eyeless, earless head mounted on it. Some were mentally retarded. . . . It is thought that all together more than ten thousand babies around the world were affected by thalidomide. (Raines, 1994: 468)

We have described other technological disasters involving women's health issues: the Dalkon Shield (see Chapter 11) and silicon breast implants (see Chapter 5).

The DES, thalidomide, and Dalkon Shield disasters and cases such as the silicon breast implant controversy demonstrate the extent to which improperly assessed technological innovations often have horrendous—and sometimes inexcusable—negative impacts on women's health and well-being. Some critics go so far as to claim that the failure of reproductive technologies, rather than being isolated events, are indicative of a problem much more systemic. According to one feminist critic:

Sexist biases permeate the new reproductive technologies and genetic engineering at all levels. In general they imply that motherhood, the capacity of women to bring forth children, is changed from a creative process, in which a woman cooperated with her body as an active human being, to an industrial production process. (Mies, 1987: 332)

More generally, critics allege that such reproductive and other technology disasters that affect women are the result of:

> . . . a style of medical practice quick to intervene with drugs and other forms of technology in an environment where profit can outweigh human concern . . . [resulting in] . . . particularly exploitative and harmful consequences for women who have little voice or direct power over their health care. (Zimmerman, 1987: 446)

All of these issues serve to point out how the combined effects of overzealous medical intervention and the subordinate position of women in society may continually create potential technological hazards to women's health and well-being. They also demonstrate why women's organizations continue to be actively involved in the technology assessment process, especially concerning technologies that directly affect women.

Women's organizations have had a positive political impact on the technology policy process. One way these organizations have been able to strengthen their impact has been by finding and promoting women candidates to run for political office. In this way, the needs and concerns of women, particularly with respect to technologies that may have negative effects, can be articulated, and women can make their demands heard, in the political arena.

Civil Rights Organizations

Activists during the civil rights movement of the 1950s and 1960s worked hard to gain racial equality for African-Americans and other minorities. The activists used a variety of tactics—community empowerment, grassroots activism, strategic use of litigation, lobbying for new legislation protecting equality of education and equal employment—to fight against Jim Crow laws, school segregation, racial discrimination in the workplace, and unequal representation in the political process. Such grassroots activism culminated in the passage of the Civil Rights Act of 1964 and the Voting Rights Act of 1965.

Nowadays, civil rights activists struggle with quite different problems, such as where to site a toxic waste dump or industrial incineration plant. The civil rights activists of the new millennium have, however, learned much from their predecessors about the importance of community activism and political protest and have applied those lessons to issues of environmental justice. Their political effectiveness could be brought to bear on technology decisions that unfairly disadvantage the poor and minorities. This is the focus of those who are advocating "environmental justice" (Novotny, 2000; Szasz, 1994). When decisions are made to site nuclear power plants, toxic waste dumps, and industrial incineration plants, they are often located in areas inhabited by the poor and minorities.

When NIMBY ("Not in My Backyard") protests—usually the work of well-organized citizens from predominantly white, middle-class communities—are successful, developers and politicians are forced to turn their attention to communities where opposition is less well organized and less powerful. This has led to what critics call the *PIBBY principle*: ("Place in Black's Backyard") or the *PIMBY Syndrome*: ("Place in Minorities Back Yard") (Roberts, 1998). Such concerns about environmental injustice date back to 1987 when the Commission for Racial Justice published an authoritative study titled *Toxic Wastes and Race* (Commission for Racial Justice, 1987). The study showed that "three out of every five Black and Hispanic Americans lived in communities with uncontrolled toxic waste" (p. xiv), and that "race proved to be the most significant variable tested in association with the location of commercial hazardous waste facilities" (p. xiii). Can it be that companies that produce hazardous substances—either as a product or as a byproduct—have chosen to site their plants in impoverished communities on the assumption that poor people are unlikely to exert strong pressure against their activities?

Poor and minority communities have since tried to fight back, and their struggles have given rise to the environmental justice movement. Politically-weak communities worried about the potential dangers of living next to dangerous technologies have sometimes built coalitions with civil rights groups concerned with social justice, labor unions concerned with worker rights and safety, and women's groups concerned with the effects of pollution on their children and on themselves, such as breast cancer.

In one well-known case, the Tulane Environmental Law Clinic filed a petition on behalf of St James Parish, Louisiana, with the U.S. Environmental Protection Agency (EPA), requesting the agency to revoke its previously issued building permit to allow Shintech, Inc., to construct a polyvinyl-chloride (PVC) plant in the community (Coyle, 2000: B1). Situated between Baton Rouge and New Orleans, St James Parish is in the heart of what is known as "Cancer Alley." This is an area that is home to approximately 130 oil and chemical companies and waste dumps that release annually more than 900 million pounds of toxins into the air, ground, and water (Roberts, 1998).

The core of the Shintech dispute was an interpretation of the Civil Rights Act of 1964, which argued that states could be held responsible if minority communities are disproportionably exposed to pollution. Title VI of the Civil Rights Act of 1964 prohibits discrimination by federally-funded state agencies. This provision forbids such agencies "to use policies that have an effect of subjecting individuals to discriminations." Seizing on this idea, community groups have filed more than 100 civil rights complaints with the EPA, accusing state and local governments of violating their civil rights by putting dangerous technologies in their back yard.

A New Jersey Federal Court seemed to concur with such sentiments when it ruled that state and local governments risk violating the Civil Rights Act when they allow dangerous and polluting industries into minority neighborhoods, even if the projects violate no environmental laws (McQuaid, 2001: 5). U.S. District Judge Stephen Orlofsky found that the New Jersey Department of Environmental Protection had violated Title VI of the Civil Rights Act of 1964, when it issued a permit to a company to build a $450-million cement plant in a predominately African-American and Hispanic section of Camden, New Jersey (McQuaid, 2001:5).

However, the environmental justice movement received a major setback from a U.S. Supreme Court ruling, holding that an individual cannot sue federally funded state and local agencies for violating federal regulations under Title VI of the Civil Rights Act. This decision is expected to have a major impact on dozens of cases pending in the courts, because Title VI has been a major policy tool civil rights activists have used in environmental justice cases (Lane, 2001: A1).

Another public policy rationale that civil rights organizations, concerned with environmental justice, could appeal to is President Clinton's Executive Order on Environmental Justice, which requires agencies to make "environmental justice part of their mission by identifying and addressing, as appropriate, disproportionately high and adverse human health or environmental effects on minority populations and low-income populations." Mayor Thomas Menino of Boston, in a document sent to Federal Aviation Administrator Jane Garvey protesting the expansion of Logan International Airport, claimed that the expansion would violate the "Clinton Administration's executive order on environmental justice" because it would triple the air traffic over some of the city's disadvantaged Federal Empowerment Zones (Abraham, 1999).

Environmental Organizations

The passage of national environmental laws, beginning with the National Environmental Policy Act in 1969, followed in 1970 by the Clean Air Act Amendments, in 1972 by the Federal Water Pollution Control Act, and in 1976 by the Toxic Sanitation Control Act, have had the effect of significantly improving the environment. They did not, however, deter some industrial polluters from despoiling the environment. Considering the relative laxness of enforcement and the relatively low probability of being apprehended, some firms have deliberately pursued a policy of violating environmental laws.

A case in point is Allied Chemical Corporation. In its application form filed with the EPA in 1971, seeking permits to discharge effluents into waterways, it deliberately omitted mentioning three toxic chemicals: Kepone, THEIC (trihydroxyethl isocyanurate) and TAIC (triallyl isocyanurate). In May 1976, Allied Chemical was indicted and charged with a total of 1,094 counts for polluting the James River (*United States v. Allied Chemical Corporation*, 1976). A federal district court in Richmond, Virginia, found Allied Chemical guilty of violating environmental laws and fined it $13.24 million—until then the largest fine ever reported in a pollution case.

Citizen concerns over the extent of dumping of chemical waste and its effects on public health and the environment prompted Congress to pass in 1980 the Comprehensive Environmental Response, Compensation and Liability Act (also known as Superfund). In a report to Congress, EPA estimated the Superfund program would require at least $11.7 billion—and perhaps as much as $22.7 billion—to clean up hazardous waste around the country. It based its spending estimates on the cleanup of 1,800 identified sites (Anonymous [a], 1984).

Although the Superfund has accomplished much since its enactment two decades ago, formidable tasks remain for years to come. To illustrate the scope of the problem, consider the following criminal convictions:

- Indian tribes across America are grappling with some of the worst of its pollution: uranium tailings, chemical lagoons and illegal dumps. Nowhere has it been more troublesome than at this Mohawk reservation the Indians call Akwesasne— "land where the partridge drums . . ."

  The Mohawks have fought a long war with GM, Reynolds Metals Co. and Aluminum Co. of America, whose factories have fouled the river that the tribe once relied on for food, income and spiritual sustenance. "When all else failed," a Mohawk says, "the river provided for us." Mohawk leaders still offer prayers of thanks to the St. Lawrence, but Akwesasne's 9,000 residents no longer can eat the perch and pike from its waters. And fluoride poisoning has decimated cattle herds. . . . Tribes had little legal access to federal sanctions against polluters. Mohawk chiefs governing the Canadian side of the Akwesasne filed a $150 million suit against Alcoa and Reynolds over fluoride, settling for $650,000 after legal fees nearly bankrupted the tribe. . . . In 1983, the EPA added the GM site to its Superfund cleanup list, estimating that the area contained 800,000 cubic yards of PCB-contaminated sediments. Akwesasne residents were warned to avoid lettuce and tomatoes from their own gardens. Women of childbearing age and children were advised to stop eating fish, their main source of protein. . . . Emboldened members of the Warriors Society have threatened to occupy factories to stop pollution. (Tomsho, 1990)

- Dexter Corp. will pay $13 million in criminal and civil penalties after pleading guilty to Clean Water Act and Resource Conservation and Recovery Act violations over four years at its specialty paper products plant in Windsor Locks, Connecticut. (Anonymous [b], 1992)

- After seven years of litigation, Ciba-Geigy officials pled guilty and agreed to pay more than $62 million in fines and cleanup costs for charges of illegally dumping toxic waste at its Toms River Plant. . . . Outraged environmentalists thought nothing but jail terms would be sufficient punishment, as fines are written off as "business expenses" and thus no deterrent to violators. (Anonymous [c], 1992)

- In a federal court, PureGro. Co., a major distributor of pesticides, fumigants and fertilizers, pleaded guilty of illegal dumping and agreed to pay a $100,000 fine and full cost of cleaning up the dump, which could run as much as $2 million. (Perry, 1993)

Notwithstanding the Superfund law and its vast budget for cleaning up hazardous wastes, violations continue, in part because of the inadequacy of the enforcement program. Reforming the provisions of the law and strengthening the system of sanctions to deter environmental crime would indeed help. This is a necessary strategy for coping with public health hazards associated with toxic wastes, but it is not sufficient. To some extent, community watchdog organizations are being formed to provide informal and supplementary enforcement capabilities for EPA. There is a dynamic that tends to promote the formation of local chapters adhering to a national or international advocacy group. Organizations such as the Sierra Club, Greenpeace, the Union of Concerned Scientists, and other national environmental organizations would be well advised to establish community-based chapters to assist in the mammoth undertaking of monitoring the environment on behalf of the citizenry.

Consumer Organizations

The idea that consumers need to be protected from the potential hazards of technology has its origins in President John F. Kennedy's famous "Consumer Bill of Rights" of 1962. As Kennedy (1962) put it: "All of us are consumers. All deserve the right to be protected against fraudulent or misleading advertisements and labels, the right to be protected against unsafe or worthless drugs and other products, the right to choose from a variety of products at competitive prices." Such a lofty proclamation prompted Ralph Nader and other activists to begin the long struggle to fashion policies aimed at regulating—on behalf of consumers—various products, services, methods, and standards for manufacturers, sellers, and advertisers.

In the early 1970s, consumer advocates charged that consumer abuses were not sporadic but systemic. As Nader put it: "What most troubles the corporations is the consumer movement's relentless documentation that consumers are being manipulated, defrauded, and injured not just by marginal businesses or fly-by-night hucksters, but by the U.S. blue-chip business firms whose practices are unchecked by the older regulatory agencies" (Nader, 1968). In order to confront such systemic abuses, Nader argued for a "countervailing force," which he envisioned would become a decentralized, loosely affiliated corps of "public citizens" with their own institutional resources and expertise. To this end, Nader and his committed associates founded the Public Citizen organization in 1971, which, in the year 2001, had a membership of more than 150,000. This loosely affiliated group of public citizens' organizations boasts a legacy of hundreds of independent congressional, judicial, and media inquiries, either convened on their behalf or which were the direct result of their persistent pressure.

Some of the inquiries by Public Citizen over the decades include: unsafely designed automobiles, adulterated foodstuffs, junk foods, ineffective or hazardous drugs, product obsolescence, faulty and unsafe appliances, deceptive packaging and false advertising, overselling of credit, and alleged insurance and health-care fraud. These inquiries have led to many consumer protection policies, including comparative product information such as auto crash safety ratings and warning labels on foods. Other Public Citizen inquiries have led to legal and procedural rights such as the right to know about toxic hazards in foods and the right to initiate class action law suits. Victims in asbestos litigation have benefited greatly from the latter right (See Chapter 1).

One of Nader's earliest crusades was against the automobile industry. His groundbreaking book, *Unsafe At Any Speed* (1965), exposed the benightedness of the American automobile industry. It argued that American automakers were more concerned with speed and chrome than they were with safety and fuel efficiency. Partly as a result of the nationwide attention Nader's book received, the National Highway Traffic Safety Administration (NHTSA), which is concerned with all aspects of automobile safety, was formed in 1970. In fact, the formation of a governmentally established regulatory agency to protect consumers is not new. For example, the Federal Trade Commission (FTC) was established in 1914 and is authorized to prevent deceptive practices in commerce and to regulate the packaging and labeling of consumer products. The Food and Drug Administration (FDA), originally established in 1906 and reorganized and renamed in 1927, administers a program of consumer protection of foods, pharmaceuticals, cosmetics, and other substances. Further evidence of the public's concern with consumer protection is the passage by Congress in 1972 of the Consumer Products Safety Act, which established the U.S. Consumer Products Safety Commission (CPSC), an independent regulatory agency. CPSC has jurisdiction over 15,000 products, "from automatic-drip coffee makers to toys to lawn mowers," according to the commission's Web site (*www.cpsc.gov*). However, despite the relatively long history of governmental regulation of consumer goods in the marketplace and despite the concerted efforts of various consumer organizations, the public continues to be persuaded to use commodities that are unsafe and unreliable and have the potential for harming innocent consumers.

Consider, for example, the alarming statistics on annual automobile fatalities and injuries. According to official NHTSA reports, more than 40,000 people are killed on U.S. highways and more than 3 million people are injured each year (National Highway Traffic Safety Administration, 2000). While not all fatalities are the result of poorly engineered, unsafe automobiles, there are enough examples in this category to cause concern to motorists (see Chapter 6). The Ford-Firestone disaster is a case in point (see Chapter 11). While federal agencies such as the NHTSA undertook investigations, and corporate executives of Ford Motor Company and Firestone-Bridgestone pointed fingers at each other over who is ultimately responsible, 203 people lost their lives and more than 700 people have been injured.

Why has there not been an increasing public demand for safer foods and safer products, including automobiles? One of the central ideas of President Kennedy's Consumer Bill of Rights is that consumers constitute an all-inclusive group. Since we are all consumers, we all *could* exert considerable influence on public policy through our purchasing power in the marketplace. For example, the high safety and reliability ratings of foreign cars (especially German, Scandinavian, and Japanese) draws American consumers to the foreign car market. Foreign car companies have captured about one-third of the market share on new automobile sales in the United States (Buss, 2001: 24). As one automobile marketing consultant put it in reference to the large market share garnered by the Japanese automotive companies: "The biggest thing they've got going for them is their . . . history of quality and reliability and dependability. People don't necessarily like the way Toyota looks, for example, and it's not an exciting performer— but, boy, is it reliable" (Buss, 2001: 25). Such a sentiment is a testament to the fact that American consumers are willing to pay for safety, reliability, dependability, and efficiency.

Consumer boycotts are another way organized consumers can express their dissatisfaction with dangerous consumer products. Consumer boycotts were a very important nonviolent means adopted by African-Americans during the civil rights era, when disadvantaged minorities used their collective buying power to bring about social changes (Weems, 1995). Labor-sponsored boycotts of a company's product have also long been a way the American labor movement has protested unfair salary and benefit programs and unsafe working conditions. Cesar Chavez underscored this fact when, through a boycott campaign, he won recognition for the plight of United Farm Workers in California's grape vineyards (Sanderson, 1974). A recent boycott campaign aimed at protesting the deplorable working conditions is a boycott of Nike Corporation over its use of "sweatshops" and child labor in the manufacturing of its popular line of running shoes (Mason, 2000).

Increasingly, we are witnessing the use of boycotts beyond the issues of racial inequality and economic exploitation to quality-of-life issues. One example is the boycott of Exxon Corporation's products because of its denial of the relationship between CO_2 emissions and global warming (Friedman, 2001: 3).

Researchers have shown that, between 1970 and 1980, 90 major consumer boycotts have occurred. Results show that a wide variety of organizations sponsored the boycotts, with labor groups and racial minorities being responsible for more than half of them (Friedman, 1985). During the 1990s, consumer boycotts skyrocketed. For example, in 1992 alone, it was reported that about 100 consumer boycotts of products and corporations were underway in the United States (Miller, 1992: 58). The major finding of the study is the pervasiveness of boycott actions and the strength of the boycotting tradition among citizens' organizations in the United States.

Consumer boycotts have also been launched by animal rights activists against fur companies and cosmetics companies such as Revlon, Avon, and Benetton that test their products on animals. Antiwar boycotts target manufacturers of military weapons such as General Electric, and boycotts have been led by environmental groups against companies that import tropical timber from rapidly vanishing rainforests, such as Mitsubishi. Environmental groups have also sponsored boycotts of companies producing dangerous pesticides and other toxic hazards—companies such as Monsanto, DuPont, Dow, Bayer, and others—since the activist groups have judged such practices to have wreaked havoc on the environment.

According to one researcher, "boycotts of businesses are reaching epidemic proportions in America largely because they are so successful" (Anonymous [d], 1990). Consider the following four striking examples of successful consumer boycotts:

- October 1984 marked the end of a seven-year boycott of Nestlé products. The boycott—organized by the "Nestlé Boycott Committee"—was carried out because of alleged marketing abuses in the promotion and sales of infant formula products in developing countries. Such consumer action led the World Health Organization to adopt a code of marketing for infant formula products. The boycott had several positive results, including: (1) the setting of precedents in the area of corporate accountability, (2) the opening up of new vistas for the discussion of human rights and commercial interests, and (3) the WHO code provides a way to define and evaluate corporate responsibility in infant formula marketing. (Post, 1985)

- Attacking brand-name products . . . the most visible example has been this year's successful campaign to stop canners selling tuna fish caught in nets that also trap and kill dolphins. The boycott zeroed in on three brands that account for about 70 percent of the canned tuna in the United States: Star-Kist, Chicken of the Sea, and Bumble Bee. When Heinz, the owner of Star-Kist, announced . . . that it would buy only "dolphin-safe" tuna, the other beleaguered companies capitulated a few hours later. (Anonymous [d], 1990: 69)

- The boycott of products of companies that experiment on animals is another example. In 1985 an organization called the

People for the Ethical Treatment of Animals (PETA) had the names of only 50 companies on its "cruelty-free shopping" list. It now lists more than 250 companies that have foresworn testing their products on animals and the list is still growing. (Anonymous [d], 1990: 70)

- Women wearing furs on New York's Fifth Avenue . . . were barracked by animal-rights advocates who held them indirectly responsible for the cruel deaths suffered by furry animals. . . . So many women have stopped wearing furs that the trade is in serious trouble. After seeing a net profit of $5 million in 1987 turn into net losses in 1988 and 1989, Fur Vault of New York agreed to sell its retail fur operations. . . . Another big New York furrier, Antonovich, has filed for protection from its creditors under federal bankruptcy laws. (Anonymous [d], 1990: 71)

It is to be expected that consumer organizations will become more active at the grassroots level and recruit citizens to lobby for safer consumer products such as food and automobiles, and more reliable and cost-effective services such as health care.

An alliance of the four types of organizations—women's groups, civil rights, environmental, and consumer—can become a powerful force for protecting citizens from dangerous technologies. Such an alliance can deliberately use the five mechanisms of citizen participation we have described to advance their interests. When an alliance of the four types of citizen organizations concerts their efforts, they can achieve a synergistic effect on technology policy decisions. One way of summarizing the potential effects of the four types of NGOs in mitigating the harms of technological disasters is as follows: when people face a problem in common, they can influence the powers that be to act for the benefit of the victims rather than for the protection of the perpetrators.

CONCLUSION

An alliance of the four types of citizen organizations, such as we have proposed, could activate the five participatory mechanisms. Critics often point out that most public hearings are not participatory in the full sense because such hearings usually consist of a strategy of one-way communication from experts to citizenry with little or no real citizen influence over the policy decisions (Renn, Webler, and Wiedemann, 1995: 10). Most decisions are already made by the experts before convening the public hearing. The public hearing, more often than not, reduces to a "decide-announce-defend" (DAD) strategy that does more to disenfranchise and anger the public than it does to create a fully participatory framework. Can such procedures permit citizens to have an input into the decision-making process?

Referenda are often criticized for delegating authority to an "uninformed/unqualified electorate" (Fiorino, 1990). Furthermore, referenda, along with public opinion surveys, offer little access to resources and information necessary to enable the public to make informed decisions.

While researchers point out that citizen advisory groups and consensus conferences also have their shortcomings, these mechanisms more fully embody the democratic ideals necessary for citizen participation in technology and science policy. For example, Rowe and Frewer (2000) point out that citizen advisory groups and consensus conferences provide citizens with technical information to enable them to make informed contributions.

The most appropriate mechanism for public participation is likely to be a hybrid of the five mechanisms discussed. Even if full consensus is not always possible, mechanisms for citizen participation provide forums for a more democratic governance when assessing technological risk. As Nelkin and Pollak (1979) put it:

> What can be generalized are . . . the conditions that will allow dissenting groups to express their concerns and to communicate effectively with administrative agencies. These conditions include: a "formula" that gives due weight to social and political factors; appropriate involvement of affected interests; an unbiased management; a fair distribution of expertise; and a real margin of choice. Actually, such procedural conditions are not likely to produce consensus, but they may reduce public mistrust and hostility toward political and administrative institutions. . . . (p. 64)

Hence, participatory mechanisms could, indeed, have a constructive influence on the reduction of technological disasters that continue to plague society. Wenk (1984) argues for such a position when he states "participatory technology . . . may be the most potent countermeasure to a stream of technologically-laden decisions leading to catastrophe" (p. 129). In the United States, Canada, and much of Europe, policy makers and concerned technologists can no longer simply ignore the problem of citizen participation in science and technology policy. It is no longer possible simply to leave the public and the experts to their own devices in an effort to maintain the status quo. Doing nothing to bridge the gap between experts and policy makers on the one hand and the public on the other hand will only arouse negative reactions on the part of a disenfranchised citizenry. Raising no questions about the judgmental fallibility of experts creates an unduly esteemed technological elite, a mandarin class. Leaving citizens to their own devices when they are known to have poor information or misinformation or to misinterpret the right information renders them less able to act in their own best interests or the best interests of the world in which they live (Daddario, 1971).

In a democratic society, it is incumbent on politically-elected leaders to ensure that there is a continuing dialogue between policy makers, experts, and the citizenry—a small price indeed to pay for mitigating the harms produced by ever more costly technological disasters.

References

Abraham, Yvonne. (1999). "Menio urges FAA to closely study problems with proposed runway," *The Boston Globe*, December 1: C17.

Anonymous (a). (1984). "EPA estimate for cleaning up dumps surges to $22 billion," *The Christian Science Monitor,* December 14: 2.

Anonymous (b). (1992). "Dexter assessed $13 million in fines for water and toxic waste violations," *Toxic Materials News Business Publishers,* September 16.

Anonymous (c). (1992). "Ciba-Geigy pleads guilty," *New Jersey Industry Environmental Alert Environmental Compliance Reporter,* March 1.

Anonymous (d). (1990). *The Economist,* May 26: 315 (7656): 69–71.

Barus, Carl. (1982). "On costs, benefits and malefits in technology assessment," *IEEE Technology and Society Magazine,* March: 3–27.

Belsten, Laura. (1996). "Environmental risk communication and community collaboration." In Star Muir and Thomas Veenendall (Eds.), *Earthtalk: Communication empowerment for environmental action.* Westport, CT: Plenum Books: 27–41.

Bradbury, Judith. (1989). "The policy implications of differing concepts of risk," *Science, Technology, and Human Values* 14 (4): 380–399.

Buss, Dale. (2001). "East meets west," *Brandmarketing* 8 (4): 24–29.

Carroll, J. (1971). "Participatory technology," *Science* 171 (19): 647–653.

Cerezo, Jose, and Garcia, Marta. (1997). "Lay knowledge and public participation in technological and environmental policy," *Research in Philosophy and Technology* 16: 33–48.

Commission for Racial Justice, United Church of Christ. (1987). *Toxic wastes and race in the United States.* New York: United Church of Christ Press.

Covello, Vincent. (1983). "The perception of technological risks: A literature review," *Technological Forecasting and Social Change* 23: 285–297.

Coyle, Pamela. (2000). "Tulane law clinic honored for work; communities got help in environmental cases," *The Times Picayune,* July 5: B1.

Daddario, Emilio. (1971). "Technology and the democratic process," *Technology Review,* July/August; 73: 12–19.

Fiorino, Daniel. (1989). "Technical and democratic values in risk analysis," *Risk Analysis* 9 (3): 293–299.

Fiorino, D. (1990). "Citizen participation and environmental risk: A survey of institutional mechanisms," *Science, Technology, and Human Values* 15 (2): 226–243.

Fischer, Frank. (2000). *Citizens, experts, and the environment: The politics of local knowledge.* Durham, NC: Duke University Press.

Fischoff, Baruch. (1979). "Informed consent in societal risk-benefit decisions," *Technological Forecasting and Social Change* 13: 347–357.

Friedman, Monroe. (1985). "Consumer boycotts in the United States, 1970–1980: Contemporary events in historical perspective," *The Journal of Consumer Affairs* 19 (1): 96–124.

Friedman, Thomas L. (2001). "Bush/Cheney will rue day U.S. dumped Kyoto," *The Houston Chronicle,* June 3: 2 STAR 3.

Funtowicz, Silvio. (1983). "Three types of risk assessment: A methodological analysis." In Chris Whipple and Vincent Covello (Eds.), *Risk analysis in the private sector.* New York: Plenum Press: 217–231.

Grabill, Jeffery, and Simmons, Michele. (1988). "Toward a critical rhetoric of risk communication: Producing citizens and the role of technical communicators," *Technical Communication Quarterly* 7: 415–439.

Hunold, Christian, and Young, Iris Marion. (1998). "Justice, democracy, and hazardous siting," *Political Studies* 46 (1): 82–95.

Johnson, Branden. (1987). "Accounting for the social context of risk communication," *Science and Technology Studies* 5 (4): 103–111.

Kasper, Raphael. (1980). "Perceptions of risk and their effects on decision-making." In R. Schwing and A. Albers (Eds.), *Societal risk assessment: How safe is safe enough?* New York: Plenum Press: 71–81.

Kennedy, J.F. (1962, March 15). Comments on presenting the Consumer Bill of Rights. Available on the World Wide Web at: www.netspace.net.au/~moloney/consumer.htm

Lane, Charles. (2001). "Justice limits bias suits under Civil Rights Act," *The Washington Post,* April 25: A1.

Lathrop, John. (1982). "Evaluating technological risk: Prescriptive and descriptive perspectives." In Howard Kunreuther and Eryl Ley (Eds.), *The risk analysis controversy: An institutional perspective.* Berlin: Springer-Verlag: 165–180.

Marcus, Ruth. (1989). "Nuclear-free zone challenged: U.S. says Oakland's proclamation is unconstitutional," *The Washington Post,* September 7: A2.

Mason, Tania. (2000). "Nike axes 'sweatshop' after BBC investigation," *Marketing,* October 19: 5–7.

McQuaid, John. (2001). "Industries lose civil rights case; Minorities get tool to fight pollution," *The Times Picayune,* May 27: 5.

Mies, Maria. (1987). "Sexist and racist implications of new reproductive technologies," *Alternatives* 12: 323–342.

Miller, Annetta. (1992). "Do boycotts work?" *Newsweek,* July 6; 120 (1): 58–61.

Mitcham, Carl. (1997). "Justifying public participation in technical decision-making," *IEEE Technology and Society Magazine* (Spring) 15: 40–46.

Montgomery, David. (1995). "Takoma park faces slow meltdown of its nuclear-free ideals," *The Washington Post,* June 25: B1.

Nader, Ralph. (1965). *Unsafe at any speed.* New York: Grossman Publishers.

Nader, Ralph. (1968). "The Great American gyp," *New York Review of Books.* Cited in Craig Smith, *Morality and the market: Consumer pressure for corporate accountability.* New York: Routledge: 211.

National Highway Traffic Safety Administration. (2000). *A compilation of motor vehicle crash data from the Fatality Analysis Reporting System.* Washington, DC: National Center for Statistics and Analysis, U.S. Department of Transportation.

Nelkin, D. (1975). "The political impact of expertise," *Social Studies of Science* 5 (1): 67–79.

Nelkin, Dorothy. (1994). *Controversies: Politics of technical decisions* (2nd ed.). Beverly Hills, CA: Sage Publications.

Nelkin, D., and Pollak, M. (1979). "Public participation in technological decisions: Reality or grand illusion?" *Technology Review,* August/September, 55–64.

Nelkin, D., and Pollak, M. (1980). "Problems and procedures in the regulation of environmental risk." In R. Schwing and A. Albers (Eds.), *Societal risk assessment: How safe is safe enough?* New York: Plenum Press: 233–246.

Novotny, Patrick. (2000). *Where we live, work and play: The environmental justice movement and the struggle for a new environmentalism.* Westport, CT: Praeger.

Perry, Tony. (1993). "Company pleads guilty to dumping of toxic waste environment," *Los Angeles Times,* January 1: B1.

Piller, Charles. (1991). *The fail-safe society: Community defiance and the end of American technological optimism.* Berkeley, CA: University of California Press.

Post, James. (1985). "Assessing the Nestlé boycott: Corporate accountability and human rights," *California Management Review* 27 (2): 113–132.

Public Law 89-209, 1965. (1966). United States Statutes at Large, Vol. 79. Washington, DC: U.S. Government Printing Office.

Raines, Jeffrey. (1994). "Thalidomide." In Neil Schlager (Ed.), *When technology fails.* Detroit: Gale Research.

Renn, Ortwin. (1990). "Risk perception and risk management," *Risk Abstracts* 7 (1): 1–9.

Renn, O., Webler, T., and Wiedemann, P. (1995). *Fairness and competence in citizen participation: Evaluating models for environmental discourse.* Dordrecht, The Netherlands: Kluwer Academic Publishers.

Robbins, Carla. (1998). "Army's huge supply of nerve gas poses unnerving questions," *The Wall Street Journal,* June 1: A1.

Roberts, R. Gregory. (1998). "Environmental justice and community empowerment: Learning from the civil rights movement," *The American University Law Review* 48: 229–269.

Rowe, G., and Frewer, L. (2000). "Public participation methods: A framework for evaluation," *Science, Technology and Human Values* (Winter) 25 (1): 3–29.

Ruckelshaus, William. (1985). "Risk, science, and democracy," *Issues in Science and Technology* (Spring): 19–38.

Sanderson, George. (1974). "The product boycott—Labor's latest tool," *The Labour Gazette* 7 (74): 477–484.

Sclove, Richard. (1996). "Consensus on technology," *Technology Review* (9) 99 : 25–31.

Shrader-Frechette, Kristin. (1985a). *Risk analysis and scientific method.* Dordrecht, The Netherlands: D. Reidel Publishing Company.

Shrader-Frechette, Kristin. (1985[b]). "Technology assessment, expert disagreement, and democratic procedures," *Research in Philosophy and Technology* 8: 103–129.

Slovic, Paul, Fischoff, Baruch, and Lichtenstein, Sarah. (1980). "Facts and fears: Understanding perceived risk." In R.C. Schwing and W.A. Albers (Eds.), *Societal risk assessment: How safe is safe enough?* New York: Plenum Press.

Spangler, Miller. (1982). "The role of interdisciplinary analysis in bridging the gap between the technical and human sides of risk assessment," *Risk Analysis* 2 (2): 101–114.

Szasz, Andrew. (1994). *Ecopopulism: Toxic waste and the movement for environmental justice.* Minneapolis: University of Minnesota Press.

Tomsho, Robert. (1990). "Dumping grounds: Indian tribes contend with some of worst of America's pollution," *The Wall Street Journal,* November 29: A1.

United States of America v. Allied Chemical Corporation (Crim. A. No. 76-0129-R, 420 F. Supp. 122, 1976 U.S. Dist).

Verhovek, Sam Howe. (2000). "Oregon ballot full of voter initiatives becomes issue itself," *The New York Times,* October 25: A24.

Weems, Robert. (1995). "African-American consumer boycotts during the civil rights era," *The Western Journal of Black Studies* 19 (1): 72–80.

Wenk, Edward. (1984). "Civic competence to manage technology," *Technological Forecasting and Social Change* 26: 127–133.

Williams, Ellen. (1994). "Diethylstilbestrol (DES)." In Neil Schlager (Ed.), *When technology fails.* Detroit: Gale Research: 462–467.

Wynne, Brian. (1989). "Sheep farming after Chernobyl," *Environment* 31 (2): 10–15, 33–38.

Zimmerman, Mary. (1987). "The women's health movement: A critique of medical expertise and the position of women." In Beth Hess and Myra Feree (Eds.), *Analyzing gender: A handbook of social science research.* Beverly Hills, CA: Sage.

Name Index

A

Abelin, T., 41
Abraham, Yvonne, 454
Ackoff, Russell, 104, 374
Aeppel, Timothy, 450
A.H. Robins Company, 355, 363, 368, 369, 373
Aiken, Howard, 182
Akgerman, A., 320
Albrecht, Steve, 365
Alevizatos, Dorothy, 388
Alford, Fred, 352
Allen, Allan, 254
Alsop, Ronald, 360
Alyeska Pipeline Service, 359
Anderson, Elizabeth, 425
Anderson, Robert, 130
Ansoff, H.L., 374
Anthony, R.G., 320
Appel, Adrianne, 358
Applegate, F.D., 240
Aronoff, C., 126
Arquilla, John, 142
Asimov, Isaac, 190, 191
Averkin, Y., 41
Aydintasbas, Asla, 35

B

Baase, Sara, 185
Babbage, Charles, 181

Babcock, G.L., 245
Bacon, Francis, 333, 336
Baldauf, Scott, 35
Baram, Michael, 420
Barash, D., 44, 46
Barbour, Ian, 424, 426
Bardeen, John, 182
Barnes, Barry, 102
Baron, S., 119
Barton, Lawrence, 354, 369, 372
Barus, Carl, 434
Bates, Christine, 319
Baskin, O., 126
Bay Area Rapid Transit System (BART), 130, 301, 321, 329
Bayer Corporation, 460
Bayless, Michael, 423
Baum, A., 49, 50, 52
Beauchamp, T. 236
Beck, U., 21
Beckman, Peter, 333
Bell, Daniel, 165
Belsten, Laura, 441
Ben-David, S., 420
Bensman, J., 252
Benz, Karl, 173
Berkowitz, Bruce, 141
Bernard, Chester J., 246
Bernstein, Jeremy, 336
Bethe, Hans, 336
Betts, Mitch, 115

Bidwai, P., 42
Bier, Vicki, 412, 417
Bierly, P., 93, 98, 99
Bignell, V., 82, 118
Bijker, Wiebe, 195
Bimber, Bruce, 388
Binkley, Christina, 73
Birsch, D., 9, 10, 16
Blaikie, P., 34
Blair, B.G., 45
Blankenzee, M., 327
Bleuer, J.P., 41
Blight, James, 47
Bloor, David, 102
Boeing Company, 319
Boggs, Danny, 392, 403
Bohr, Niels, 333, 335
Boisjoly, Roger, 279, 280, 281, 285–286
Boisoly, Russell, 279
Bondurant, S., 136
Borman, F., 354
Bougumil, R., 418
Bowers, James, 389
Bowie, Norman, 375
Bowonder, B., 148, 289
Bradbury, Judith, 438
Brandeis, Justice Louis, 381
Branscomb, Lewis, 402
Brick, Michael, 404, 405
Bridis, Ted, 61
Brittian, Walter, 182

468

Subject Index

M

N